2ª

The Dynamics of Evolution

The Dynamics of Evolution

The Punctuated Equilibrium Debate in the Natural and Social Sciences

Edited by
ALBERT SOMIT
and
STEVEN A. PETERSON

Cornell University Press
Ithaca and London

First pubiished 1992 by Cornell University Press.

International Standard Book Number 0-8014-2531-x (cloth)
International Standard Book Number 0-8014-9763-9 (paper)
Library of Congress Catalog Card Number 91-55569

Printed in the United States of America

Librarians: Library of Congress cataloging information appears on the last page of the book.

♾ The paper in this book meets the minimum requirements of the American National Standard for Information Sciences—Permanence of Paper for Printed Library Materials, ANSI Z39.48-1984.

Contents

Acknowledgments

We have many people to thank. We are profoundly indebted to Harvey Wheeler, then editor of the *Journal of Social and Biological Structures,* who encouraged us to use the promise of publication in his journal as a lure for potential contributors. The resulting collection of essays, in an earlier version, appeared as a special issue of the *Journal of Social and Biological Structures* (Academic Press) in July 1989. We have incurred at least an equal obligation to Holly M. Bailey, formerly editor at Cornell University Press, whose suggestions and encouragement led to the present volume.

We would also be remiss not to thank the College of Liberal Arts and Sciences at Alfred University for logistical and psychological support that helped make this volume possible. And a parallel expression of appreciation is certainly owed to the College of Liberal Arts, the School of Law, and the Department of Political Science of Southern Illinois University at Carbondale.

Karen Mix labored over the manuscripts to have them meet Cornell University Press specifications. She and Michael Yehl, an undergraduate student at Alfred University, were forced to learn how to use scanner software in order to read the hard copy manuscripts into word processing files. Regrettably, several essays refused to scan and had to be typed, the old-fashioned way, into word processing files.

Finally, we express our appreciation to our contributors. It is their ideas that give this book its intellectual substance and their unstinting cooperation that, in the final analysis, made its publication possible.

A. S. AND S. A. P.

The Dynamics of Evolution

Punctuated Equilibrium: The View from the Elephant

ALBERT SOMIT
STEVEN A. PETERSON

How does change occur in the biological world? Does it happen gradually, as classical evolutionary theory has long assumed, or does it occur by rapid transformations of structure and behavior? Is the nature of biological change relevant for our understanding of human behavior and, if so, how? These are the questions posed by a "new" conceptualization of the evolutionary process—the perspective most commonly referred to as punctuated equilibrium. And these are the questions this book addresses.

Punctuated equilibrium theory makes two contentions: that evolutionary change (or at least very significant proportions thereof) occurs in rapid bursts over (geologically) short periods of time, and that there is relative stasis after the punctuational burst. (As we shall see, even these two propositions are subject to varying interpretations.) In 1971, paleontologist Niles Eldredge published a brief essay outlining this "new" evolutionary model (some critics, challenging the claim of originality, point to a much earlier volume that, they maintain, adumbrates punctuated equilibrium. See Simpson, 1944). Eldredge's 1971 paper and a subsequent article by Eldredge and Stephen Jay Gould in 1972 unleashed a controversy that still roils the scientific journals and has divided evolutionary theorists into often angrily contesting camps. It seems only sensible, after almost two disputative decades, to attempt an assessment of its implications for the neo-Darwinian synthesis.

In structuring this volume, we had three goals. First, we wanted to identify the component elements of punctuated equilibrium. What are the basic tenets of punctuated equilibrium theory? From what sources did it draw? What are its specific claims? Our second objective was to assess the present status of punctuated equilibrium among evolutionary theorists. Which punctuationist concepts have earned general acceptance? Which have been rejected? Which have been modified? Which are still being controverted?

Third, we sought to weigh the scientific implications of punctuated equilibrium. This undertaking had two aspects. Most immediately and obviously, what are its implications for biology (i.e., evolutionary theory)? We were also interested, however, in its implications for the social and behavioral sciences. Over the past twenty-some years, more and more social scientists have turned to biological theory (and to biologically derived research techniques) in search of a better understanding and explanation of the phenomena with which their own fields are concerned. As a result, biologically oriented subfields have emerged in sociology, economics, political science, archaeology, and, most recently, law; there has long been, of course, a strong "biological" component in psychology and anthropology. In borrowing from biology, these groups relied heavily on neo-Darwinian theory; the significance to them of a possible major change in that paradigm needs no elaboration.

As has been the case with many other theories, ideological issues have become intertwined with scientific ones in the debate over punctuationism. There have been allegations (see the chapters by Gould and Ruse herein) that the proponents of punctuated equilibrium were influenced, if not motivated, by Marxian ideological considerations, an allegation not discouraged by Gould's active involvement in the Sociobiology Study Group of Science for the People and that group's ideological criticism of Edward O. Wilson's *Sociobiology*. On the other side of the scientific fence, some of those who led the opposition to punctuationism made no secret of their distaste for its potential social and political overtones. For instance, some have noted that punctuational theory may provide at least metaphoric support for advocacy of revolutionary change (Peterson and Somit, 1983). Combine this with Gould's candid comment about learning Marxism at his father's knee, and one can understand the ideological suspicions the theory has elicited.

The formulation of punctuated equilibrium also coincided with the rise of sociobiology and the fascinated attention given to Edward O. Wilson and his followers by the popular press and media. Sociobiology proposed to explain human behavior—or at least most of it—in terms of natural selection; more specifically, in terms of "inclusive fitness." Wilson rather rashly announced that it was only a matter of time before sociobiology would literally subsume sociology and the other social sciences. Small wonder that social scientists, faced with this happy prospect, have been attentive to punctuationism's apparent challenge to the neo-Darwinian synthesis on which sociobiology so heavily relied.

As our work progressed, the book took unexpected turns. We had assumed that, in the pro-and-con punctuated equilibrium pieces, our contributors would deploy the most recent evidence to bolster their respective positions. This they did, where appropriate, but they also did a great deal more. Michael Ruse, for instance, reverses his earlier negative assessment of punctuated equilibrium with the refreshing admission, "Now, I think I was wrong; at least I think I was altogether too quick." Ernst Mayr provides a magisterial review of punctuated equilibrium's intellectual antecedents, as well as a thoughtful critique of the theory itself. And Stephen Gould, in a fascinating *apologia pro vita sua*, places the emergence of punctuated equilibrium against the broader panoply of thought in evolutionary theory generally and paleontology in particular.

A second surprise was the range and depth of disagreement that still exists, after nearly twenty years of debate, on two closely related and very basic issues: What are punctuated equilibrium's constituent concepts and contentions? What specific claims does it make with regard to evolution and the mechanisms and/or forces that drive the evolutionary process?

The Nature of Punctuated Equilibrium

Though we expected some differences between proponents and opponents, they proved to be much wider than we had thought. We discovered, too, that supporters, critics, and even "neutralists" differ among themselves as to punctuated equilibrium's explicit doctrinal content and implicit claims. (Seeking to explain this phenomenon, some argue that Eldredge and Gould have shifted positions repeatedly, and rather substantially, since 1972, an allegation that Gould emphatically rejects.)

What is punctuated equilibrium? In the metaphor of the ancient fable, only the elephant knows. Each of our contributors in the first part of this book has his own conception, and the (sometimes striking) differences among them are clearly at the heart of their disagreements over the validity of the doctrine and, consequently, of its implications for evolutionary theory. This situation posed something of an editorial dilemma: if distinguished scientists themselves dispute this matter, would it not be difficult for our readers, unless they are both knowledgeable and persistent, to sort out the several issues and positions? To facilitate and simplify that effort, we have sought to identify, for each of our contributors to the first half of the volume, that individual's understanding of punctuated equilibrium and to present that understanding in what we hope will be more readily grasped summary form.

As Table I.1 shows, almost everyone agrees that, at its most basic, punctuated equilibrium involves two key propositions: first, that species undergo long periods of little or no evolutionary change; second, that these lengthy intervals of stasis (i.e., equilibrium) are broken (i.e., punctuated) by relatively rapid speciation events. Even this basic agreement erodes, however, with regard to the relative importance of natural selection in accounting for speciation or with regard to other claims that, depending on one's perspective, were or were not made, implicitly or explicitly, by the punctuationists.

A quick look at each author's conception of punctuated equilibrium may be helpful. Given the purposes of his essay, Steven M. Stanley, a supporter, needed to identify only the aforementioned propositions that the general pattern found in the fossil record is one of "relative stasis" and that "events of rapid evolutionary transformation . . . account for most evolution in the history of life" (p. 85). Stanley's formulation leaves open the possibility that natural selection may play a dominant role in the process of speciation.

Ernst Mayr, the dean of contemporary evolutionary theorists, recognizes the same two basic tenets, which he summarizes in essentially similar language. But Mayr, on whose ideas Gould and Eldredge drew quite heavily, also calls attention to other claims that have been associated with punctuated equilibrium—namely, that practically all change is due to these rapid speciation events; that speciation can and does occur via saltation, as well as punctuation; and that, at least in some versions, punctuated equilibrium theory necessarily implies some type of selection above the level of the organism. The latter two contentions, in particular, go well beyond the original emphasis on lengthy periods of stasis punctuated by rapid episodes of speciation.

Implicitly concurring with Mayr's thesis that punctuated equilibrium subsequently "expanded" to include ideas not initially voiced, Antoni Hoffman and Michael Ruse both seek to resolve the problem by postulating a multistage development. According to Hoffman, there was the "weak" version advanced in the 1972 paper; this was followed by

Table I.1. Basic tenets of punctuated equilibrium

Stanley	First, that "species usually evolve very sluggishly" and the general pattern is one of "relative stasis"; second, that "rapid [evolutionary] events account for most evolution in the history of life" (p. 85).
Mayr	"The theory of punctuationism . . . consists of two basic claims: that most or all evolutionary change occurs during speciation events, and that most species enter a phase of total stasis after the end of the speciation process" (p. 30). But Mayr also identifies other "specific claims" made by or attributed to Eldredge and Gould. Among these are (1) that speciation was followed by "total stasis" (p. 27); (2) that "'if Mayr's (1963:586) characterization of the synthetic theory is accurate, then that theory as a general proposition is effectively dead' (Gould 1980:120)" (p. 28); (3) that Goldschmidt's ideas "were akin to punctuationism" (p. 28).
Hoffman	*Weak Version* "Almost all evolutionary change should be concentrated in speciation events, whereas the change taking place within particular species, between successive speciation events, is insignificant and virtually nonexistent" (p. 124). "Speciation [is] a necessary condition for any significant evolutionary change of organisms" (p. 124). *Strong Version* "Gradual phenotypic change is almost absent from the evolution of phyletic lineages and . . . periods of complete phenotypic stasis of species are interrupted solely by speciation events" (p. 125). "Proposed that there is a fundamental biological difference between the processes of microevolution and speciation" (p. 127). "Speciation is a macroevolutionary phenomenon caused by a macroevolutionary process, irreducible to the processes of microevolutionary change operating within the framework of a species' environment and biological constitution" (p. 127). *Moderate Version* Origin of new species attributed to microevolution: "the evolutionary history of the majority of species consists mainly of a very long period of evolutionary stasis, when the species remains in homeostasis with its environment and essentially does not evolve" (p. 129).

Table I.1. (continued)

Ruse	*Phase 1 (1972)* Fossil record does not show gradual change; most change at time of speciation; speciation as allopatric; Mayr's "Founder's Principle" as a major causal element; change in species is change in organisms; stressed conventionality of view of natural selection. *Phase 2 (circa 1980)* Deemphasis of organic adaptation; downplaying of natural selection; notion of macromutations; father figure changes from Mayr to Goldschmidt; view that the "synthetic theory of evolution [is] effectively dead" (p. 142). *Phase 3 (1982–)* Macromutation disassociated from punctuated equilibrium; hierarchical view of evolution; possibility of species selection; adaptation "counts" but is only one factor.
Eldredge	Stasis, i.e., long period of little or no change; species selection with species as real individuals.
Gould	"Species almost always arise in small populations isolated at the periphery of parental ranges" (p. 56). "Geologically instantaneous origin and later stasis of species" (p. 66). "The predominant amount of morphological change accumulated in evolutionary trends must be generated in punctuational events of cladogenesis" (p. 74). "Evolutionary trends as a sorting among species, and not as a simple extrapolation of processes occurring within single populations" (pp. 59–60). This led to "species selection" and "species sorting."

both "strong" and "moderate" versions. The basic elements of each are outlined in Table I.1.

Like Hoffman, Ruse discerns three versions or, more precisely, phases. In Phase 1 (1972), the claims are rather modest, emphasizing primarily the pervasiveness of stasis and the importance of relatively rapid speciation events, rather than the gradual, slow changes usually associated with natural selection. Phase 2 (1980–82) is characterized by much more sweeping pronouncements, à la Gould's statement that the synthetic theory of evolution is "effectively dead." Phase 3 (1982 and later), in contrast, reverts to the more modest contentions of the 1972 paper. (Here, again, Table I.1 may be helpful.) But, as Ruse emphasizes, this "history" rests primarily on his analysis of Gould's writings and does not necessarily apply, or applies less directly, to Eldredge. Ruse explores, furthermore, the factors—personal, philosophical, possibly political, some even metaphysical—that may have influenced Gould's scientific thinking. (We share Ruse's regret, and so we think will our readers, that he was unable to provide a similar discussion of Eldredge's ideas.)

Neither Eldredge nor Gould (in their respective contributions to this book) directly addresses the charge that punctuated equilibrium has been a moving target. Aside from rather perfunctory references to stasis and to species selection and species sorting, Eldredge does not attempt to delineate, in any concrete fashion, what he views as the central propositions of punctuated equilibrium, either as originally stated or as (possibly) subsequently modified. That omission is understandable, given the subject of his essay—the implications of punctuated equilibrium for the study of large-scale biological and cultural systems.

Gould, however, explains at length just what he and Eldredge had in mind in 1972 and traces the intellectual journey that led to their now classic paper. That paper, he insists, was completely consistent with the reigning neo-Darwinian synthesis; neither he nor Eldredge then had more than a "groping and tentative" perception of the larger implications of their argument, implications that led eventually to what was widely perceived as an argument for species selection. He confesses that "we made our biggest conceptual mistake and unfortunately sowed much subsequent confusion" (p. 61) by failing to distinguish initially between "species selection" and "species sorting."

It was this confusion, compounded by a conjunction of avid public interest in creationism, circa 1980, plus "egregiously bad and harmful" press accounts (p. 58), which spawned the accusation that punctuated equilibrium contradicted, or was at least incompatible with, the neo-Darwinian synthesis. That charge, Gould insists, was untrue: punctuated equilibrium is not a theory of change via saltation; it does not deny that natural selection might account for speciation in some instances or insist that all change is due to speciation events; and, contrary to the readings of its critics, it consistently holds that natural selection (as distinct from sorting) operates primarily, although not exclusively, on the individual organism. In short, Gould maintains, punctuated equilibrium has remained solidly within the Darwinian tradition, although it may eventually require "an important revision and extension of Darwinism" (p. 62).

Nearly twenty years into the fray, then, there remain at least two levels of controversy among the proponents and opponents of punctuated equilibrium. The first centers on the respective roles of natural selection and punctuated speciation in accounting for evolutionary change. Paralleling or associated with this question is that of the type of evidence—or the kind of investigation (see Gould, herein)—which might satisfactorily resolve the matter.

A second level of controversy relates to "species selection" or, as Gould now says, what should more properly be called "species sorting." Does punctuated equilibrium necessarily imply a "higher order" macroevolutionary process operating on other than the individual organism? If so, what evidence is there to support this contention? And, perhaps most important, is the notion of "hierarchical selection" compatible with the neo-Darwinian synthesis?

As the first half-dozen essays reveal, and as this introduction has sought to explain, our several distinguished biologists, reflecting the diversity of opinion among evolutionary theorists, are unable to agree on a common definition of the constituent elements of punctuated equilibrium. That being the case, it is almost inevitable that we also encounter, in these essays, profoundly different readings of the implications of punctuated equilibrium.

Implications of Punctuated Equilibrium Theory

As mentioned earlier, we wanted to examine two dimensions of this question: What are the implications of punctuated equilibrium for evolutionary theory, that is, for the neo-Darwinian synthesis, and what are its implications for the social sciences?

With regard to the implications for evolutionary theory, we encounter practically polar differences of opinion. As might be expected, these correlate rather closely with assessments of punctuated equilibrium's scientific merits.

The most severe judgment is that of Antoni Hoffman, who identifies three versions of the theory—weak, strong, and moderate. Right or wrong, Hoffman nonetheless elevates brevity to an art form as he characterizes these, in order, as "entirely trivial," "blatantly false," and "empirically untestable."

Against Hoffman's views we can juxtapose Stanley's somewhat cautious conclusion, manifestly based on quite another interpretation of punctuated equilibrium, that "in calling for rapidly divergent speciation, the punctuational model is not demanding the impossible" (p. 100).

We get a much more positive reading from Ruse who, writing as a philosopher and a historian of science, asks, "Is the theory of punctuated equilibria a new paradigm?" Carefully distinguishing among the several meanings of "paradigm" (sociological, psychological, epistemological, and ontological), Ruse acknowledges that his earlier negative judgment was mistaken. In the psychological and epistemological senses of the term, the latter two phases of punctuated equilibrium do constitute a new paradigm—or very nearly so. Biologists will now have to choose, Ruse predicts, between two "metaphors" or "traditions."

On this issue, Gould steers a course between Ruse's "new paradigm" and Hoffman's brusque, near total dismissal—a dismissal tempered by Hoffman's judgment that, as a heuristic tool, punctuated equilibrium has been successful. According to Gould, punctuated equilibrium was from its beginning, and has remained, within the mainstream of neo-Darwinian evolutionary thought. But Gould is also confident that punctuated equilibrium will have a significant impact on the future course of that thought. We can do no better than quote his two-sentence assessment:

> We may say at least that punctuated equilibrium directed the attention of macroevolutionists to the causally distinct status of species and that this insight helped to lead the larger discipline of evolutionary biology toward the expanded Darwinism of hierarchy theory. I think that I speak for Niles Eldredge as well in stating that we view the elaboration of punctuated equilibrium into a macroevolutionary basis for hierarchy theory as its major potential contribution, although we certainly take pride in the achievement of our original goal—the removal of the gradualist filter from paleontological perspectives and the recognition of punctuated equilibrium as a real and basic empirical pattern in life's history, not a feature of the record's imperfection, worthy only of our inattention or embarrassment. (Pp. 68–69.)

Nonetheless, Gould concludes, important as are these new emphases and insights, they represent "an expansion, not a refutation of Darwinism" (pp. 67–68).

Ernst Mayr ends his essay with a section titled "The Contributions of Punctuationism to Evolutionary Theory," and his balance sheet is not too unlike Gould's. Above all, says Mayr, punctuationism has "shown how one-sided has been the myopic focusing of paleontologists and population geneticists on the one-dimensional, transformational, upward movement of evolution" (p. 48). The punctuationists, and the punctuated equilibrium debate, have led to a belated general realization that variational evolution takes place not only at the level of the individual and of the population but at the level of species as well. In short, he concludes that "speciational evolution is Darwinian evolution at a higher hierarchical level. The importance of this insight can hardly be exaggerated" (p. 48).

We have held Eldredge's appraisal to the last because, given his interest in large-scale cultural, as well as biotic, organisms, we also deal with his views in our later discussion of the implications for large social systems. For Eldredge, too, the main legacy of punctuated equilibrium is that "large-scale, interactive biotic entities of which organisms are parts do in fact exist; that their natures can be specified; and that their causal interactions can thus be elucidated" (p. 104).

Thus, there has been much debate among students of evolution. However, there may be implications of this debate for the social and behavioral sciences, too. Social scientists have become increasingly sensitive to the role of biological factors in shaping human behavior, social and individual, and have adopted an "evolutionary perspective" in their efforts to explain that behavior. What, then, are the implications of punctuated equilibrium for the social sciences? That is the question to which the second half of this book is addressed.

There has long been interest in examining human behavior as, in part, a reflection of natural forces. To stretch the point, one can go back to Plato and Aristotle for examples (Peterson, 1976; Somit and Peterson, 1986). Perhaps the major promise of this approach has been that the biological study of *Homo sapiens* can help us to delineate what the essence of human nature might be. This enterprise is critical because any theory of human behavior rests on a vision of the nature of the beast (Somit, 1981). Generally, biology is currently viewed as one of the disciplines most likely to inform the effort to understand better the nature of humans.

Over time, students of human behavior who have been interested in the potential applications of biological theory and research have been adept at sensing changing currents in biology and exploring the implications for human behavior. Sometimes this borrowing has been profitable; at other times, it has not (e.g., see Peterson, 1976; Somit and Peterson, 1986). At this point, we can only say that those scholars interested in the linkage between punctuated equilibrium theory and social phenomena are still in the early stages of examining it. The chapters that explicitly explore the punctuated equilibrium–social behavior nexus should be read and evaluated in this light. These are among the first steps at assessing the complex interplay between evolutionary forces as described by punctuated equilibrium theory and human social behavior.

Individual chapters examine the extent to which this theoretical perspective informs economics, sociology, anthropology, psychology, and political science. One approach to organizing the arguments contained therein is to realize that there are two distinct uses to which punctuational theory can be and has been put—as direct application of the actual theoretical framework and as metaphor. The former is more often the target of criticism in the following pages than the latter.

The direct application of the theory is the point of departure for several of the following chapters, especially those, respectively, by Susan Cachel, Allan Mazur, and Brian A. Gladue. Each author seeks to establish whether "hard-core" punctuationism, as expounded by Gould, Eldredge, Stanley, and others, is directly applicable to a given discipline or subfield (such as physical anthropology, cultural anthropology, archaeology, sociology, and psychobiology). All three address the same question: Is human behavior the result of punctuational evolutionary processes? For diverse reasons, all three conclude that punctuated equilibrium theory has very modest explanatory value. Mazur believes that we do not have the appropriate data base to test predictions from punctuational theory about the evolution of human social behavior; Gladue argues that the time frame involved in evolution is so long that punctuationism is simply not applicable to an understanding of ongoing behavior; and Cachel contends that inherent difficulties with punctuated equilibrium theory make it unlikely that it will have much impact in anthropology.

These chapters stand, obviously, as a challenge to those who would apply punctuated equilibrium—or other evolutionary approaches—directly to the study of human behavior. This is, however, only one of two ways in which the theory can have value for our understanding of behavioral phenomena. Theories as metaphors or models can also be substantial aids for advancing our comprehension of social reality. E. F. Miller has defined metaphor as (1979:156) "all types of transference of words and meaning from one kind of thing to another." Metaphors as models have had an enormous impact and, in this guise, punctuationism may also have considerable heuristic value. As Miller has remarked (1979:162), "Once we see the world from the point of view of a particular metaphor, the face of it is changed. Adopting a new metaphor changes our attitude toward the facts."

Although Miller is speaking of politics in the following passage, the message that he is presenting is much more broadly applicable. He says (1979:169): "Metaphor is necessary to political knowledge precisely because the meaning of reality of the political world transcends what is given to observation. . . . Metaphors can take us beyond the observable and also make manifest the intelligible structure of the unobservable." This appreciation is reflected in the contributions by Eldridge, Kenneth E. Boulding, and, to some extent, Glendon Schubert and Roger D. Masters. Although there are serious and substantial criticisms of the "hard-core" use of punctuational theory, there is little criticism of its metaphoric usage in the chapters to follow, with Gladue as one exception.

There already have been some attempts to apply the metaphor in political science. Chong Phil Ra (1988) has argued, for instance, that political socialization is discontinuous, not a smooth process of gradual learning. At certain points in one's life, one acquires a great deal of political knowledge—and this may lead to a rapid transformation in political views; at other times, one's political values remain stable. All in all, Ra concludes, punctuated equilibrium provides an apt metaphor. Peterson and Somit (1983) have speculated that punctuationism may bear on the age-old question posed by political philosophers about the nature of change. Are desired political changes best accomplished by evolutionary or revolutionary measures? Is revolutionary change inherently bad, as Edmund Burke insisted, because it entails the overthrow of adaptive, stabilizing forces and institutions in society? We suggested very briefly that, as metaphor, punctuationism implies that gradualism is not the only or even the most probable mode of change, and it is perilous to assume a priori that existing social forms are ipso facto adaptive and that those brought about by revolutionary change are not. Finally, Edward G. Carmines and James Stimson (1989) make brief reference to punctuational theory as a device to explain changes in American electoral behavior.

It may well be that this is the manner in which punctuated equilibrium theory renders its greatest contribution to the behavioral sciences. By providing a different metaphor for explaining social phenomena, the theory may assist us in better understanding human behavior in all of its manifestations.

The chapters in Part II point the way to such an understanding.

In a broad-ranging and provocative discussion, Kenneth Boulding examines the relevance of punctuationism not simply for economics but for the study of biological systems and social systems as well. The two have many similarities. The "biosphere" and the "sociosphere" closely interact; both are subject to many of the same regularities and constraints; both evince much the same behavioral patterns and the same "strange mixtures of stability and instability, continuity and discontinuity" (p. 178).

Both evolve, too, by some as yet undetermined mix of mutation and selection. Here punctuated equilibrium becomes important because it calls attention to the possibility that "certain improbable events may produce large consequences very rapidly" (p. 179). This likelihood (Boulding terms it "Browning's Law") may well explain "the extraordinary phenomenon of acceleration in social and economic systems, especially in the last few centuries" (p. 183). It also raises the question "as to whether the phenomenon of punctuation in biological evolution may not show some parallels"—that is, similarly accelerated change (p. 183).

Eldredge's essay in Part I is also concerned with the implications of punctuationism for the understanding of large-scale cultural, as well as biological, systems. Toward this end, he calls attention to the disciplinary parallels between archaeology, based largely in the social sciences, and paleontology, based in biology. Both have traditionally served an essentially historical role. Now, however, punctuated equilibrium has forced a sweeping reconceptualization of the nature of species. "To what extent," he asks, "are these changes in paleontological thinking relevant to a consideration of the histories and evolution of social systems in general and to the study of human history and archaeology in particular?" (p. 111).

Just as biological species can correctly be conceptualized as ontologically real and hierarchically structured so, too, "social systems may profitably be viewed as hierarchically arrayed stable entities—large-scale spatiotemporally bounded entities" (p. 118). When this is done, the "intriguing possibility emerges that such social entities may themselves be susceptible to a form of selection directly analogous to large-scale natural selection" (p. 118). The additional possibility that "lower-level events—the invention, maintenance, modification, and eventual disappearance of both biological and cultural traits . . .—may also be largely determined by interactions and behaviors of large-scale entities [Teggard's downward causal influence] makes the recognition of the importance of such systems all the more critical in the further elaboration of both biological and cultural evolutionary theory" (pp. 118–119). That recognition, Eldredge believes, could transform both eology and paleontology from "historical" to "functional" science.

Susan Cachel looks at punctuated equilibrium from the perspective of anthropology and one of its major subdisciplines, archaeology. Despite Gould's (1984) claim, she believes that it has had little impact on social or cultural anthropology. The most likely influence, she feels, will be on physical anthropology, in which such issues as "rates of evolution, species discrimination in the fossil record, and longevity and stasis of species" will affect nonhuman primate evolutionary studies and paleoanthropology (p. 212). If so, some of the possible consequences might include the virtual abandonment of the study of human adaptability and the effect of natural selection on human evolution. These, conceivably, would be replaced by "description of human variation, typology, and the construction of phylogenies"—a change signifying, in effect "a return to the pre-1950s state of affairs in physical anthropology" (p. 206). Cachel also provides a careful analysis of the theoretical and practical interplay among punctuationism, cladistic systematics, and what Stephen Jay Gould and R. L. Lewontin have called the "adaptationist program"; and a critique of attempts to apply punctuationist concepts to the interpretation of human evolution.

With regard to archaeology, she is more cautious than Eldredge. Punctuationism has begun to affect archaeological interpretations of Pleistocene and Holocene extinctions, but there has been no major change in archaeological thinking.

Anthropologists who study human evolution have been primarily concerned with the changes in physical characteristics during the transition from *Homo erectus* to *Homo sapiens*; sociologists tend to be more interested in possible behavioral changes and especially in evolving social behavior. It is from this latter vantage point that Allan Mazur examines the implications of punctuated equilibrium.

Mazur begins with a cautionary question: "With so much controversy over the place of punctuation in physical evolution, which is relatively well documented, what can be said for social evolution, which is not?" (pp. 222–223). Regrettably, not too much. Both methods of tracing the evolution of human social behavior (the comparison of observable behavior across living species and the use of datable prehistoric cultural artifacts) have serious shortcomings, given our present state of knowledge. Although Mazur's essay is devoted primarily to the application of these two techniques, he repeatedly reminds us of their limitations.

There have been, Mazur tentatively concludes, three great punctuational events in the evolution of human culture and behavior. The first was the abrupt break from the million-year-old Acheulean tradition, which occurred when *Homo eructus* "left the earth" to early *Homo sapiens*. Mazur views the cultural and behavioral changes associated with the subsequent appearance of *Homo sapiens* largely as a preliminary to the second punctuational event—the transition, beginning about ten thousand years ago, from a hunter-gatherer society to an agrarian mode of existence. "In one of the most remarkable coincidences of history," he observes, "this transformation to agrarian civilization occurred not once but in at least six places at nearly the same time: Mesopotamia, Egypt, India, China, Mexico, and Peru" (p. 231). The third great change was, of course, the recent emergence of industrialized society.

But what caused the first two bursts of social innovation, if not the third as well? Here, punctuationism provides little help. The "answers to these questions," Mazur tell us, "will remain at the level of untestable speculation . . . unless some new empirical method that is not yet apparent can be found for probing deeper into social evolution than we can now" (p. 233).

Donald Campbell's 1974 presidential address at the American Psychological Association meeting urged a more Darwinian approach to psychological phenomena. Brian Gladue begins his essay from that perspective, emphasizing the importance of biological concepts and theory for psychobiology. He addresses the issue of whether the debate in evolutionary circles over punctuational change versus gradual change has any relevance for psychologists. Like Mazur, his answer is no—but for different reasons.

First, Gladue contends, the time frame within which even rapid punctuational change takes place is too long to have any practical significance for psychology. "As a species," he observes, "we are relatively undifferentiated from our ancestors, and as a species, it really does not matter whether we got here via gradualism or punctuated equilibrium" (p. 242). Second, he argues, the application of punctuational theory to human behavior has often been at a metaphorical level only. Furthermore, "metaphors, however elegant and intuitively appealing, do not predict in the manner of theories or hypotheses" (p. 244). Gladue concludes that "punctuated equilibrium or gradualism, whichever scheme best explains an expanding collection of data from the fossil record, will have no direct altering effect on advances in the area of real-time proximal psychobiology, the 'how' aspects of explaining the biological mechanisms influencing human behavior" (p. 245).

Two of our contributors, Glendon Schubert and Roger Masters, look at punctuated equilibrium from the perspective of political science. Schubert is particularly concerned with its implications for our approach to the survival of *Homo sapiens*. Having speedily wiped out vast numbers of other species, says Schubert, we ourselves are well on our way to extinction, which we will soon accomplish, even in the absence of some massive extraterrestrial impact, by our inexorable pollution of the environment or via nuclear holocaust, deliberate or accidental. Not unreasonably, Schubert believes that these threats "demand revolutionary political change right now, in the immediate present" (p. 268).

But the traditional political science approach sees "slow, moderate, and gradual change . . . as the expected (and desired) norm" (p. 268) and presumes that rapid change is both abnormal and undesirable. It is here that the practical political consequences of punctuationism and gradualism sharply diverge. To accept "biological gradualist evolutionary theory as the model for human politics," Schubert warns, "is a blueprint for the Doomsday of the human species by reinforcing the politics of certain failure to change either enough or in time" (p. 268).

The adoption by political and social science of catastrophic evolutionary theory would constitute, he grants, a major paradigm shift; nonetheless, that "shift is itself the first step to be taken in trying to do something to postpone the extinction of our species" (p. 275). Why? According to Schubert, "The great advantage of catastrophism in political theory is that it proffers an ideology, premised in the biology of our species, that can be used to legitimize the revolutionary solutions demanded by the catastrophic changes already well advanced in our biospheric niche" (p. 274).

As both Ruse and Gould briefly mention, the advocates and the opponents of punctuationism have accused each other of permitting ideological considerations to color, if not shape, their scientific views. Is there any logical or necessary relationship between the advocacy of rapid or gradual biological change, on one hand, and the advocacy of gradual or rapid political change, on the other? That is the problem Roger Masters poses.

In search of an answer, Masters surveys a spectrum of philosophers who took positions on both issues, focusing especially on Aristotle, Lucretius, Hobbes, Rousseau, and Marx. Some, like Aristotle and Marx, held congruent views about rates of change in biology and in politics; others, such as Lucretius, had no difficulty in combining a belief in discontinuous natural processes with gradualist models of human change or, à la Rousseau, of simultaneously espousing biological gradualism and revolutionary political change. Since all possible combinations of views toward rates of change—both evolutionary and political—can be identified in the Western tradition, Masters concludes that concepts of change in biology are so independent of orientations toward political change that there is no reason to attribute inherent political bias to either model.

To which sensible finding, we suspect, the ideologists on both sides will promptly respond, "Yes, but . . ."

This volume, then, has set ambitious goals. Our hope is that it will clarify the debate over exactly what punctuated equilibrium *qua* theory really says; indicate the nature of the scientific debate over the claims of this theory; examine the implications of punctuationism for evolutionary theory; and lay out the implications for the study of human behavior. Each of these topics generates controversy in the following pages; in the process, we believe, the nature of the issues involved, if not resolved, will at least become much clearer.

References

Carmines, E. G., and J. A. Stimson. 1989. *Issue Evolution*. Princeton: Princeton University Press.

Eldredge, N. 1971. The allopatric model and phylogeny in Paleozoic invertebrates. *Evolution* 25: 156–167.

Eldredge, N., and S. J. Gould. 1972. Punctuated equilibria: an alternative to phyletic gradualism. In T. J. M. Schopf, ed., *Models in Paleobiology*. San Francisco: Freeman Cooper, pp. 82–115.

Gould, S. J. 1984. Toward a vindication of punctuational change. In W. A. Berggren and J. A. Van Couvering, eds., *Catastrophes and Earth History*. Princeton: Princeton University Press, pp. 9–34.

Miller, E. F. 1979. Metaphor and political knowledge. *American Political Science Review* 73:155–170.

Peterson, S. A. 1976. Biopolitics: lessons from history. *Journal of the History of the Behavioral Sciences* 12:354–366.

Peterson, S. A., and A. Somit. 1983. Punctuated equilibria, sociobiology, and politics. Paper presented at Southern Political Science Association, Birmingham, Ala.

Ra, C. P. 1988. An ontogeny of political behavior: a punctuated equilibrium model of political socialization. Ph.D. dissertation, University of Hawaii.

Simpson, G. G. 1944. *Tempo and Mode in Evolution*. New York: Columbia University Press.

Somit, A. 1981. Human nature as the central issue in political philosophy. In E. White, ed., *Sociobiology and Human Nature*. Lexington, Mass.: D. C. Heath, pp. 167–180.

Somit, A., and S. A. Peterson, 1986. Biological correlates of political behavior. In M. Hermann, ed., *Political Psychology*. San Francisco: Jossey-Bass, pp. 11–38.

Part One

Evolutionary Change: The Debate and Its Scientific Implications

1

Speciational Evolution or Punctuated Equilibria

Ernst Mayr

Only recently have we understood how different are the concepts to which the term "evolution" has been attached. With wisdom of hindsight, we can now (250 years after Buffon) distinguish three very different concepts of evolution: saltational evolution, transformational evolution, and variational evolution.

Theories postulating saltational evolution are a necessary consequence of essentialism. If one believes in constant types, only the sudden production of a new type can lead to evolutionary change. That such saltations can occur and indeed that their occurrence is a necessity is an old belief. Almost all of the theories of evolution described by H. F. Osborn (1894) in his *From the Greeks to Darwin* were saltational theories, that is, theories of the sudden origin of new kinds. The Darwinian revolution (Darwin, 1859) did not end this tradition, which continued to flourish in the writings of Thomas H. Huxley, William Bateson, Hugo De Vries, J. C. Willis, Richard Goldschmidt, and Otto Schindewolf. Traces of this idea can even be found in the writings of some of the punctuationists.

According to the concept of transformational evolution, first clearly articulated by Lamarck, evolution consists of the gradual transformation of organisms from one condition of existence to another. Almost invariably, transformation theories assume a progression from "lower to higher" and reflect a belief in cosmic teleology resulting in an inevitable steady movement toward an ultimate goal, an ultimate perfection. In biology all so-called orthogenetic theories, from those of K. E. von

Baer to Osborn, L. S. Berg, and Teilhard de Chardin are in this tradition.

As R. C. Lewontin (1983) has correctly pointed out, Darwin introduced an entirely new concept of evolution: variational evolution. New gene pools are generated in every generation, and evolution takes place because the successful individuals produced by these gene pools give rise to the next generation. Evolution thus is merely contingent on certain processes articulated by Darwin: variation and selection. No longer is a fixed object transformed, as in transformational evolution, but an entirely new start is, so to speak, made in every generation. Evolution is no longer necessarily progressive; it no longer strives toward perfection or any other goal. It is opportunistic, hence unpredictable.

What Darwin did not fully realize is that variational evolution takes place at two hierarchical levels, the level of the deme (population) and the level of species. Variational evolution at the level of the deme is what the geneticist deals with. It is effected by individual selection and leads minimally to the maintenance of fitness of the population through stabilizing selection.

The second level of variational evolution is that of the species. Owing to continuing (mostly peripatric) speciation, there is a steady, highly opportunistic production of new species. Most of them are doomed to rapid extinction, but a few may make evolutionary inventions, such as physiological, ecological, or behavioral innovations that give these species improved competitive potential. In that case they may become the starting point of successful new phyletic lineages and adaptive radiations. Such success is nearly always accompanied by the extinction of some competitor. This process of succession of species is often referred to by the term "species selection," but to prevent misunderstandings it may be better to call it "species turnover" (see below).

The transfer from transformational to variational evolution required a conceptual shift that was only imperfectly carried through by most Darwinians. As a consequence, geneticists described evolution simply as a change in gene frequencies in populations, totally ignoring the fact that evolution consists of the two simultaneous but quite separate phenomena of adaptation and diversification. The latter results from a process of multiplication of species, a process almost totally ignored in the writings of R. A. Fisher, J. B. S. Haldane, Sewall Wright, and other leading evolutionary geneticists.

Transformational thinking likewise continued to dominate paleontology, expressed in the concept of phyletic gradualism. Since most paleontologists were typologist (in an almost Platonian sense), they subconsciously assumed that species were everywhere the same and, thus, at any given time essentially uniform. Speciation consisted of the gradual transformation of such species in geological time. Since the gradualness of such phyletic transformation could be documented in the geological record only in the rarest cases, it was postulated that the absence of intermediates was a consequence of the notorious incompleteness of the fossil record. The so-called evolutionary species definition adopted by most paleontologists (Simpson, 1961; Willmann, 1985) reflects the same focus on the vertical (i.e., time) dimension. If adopted, it leaves only two options: speciation is explained either by gradual phyletic evolution, with the gaps between species being due to the deficiency of the fossil record, or by sympatric saltational speciation. Indeed, most paleontologists adopted both options. Acceptance of phyletic gradualism does not require the acceptance of a constant rate of evolutionary change. The rate may accelerate or slow down, but change leads inexorably to the steady transformation of a lineage.

Even Darwin, for reasons that relate to his struggle against creationism, stressed the transformational aspect of evolution. He was, however, fully aware of highly different rates of evolution, from complete stasis to rates of change so fast that intermediates could not be discovered in the fossil record (Gingerich, 1984; Rhodes, 1983; and others). Owing to his adoption of sympatric speciation, however, Darwin never needed to consider the geographical component in his theorizing. When he said that a new species might originate as a local variety, he did not claim that it was an isolated population. It seems to me that for Darwin the pulsing of evolutionary rates was a strictly vertical phenomenon.

The geneticists, with the exception of a few saltationists such as DeVries and Bateson, usually ignored the problem of speciation altogether. The only geneticists who showed an interest in the multiplication of species were those who had been educated as taxonomists, like Theodosius Dobzhansky and G. L. Stebbins. The problem of relating speciation to macroevolution occupied primarily three zoologists, Julian Huxley (1942), Mayr (1942, 1954), and Bernhard Rensch (1947), who were neither geneticists nor paleontologists. Since these three were among the architects of the evolutionary synthesis, one can state that the problem of the relation between speciation and macroevolution was not entirely ignored by the evolutionary synthesis.

The widespread neglect of the role of speciation in macroevolution continued until Niles Eldredge and Stephen Jay Gould (1972) proposed their theory of punctuated equilibria. Whether one accepts this theory, rejects it, or greatly modifies it, there can be no doubt that it had a major impact on paleontology and evolutionary biology.

The gist of the theory was that "significant evolutionary change arises in coincidence with events of branching speciation, and not primarily through the transformation of lineages" (Gould, 1982a:83, 1983). The contrast between the previously dominant view of evolutionary change was as follows. Traditionally, evolution had been seen as a single-phase phenomenon of gradual change, albeit sometimes more slowly, sometimes more rapidly. Now evolution was seen as an alternation between speciation events during which the major evolutionary (particularly morphological) change occurred and lengthy periods of stasis.

Historical studies have since shown that the term "punctuated equilibrium" was more novel than the concept. A role for peripheral populations in speciation was already postulated by L. v. Buch (1825) and fully substantiated by Darwin for the Galapagos mockingbirds. Unfortunately, by the time Darwin published the *Origin* (1859), he had adopted sympatric speciation (Mayr, 1982a). When he said that a new species might originate as a local variety, he did not necessarily mean an isolated population. Nor are the changes in the rate of evolution to which Darwin refers brought in relation to speciation.

Before going any further in the analysis of the literature, it is important to call attention to a prevailing confusion between two distinct evolutionary phenomena, gradualism and uniformity of evolutionary rate. Darwin emphasized gradualism (Rhodes, 1983), but, as I shall show, even that term is ambiguous, allowing for two very different interpretations. What Darwin did not insist upon was a uniformity of rates (Huxley, 1982; Penny, 1983; Rhodes, 1983). The existence of so-called living fossils was known to paleontologists early in the nineteenth century, and the occurrence of different rates of evolution in different phyletic lines was paleontological dogma already in Darwin's lifetime. George Gaylord Simpson (1953) analyzed this phenomenon in great detail and even introduced a special terminology to characterize lineages with average, very rapid, and extremely slow evolutionary rates. Eldredge and Gould never claimed to have discovered this difference in rates, and the part of the ensuing polemic stressing these differences is therefore irrelevant for the evaluation of the punctuation theory.

It is not always easy to interpret Darwin's statement (Rhodes, 1983) because isolation (at least during the process of speciation) had become unimportant for him owing to his adoption of sympatric speciation. I have been unable to discover in Darwin's writings any connection between allopatric speciation and change of evolutionary rate. Gould and Eldredge correctly state that Simpson likewise failed to make such a connection. His quantum evolution was a vertical (temporal) phenomenon, as it had to be considering his evolutionary species definition (Simpson, 1944:207–217).

Paleontologists knew that new species may originate in a very circumscribed area and turn up in the fossil record only after having spread more widely (Bernard, 1895). This insight was made use of, however, only in stratigraphic research and not in studies of macroevolution. On the contrary, the importance of peripatric speciation was minimized after Fisher (1930) and Wright (1931, 1932) had asserted, although for different reasons, that evolution was most rapid in populous, widespread species, a conclusion adopted also by Dobzhansky (1937, 1951) and by most evolutionists before the 1970s.

I believe I was the first author to develop a detailed model of the connection between speciation, evolutionary rates, and macroevolution (Mayr, 1954). Although long ignored, my new theory of the importance of peripatric speciation in macroevolution is now widely recognized. "Mayr's hypothesis of peripheral isolates and genetic revolution must of necessity be a centerpiece of the punctuated equilibria theory; it is the theory, for all practical purposes" (Levinton, 1983:113). I once more presented my theory in great detail (Mayr, 1963:527–555). Under these circumstances it is most curious that the theory was completely ignored by paleontologists until brought to light by Eldredge and Gould (1972).

The major novelty of my theory was its claim that the most rapid evolutionary change does not occur in widespread, populous species, as claimed by most geneticists, but in small founder populations. This conclusion was based on empirical observations gathered during my studies of the speciation of island birds in the New Guinea region and the Pacific. I had found again and again that the most aberrant population of a species—often having reached species rank, and occasionally classified even as a separate genus—occurred at a peripheral location, indeed usually at the most isolated peripheral location. Living in an entirely different physical as well as biotic environment, such a population would have unique opportunities to enter new niches and to select novel adaptive pathways.

As I pointed out elsewhere (Mayr, 1982b), my conclusion was that a drastic reorganization of the gene pool is far more easily accomplished in a small founder population than in any other kind of population.

Indeed, I was unable to find any evidence whatsoever of the occurrence of a drastic evolutionary acceleration and genetic reconstruction in widespread, populous species.

In view of frequent recent misrepresentations of my 1954 theory I must emphasize also what I did not claim (see also Mayr, 1982b):

1. I did not claim that every founder population speciates. In the vast majority, only minor genetic reorganizations occur, and the majority of such founder populations soon become extinct or merge with the parental species.
2. I did not claim that every genetic change in a founder population is a genetic revolution. Evidently it requires a special constellation for the occurrence of a more drastic genetic reorganization. All I claimed was that when a drastic change occurs, it occurs in a relatively small and isolated population.
3. I did not claim that speciation occurs only in founder populations. Finally, I nowhere claimed that I chose the name "peripatric" (Mayr, 1982c) because the founders came from the periphery of the parental range. I chose that name because the founder populations were at peripheral locations. My interpretation throughout was very pluralistic and was naturally misunderstood in an age when singular, deterministic solutions were strongly preferred.

In 1954 I was already fully aware of the macroevolutionary consequences of my theory, saying that "rapidly evolving peripherally isolated populations may be the place of origin of many evolutionary novelties. Their isolation and comparatively small size may explain phenomena of rapid evolution and lack of documentation in the fossil record, hitherto puzzling to the palaeontologist" (p. 179).

I later supplemented my theory by pointing out (Mayr, 1982c) that peripatric speciation may occur not only in founder populations but also in any population going through a severe bottleneck such as refuge populations during Pleistocene glaciations (Haffer, 1974).

The first to pick up my theory was Eldredge (1971), who found in his study of Paleozoic trilobites that the majority of species showed no change in species-specific characters throughout the interval of their stratigraphic occurrence, whereas new species appear quite suddenly in the strata. He therefore proposed that the allopatric model be substituted in the minds of palaeontologists for phyletic transformation as the dominant mechanism of the origin of new species in the fossil record. This was followed in 1972 by the Eldredge and Gould paper, in which the term "punctuated equilibrium" was proposed. The Eldredge-Gould proposal was essentially my 1954 theory, except for a far stronger emphasis on stasis, indeed a belief that no further evolutionary change would occur after the speciation process was completed.

Questions and Objections

A modest theory of punctuationism is so strongly supported by facts and fits, on the whole, so well into the conceptual framework of Darwinism, that one is rather surprised at the hostility with which it was attacked. The controversy over punctuationism is, by now, more than twenty years old, and it is possible to distinguish different classes of objections.

There are questions that deal with the core ideas of the theory: What is stasis? How can one account for it? Do all species experience stasis? Is all evolutionary change restricted to bouts of speciation? If so, why? What are the genetic aspects of speciation? These and other questions will be analyzed in the second part of this essay.

But not all the objections raised against punctuationism deal with these core ideas. Others were raised against rather specific claims made by Eldredge, Gould, or both, or against the way they treated their evidence. It will be helpful to deal with these objections first. They relate largely to claims that are not part of a punctuationist theory of evolution. To deal with them separately and to test them for their validity will clear the field for a subsequent testing of the core ideas of punctuationism.

Four aspects of the treatment of punctuationism by Gould and Eldredge were objected to most frequently.

First to receive attention was the seemingly monolithic nature of the claims. Even though Eldredge and Gould (1972) nowhere stated that a neospecies enters a period of total stasis, this is what all their graphic presentations suggested (figs. 5-4, 5-8, 5-10). Furthermore, evolutionary trends are explained by a process of species selection of completely static species (fig. 5-10). Not surprisingly, their opponents assumed that Eldredge and Gould had postulated total stasis for all species after they had completed the process of speciation.

Professor Gould assures me that they had never adopted such an extreme position, and in their next paper they stated emphatically: "We never claimed that gradualism could not occur in theory or did not occur in fact. . . . The fundamental question is not 'whether at all,' but how often" (Gould and Eldredge, 1977:119). In their abstract, Gould and Eldredge specify: "Most species, during their geological history, either do not change in any appreciable way, or else they fluctuate mildly, with no apparent direction. Phyletic gradualism is very rare"(p. 115).

The second point of contention was the claim of novelty. Nothing incensed some evolutionists more than the claims made by Gould and associates that they had been the first to have discovered, or at least to have for the first time properly emphasized, various evolutionary phenomena already widely accepted in the evolutionary literature. G. L. Stebbins and F. J. Ayala (1981), Verne Grant (1982, 1983), and J. S. Levinton (1983) were fully justified in rejecting these claims of novelty. In particular, they showed that an insistence on gradualism by Darwin and his followers was a denial of saltationism but not a denial of different and changing rates of evolution.

Third, vigorous objection was raised to the claim that punctuationism would require a revision of Darwin's "evolutionary synthesis." "I have been reluctant to admit it, but if Mayr's (1963:586) characterization of the synthetic theory is accurate, then that theory as a general proposition is effectively dead" (Gould, 1980:120). The gist of my statement to which Gould refers was that, contrary to Goldschmidt and Schindewolf, nothing happens in macroevolution that does not happen in populations. What Gould actually attacks, and rightly so, is the completely reductionist characterization of evolution by the mathematical population geneticists. To equate these reductionist views with the theories of the evolutionary synthesis is unjustified, however, as I pointed out in a critical review of similar statements published by M. W. Ho and P. T. Saunders (Mayr, 1984b). A rejection of the axiom of most population geneticists, "Evolution is a change of gene frequencies," is not a rejection of the evolutionary synthesis. The theory of the synthesis is much broader and constitutes in many respects a return to a more genuine Darwinism. The events that take place during peripatric speciation, no matter how rapid they may be, are completely consistent with Darwinism.

Curiously, some authors also mistakenly assumed that the occurrence of stasis would refute Darwinism. Teleological thinking requires continuous evolutionary change, but Darwin rejected teleology (Mayr, 1984a) and accepted stasis (Rhodes, 1983). An evolutionary lineage may continue to vary genetically without undergoing any major reconstruction. Alternatively, a stable lineage may continue to send out founder populations, some of which, through peripatric speciation, could become more or less distinct daughter species.

The fourth reason why punctuationism faced so much opposition is that at one stage Gould pleaded for a revival of Goldschmidt's ideas and implied that they were akin to punctuationism. This claim clearly indicated that there was considerable conceptual confusion as to what punctuated equilibria really means. Before the possibility that Goldschmidt was a forerunner of punctuationism can be discussed constructively, it is necessary to discriminate among four interpretations of punctuationism:

1. An evolutionary novelty originates by a systemic mutation: the individual produced by such a mutation is the representation of a new species or higher taxon.
2. Evolutionary change is populational, but all substantial evolutionary changes take place during bouts of speciation. As soon as the process of speciation is completed, the new species stagnates ("stasis") and is unable to change in any significant way. Early statements by Eldredge and Gould (1972) and Gould and Eldredge (1977) gave the impression that this was their interpretation.
3. Phyletic lineages ("evolutionary species") can evolve slowly and gradually into different species and even genera, but the more pronounced evolutionary changes and adaptive shifts take place during speciational bouts in isolated populations. This has been all along my own interpretation (Mayr, 1954, 1982b) and is presumably that of many evolutionists familiar with geographic speciation.
4. A multiplication of species (the branching of lineages) occurs but is of no greater evolutionary importance than changes within lineages. In fact, phyletic gradualism is responsible for most evolutionary change. It was this view, held by the majority of paleontologists, that induced Eldredge and Gould (1972) to propose their theory of punctuated equilibria.

Only the first one of these four theories conflicts with Darwinism. It was Goldschmidt's theory, and because Goldschmidt has often been cited in connection with punctuationism, it is necessary to discuss his ideas in more detail.

To strengthen the punctuationism case, Gould cited Goldschmidt's views on macroevolution, indicating that "during this decade Goldschmidt will be largely vindicated in the world of evolutionary biology" (Gould, 1980:186). Goldschmidt had claimed that the differences among subspecies, and more broadly all geographic variation, was caused by minimal genetic changes, mutations of alleles, mostly being selected merely for climatic adaptation. Such changes would not permit any transgression of the ancestral type. Any genuine evolutionary novelty was due to the origin of a "hopeful monster," caused by a systemic mutation. This thesis followed from Goldschmidt's rather eccentric conception of the nature of chromosomes and of the genotype. According to him, a systemic mutation is a complete change of the primary pattern or reaction system into a new one and has the capacity to produce a strikingly different new individual that could serve as the founding ancestor of a new type of organism. As J. Maynard Smith (1983:276) pointed out, hopeful monsters, by contrast, are drastically altered phenotypes. They are possible, at least in theory, and it should be possible to discover empirically how often they occur and how often (if ever) they are selectively superior.

It entirely misrepresents Goldschmidt's theory to claim that Goldschmidt "argued that speciation is a rapid event produced by large genetic changes (systemic mutations) in small populations" (Gould and Calloway, 1980:394). The whole concept of populations was alien to his thinking. According to him, a new type is produced by a single systemic mutation producing a unique individual. Gould (1982) is also wrong in claiming that Goldschmidt never had the view "that new species arise all at once, fully formed, by a fortunate macromutation." Actually, this is what Goldschmidt repeatedly claimed. For instance, he cited with approval Schindewolf's suggestion that the first bird hatched out of a reptilian egg, and he was even clearer on this point in a later paper (1952:91–92) than in his 1940 book.

In refutation of Goldschmidt's claims I demonstrated (Mayr, 1942) that geographic variation in isolated populations could indeed account for evolutionary innovations. Such populations have a very different evolutionary potential than contiguously distributed, clinally varying populations in a continental species. As I stated (Mayr, 1954), and have reiterated (Mayr, 1963, 1982b), one can defend a moderate form of punctuationism, based on strictly empirical evidence, without having to adopt Goldschmidt's theory of systemic mutations.

Some Basic Questions about Punctuationism

The theory of punctuationism, to repeat, consists of two basic claims: that most or all evolutionary change occurs during speciation events, and that most species usually enter a phase of total stasis after the end of the speciation process. The two claims are to some extent two separate theories.

The controversy that followed the proposal of this theory revealed that there are considerable conceptual and evidential difficulties in either substantiating or refuting this theory. First, the nature of the fossil record makes it exceedingly difficult, if not impossible, to obtain irrefutable evidence either for stasis or for a very short time span speciation.

Second, throughout the controversy one encounters considerable terminological vagueness and equivocation, as for instance concerning the meaning of such words as "gradual," "stasis," "speciation," and "species selection." A careful analysis of the terms most frequently used in the punctuationism controversy is therefore indispensable.

Gradualness

Whether evolution is gradual became the focus of a heated controversy in the punctuationist argument. Darwin (1859), as everyone knew, had frequently emphasized the gradual nature of evolutionary change (pp. 71, 189, 480), largely because of his opposition to two ideologies dominant in his time, creationism and essentialism. After these ideologies lost their power, it was no longer necessary to be so single-mindedly opposed to the occurrence of discontinuities. Yet the recent controversy concerning the saltational versus gradual origin of evolutionary novelty revealed an equivocation.

Most modern authors failed to distinguish between two very different phenomena: the production of a new taxon, and the production of a new phenotype. If the production of a new taxon is gradual, it is taxic gradualism; if it is instantaneous, it is taxic saltation. Likewise, one can distinguish phenotypic gradualness and phenotypic saltation. What Darwin mostly argued against was the thesis that evolutionary novelties could originate through taxic saltation, that is, through the production of a single individual representing a new type, a new taxon. Instead, he proposed that all evolutionary innovation is effected through the gradual transformation of populations.

This distinction became important after Goldschmidt revived the essentialistic idea that a new higher taxon could be established as the product of a single systemic mutation. Even though the success of such a taxic saltation is too improbable to be endorsed by a contemporary evolutionist, it still leaves the possibility of the occurrence of phenotypic saltations. If a mutation with a drastic phenotypic change could be incorporated in a population and become part of a viable phenotypic polymorphism, it could lead to a seemingly saltational evolutionary change. Gould (1980:127) indeed envisages a "potential saltational origin for the essential features of key adaptations. Why may we not imagine that gill arch bones of an ancestral agnathan moved forward in one step to surround the mouth and form proto-jaws?" Maynard Smith (1983:276) points out that the occurrence of "genetic mutations of large phenotypic effect is not incompatible with Darwinism." Steven M. Stanley (1982) has argued quite persuasively that gastropod torsion might have originated through a single mutation. It would have had to pass through a stage of polymorphism until the new gene had reached fixation. Evidently such a process is feasible, but its importance in evolution is contradicted by the fact that, among the millions of existing populations and species, mutations with large phenotypic effects would have to be exceedingly frequent to permit the survival of the occasional hopeful monster among the thousands of hopeless ones. But this in not found. Furthermore, enough mechanisms for the gradual acquisition of evolutionary novelties are known (Mayr, 1960) to make the occurrence of drastic mutations dispensable, at least as a normal evolutionary process.

The argument, thus, is not whether phenotypic saltations are possible, but rather whether evolution advances through the production of individuals representing new types or through the rapid transformation of populations. No matter how rapid, such a populational "saltation" is nevertheless Darwinian gradualism.

Stasis

Of all the claims made in the punctuationist theory of Eldredge and Gould, the one that encountered the greatest opposition was that of "pronounced stasis as the usual fate of most species," after having completed the phase of origination (Gould, 1982a:86). Yet it was this very claim which the authors designated as their most important contribution.

The extraordinary longevity of the so-called living fossils had, of course, been known since the early days of paleontology (Eldredge and Stanley, 1984; de Ricqles, 1983). But is such stasis the usual fate of most species? Evidence supporting this claim can be found in Stanley's book (1979), some review papers (e.g., Levinton, 1983; Gould, 1982b), and recent volumes of *Paleobiology, Systematic Zoology*, and other journals. Yet the literature also reports numerous cases of seeming speciation by phyletic gradualism (e.g., Van Valen, 1982:99–112). Perhaps most convincing are the cases of significant evolutionary transformation in continuous phyletic lineages reported by K. D. Rose and T. M. Brown (1984) for Eocene primates and by J. Chaline and B. Laurin (1986) for Pliocene rodents. Such phyletic speciation seems to be more frequent in terrestrial than in marine organisms.

Two objections have been raised against the seeming cases of phyletic speciation. First, hiatuses and depositional breaks seem to occur even in the most complete sequences; second, the so-called species of these sequences may not be valid species because they usually differ only in minor characters of size and proportions. Be that as it may, Gould has recently seemed to concede that speciation by phyletic gradualism does occur.

I agree with Gould that the frequency of stasis in fossil species revealed by the recent analysis was unexpected by most evolutionary biologists. Admittedly, stasis is measured in terms of morphological difference, and the possibility cannot be excluded that biological sibling species evolved without this being reflected in the morphotype. Let us tentatively assume that some species enter complete stasis while others evolve by phyletic gradualism.

The question of what percentage of new species adopts one or the other of these two options cannot be resolved either by genetic theory or through the study of living species. It can be decided only though an analysis of the paleontological evidence, and this poses great methodological difficulties (Levinton and Simon, 1980; Schopf, 1982). For instance, in the analysis of the benthic foraminifera, the calculated average age of 20 million years was based on only 15 percent of the recent species. For all the others the fossil record was too spotty to permit any determinations. In other words, the proof of stasis was based on a highly biased sample, consisting of common widespread species, which one could expect to have longevity and which comprised a small minority of the entire fauna. It is conceivable that a considerable fraction of the remaining 85 percent underwent rapid phyletic speciation and thus became unavailable for analysis. The indications are that the vast majority of the so-called rare species are short-lived, probably not for reasons of rapid phyletic change but rather owing to extinction. The best one can do under the circumstances is to adopt an intermediate position by admitting the occurrence of some gradual phyletic speciation but pointing also to the unexpectedly large number of cases in which fossil species showed no morphological change over many millions of years.

Recent discoveries in molecular biology have raised questions about the meaning of stasis. The stasis found in morphological characters in such old genera as *Rana, Bufo, Plethodon*, or even *Drosophila* is not at all owing to the retention of an entirely unchanged genotype. Through the electrophoresis method, countless changes in quasi-neutral enzyme genes have been discovered, but numerous other nonmorphological changes have also taken place in these genera, such as the acquisition of new isolating mechanisms, as well as of numerous adaptations to changing environments. What has remained stable, however, is the morphotype, the basic *Bauplan*. The species in some lineages that can be inferred to have separated 30 to 60 million years ago are morphologically still almost indistinguishable except in size, coloration, and minor differences in skeletal dimensions. This is just one more piece of evidence for the interesting phenomenon that organisms do not evolve as harmonious types but that different characters may evolve at highly different rates. *Archaeopteryx*, with its incongruous mixture of reptilian and avian characters, illustrates this well, as does the peculiar mixture of similarities and differences revealed by a comparison of human and chimpanzee.

The discovery of highly unequal rates of evolution of different components of the genotype does not, however, eliminate the problem of stasis. Why do certain components of the genotype and phenotype remain so stable for millions or tens of millions of years?

This is a particularly puzzling question because there is an almost inexhaustible source of variation. Not only is there a steady rate of mutation at all gene loci, but various phenomena have been discovered by molecular biology in recent years that would seem to lead almost inevitably to a frequent revamping of the genotype. For instance, merely isolating stocks of *Drosophila* in laboratory populations may lead to all sorts of mutual incompatibilities. Transposons and other genetic elements may change the mutability of adjacent loci, and a host of other molecular phenomena would seem to encourage genetic changes of evolutionary significance. Furthermore, selection pressures in an incessantly changing physical and biotical environment would seem inducive to continuing evolutionary change. That we encounter the stability found in the living fossils, and to a lesser extent in the majority of species and genera, is very puzzling indeed.

On the whole, three explanations have been advanced to explain stasis. Reductionist geneticists attribute stasis entirely to stabilizing (normalizing) selection. All mutants or recombinants that deviate from the norm are eliminated by natural selection. This is, of course, no explanation because abundant normalizing selection also takes place in rapidly evolving lines. Obviously, all zygotes with lowered viability are apt to be eliminated in any population either before birth or at least before reproduction. Such selection under the name of "elimination of degenerations of the type" was acknowledged by essentialists long before the establishment of evolutionism (Mayr, 1982a:488). The large mortality of zygotes both in static and in rapidly evolving species is evidence for such elimination. The "internal selection" of certain authors is largely such elimination, as pointed out by J. Remane (1983).

This, however, does not explain why, in spite of the universality of such normalizing selection, certain lineages evolve rapidly while others remain in total stasis. Nor does invoking "normalizing selection" make a distinction between the elimination of new deleterious mutations and that of deleterious recombinants, particularly those interfering with developmental constraints. It is misleading to say that stasis is caused either because deviants were weeded out or because the developmental system prevented them from arising. Even in the latter case it is normalizing selection through which the developmental constraints operate.

Another explanation is that species in stasis had reached optimal adaptation and were no longer answering any directional selection. This is improbable for two reasons. First, selection virtually never succeeds in completely optimizing a genotype; hence a change in the degree of optimality should be recognizable during millions of years. Furthermore, the environment is known to have changed considerably in periods during which certain species displayed complete stasis. Why is this environmental change not reflected in an evolutionary response by these species?

Considering that both of these explanations are unsatisfactory, one must ask whether there is any other possible explanation. Yes, provided one gives up the atomistic view that each gene is independent both in its actions and in the effects of selection on it. If one adopts a more holistic (integrative) view of the genotype and assumes that the genes perform as teams and that large numbers of other genes form the "genetic milieu" (Chetverikow, 1926) of any given gene, one can suggest an explanation. It is that epistatic interactions form a powerful constraint on the response of the genotype to selection. Such epistatic interactions were already dimly appreciated by Darwin (1859:11, 146; 1868, vol. 2:319–335) and have since been stressed by S. S. Chetverikov (1926) (genetic milieu), M. Lerner (1954) (genetic homeostasis), K. Mather (1953) (internal balance), and Mayr (1963, 1970, 1975, 1982b) (cohesion of the genotype).

Holists claim therefore that much of macroevolution cannot be explained by atomistic gene replacements or by selection pressures on single genes, but only by a more drastic reorganization, made possible by loosening the tight genetic cohesion of the genotype found throughout widespread populous species. Mayr, H. L. Carson (1975), Eldredge and

Gould, and Stanley ascribe the stability of the phenotype, as observed in static species, to such an internal cohesion of the genotype, or parts of it. Significant evolutionary advance can take place only after a breaking up of previously existing epistatic balances.

There is, of course, no conflict whatsoever between this holistic view and Darwinism because the cohesive domains of the genotype must have come into existence through natural selection. Unfortunately, current genetic techniques seem unable to analyze such cases of restructuring. Until such techniques become available—probably in the not too distant future—it is impossible to prove conclusively the existence of such genotypic domains and a general cohesion of the genotype, or at least of parts of it.

There are numerous aspects of geographic variation that make sense only if one accepts the notion of a cohesiveness of the genotype. For instance, how else can one explain the pattern of geographic variation of the superspecies *Tanysiptera galatea* (Mayr, 1954)? *T. galatea* is distributed over all of New Guinea, where it is adapted in the northwest of the island to a purely tropical wet climate without any seasons and in the southeast to a trade-wind climate with a nine-month dry season. One would expect that two extremely different phenotypes of this species would have evolved at the two ends of New Guinea in response to two drastically different climatic selection pressures. Actually there is only minimal geographic variation on the mainland in contrast to a series of strikingly different species that have budded off *T. galatea* on nearby islands. What else could have been responsible for the unexpected stasis on the mainland, except some process akin to genetic homeostasis?

It has been claimed that this holistic view of the genotype was not within the spirit of modern Darwinism. This is erroneous. The atomistic viewpoint was defended only by mathematical geneticists from R. A. Fisher to B. Charlesworth and R. Lande. The more holistic viewpoint was promoted by numerous Darwinians from Chetverikow to Mather, Lerner, and Bruce Wallace, and was vigorously promoted by me from 1950 on. It was a strong tradition in evolutionary biology long before the introduction of the theory of punctuated equilibria.

SPECIATIONAL EVOLUTION

One of the differences between the punctuationists and the defenders of phyletic gradualism is that for the gradualists speciation is the continuing change of a phyletic series until it has become a different species (the total number of species remaining constant), while for the punctuationists speciation means a multiplication of species. If this multiplication is effected by the establishment of a founder population and its rapid genetic reconstruction, then such speciation can occur in a relatively short span of time. It is not sudden like saltational speciation, but it may be "instantaneous" as far as the fossil record is concerned. Even if it takes hundreds or thousands of years, the paleontological analysis would record it as a sudden event.

Many readers of the early punctuationist papers gained the impression that punctuationist speciation was an instantaneous event. Gould (1982a) pointed out, however, that for a geologist "instantaneous" means something very different from what it means to a student of living biota. Instantaneous is to be "defined as one per cent or less of later existence in stasis" (Gould, 1982a:84). Hence one hundred thousand years would be instantaneous for a species experiencing a 10-million-year stasis. The semantic problem is evident when we consider that all population evolution—that is, all evolution we are concerned with—is gradual. It is obvious from the recent controversy that the chronology of speciation events cannot be established by paleontological analysis. Rather, it will have to be inferred from an analysis of currently living speciating species, as I have repeatedly attempted to do (e.g., Mayr, 1963). In freshwater fishes it may take less than four thousand years (Mayr, 1984c).

That peripatric speciation is by far the most common mode of speciation is indicated not only by the pattern of distribution of incipient recent species but also by the frequency by which new species, apparently having originated somewhere else, suddenly appear in the fossil record. Such cases are reported in almost every revision of a fossil genus. For instance, in a Tertiary genus of bryozoans in at least seven cases the ancestral species persisted after having given rise to a descendant species (Cheetham, 1986).

The Genetics of Speciational Evolution

The crucial question one must ask is how peripatric speciation differs in its genetic impact from gradual phyletic speciation. Are the genetics of the two processes truly different? Does peripatric speciation speed up evolution? Is the evolutionary pulsing provided by peripatric speciation necessary for the origin of evolutionary innovations? Honesty demands that we admit a lack of concrete knowledge that would permit us to answer these questions. All that we can do at present is to hypothesize; and in that respect we have not made much progress since 1954.

The genetic interpretation I gave in 1954 (see also Mayr, 1982b) was based on the genetic views of that period. I was much impressed by the findings of Mather, Lerner, and Wallace, largely supported also by Dobzhansky, on the genetic homeostasis of the genotype and the constraints on evolutionary departures imposed by this internal balance. I postulated, therefore, that certain events in the founder population might help loosen this cohesion and liberate the founder population from the straitjacket imposed on it by the epistatic balances of its genotype. I designated such an event a genetic revolution. Curiously, several authors reporting on my papers have claimed that I had postulated macromutations or "thousands of mutations," whereas I had invoked not even a single mutation. Rather, my interpretation was based on a developmental point of view that apparently was incompatible with then current thinking. Developmental considerations were at that time ignored by most evolutionary geneticists. To understand the mechanism I proposed, one must remember that natural selection is a two-step process. My theory dealt exclusively with the first step, the generation of variation. It suggested how additional variation might be made available in a founder population. The only possibility I could see on the basis of then available genetic theory was a change of epistatic balances. Thirty-five years later, such a change is still a conceivable process and is probably involved in some cases of peripatric speciation. Yet, many different kinds of DNA have been discovered which might control a drastically speeded-up genetic reorganization in small populations, also effecting "genetic revolutions." But, these new discoveries do not weaken the basic message of my theory, the role of small founder populations in loosening up genetic cohesion and thus facilitating evolutionary change. A somewhat different version has been suggested by H. L. Carson and A. R. Templeton (1984), likewise based on the concept that recombinational rather than mutational events are the decisive factor in this restructuring and likewise based on a change in epistatic balances.

Considering our ignorance of what happens to the genotype during peripatric speciation, it is only natural that models would be proposed by reductionist geneticists that can explain everything in terms of the simplest single-gene assumptions. These models cannot be refuted, but they leave far more natural phenomena unexplained than does the theory of a genetic restructuring of founder populations. Most important, though, is that none of the recent attacks on punctuationism, particularly in its typological quasi-Goldschmidtian version, has affected the recognition of the great potential importance of founder populations. It reveals a complete misunderstanding if any author says of my 1954 theory that it was a "macromutation theory . . . an alternative to the selectionist program" (Turner, 1984:351). Nothing was stressed as much in my 1954 theory as selection.

One more point must be emphasized: there is great pluralism in speciation events. Whenever sexual selection in an isolated population leads to the origin of new behavioral isolating mechanisms, new species may evolve that differ by only a few genes (Mayr, 1984c). Such species are usually almost indistinguishable morphologically. They did not experience a genetic revolution.

The punctuationists have rightly criticized the gradualists for using the notorious imperfection of the fossil record to support gradualism. It is ironic that the punctuationists use the same argument when claiming that the populations in which peripatric speciation takes place are too localized and ephemeral to leave a fossil record. The incompleteness of the fossil record is thus as much a part of the argument of the punctuationists as it is of the gradualists.

Species Selection

As I described above, one of Darwin's brilliant insights was that evolution is variational rather than transformational, as was believed by Lamarck and by many anti-Darwinians after 1859. In variational evolution it is the survival of certain favored individuals that leads to evolutionary change; the emphasis is on the individual and the population. But variational evolution also occurs at a higher hierarchical level, that of the species, a fact particularly stressed by punctuationism. Whenever important evolutionary innovations occur, they occur during speciation. Once a species has become successful and stabilized, it will tend to change very little. The species therefore is considered the unit of evolution.

Recognition of the evolutionary importance of species long antedates punctuationism. I know of no better description of the role of species in evolution than the one I gave in my *Animal Species and Evolution* (Mayr, 1963):

> It is the very process of creating [new] species which leads to evolutionary progress. . . . Since each coadapted gene complex has different properties and

> since these properties are, so to speak, not predictable, it requires the creation of a large number of such gene complexes before one is achieved that will lead to real evolutionary advance. Seen in this light, it appears then that a prodigious multiplication of species is a prerequisite for evolutionary progress.
>
> Each species is a biological experiment. The probability is very high that the new niche into which it shifts is an evolutionary dead-end street. There is no way to predict, as far as the incipient species is concerned, whether the new niche it enters is a dead-end street or the entrance into a large new adaptive zone. . . .
>
> The evolutionary significance of species is now quite clear. Although the evolutionist may speak of broad phenomena, such as trends, adaptations, specializations, and regressions, they are really not separable from the progression of entities that display these trends, the species. The species are the real units of evolution as the temporary incarnation of harmonious, well-integrated gene complexes. And speciation, the production of new gene complexes capable of ecological shifts, is the method by which evolution advances. . . . The species, then, is the keystone of evolution. (P. 621.)

The continuous production of new species inevitably leads to competition among species and to a great deal of extinction. This process has been referred to as species selection (Stanley, 1975, 1979). The term is new, but the concept goes far back in the history of biology. Even in pre-Darwinian days, Charles Lyell postulated that the introduction of a new and better-adapted species might lead to the extinction of an inferior species, or that the extinction of a species would be followed by the introduction of a better-adapted species. Species extinction by competition was an important process also for Darwin. He illustrated it by the fate of the indigenous New Zealand fauna when it encountered species introduced from the British Isles (Darwin, 1859:201). Extinction of species caused by the appearance of better-adapted species has been frequently discussed in the post-Darwinian evolutionary literature.

The long-standing recognition of species selection refutes the claim that the recognition of the hierarchical level of species selection invalidated Darwinism. Gould (1982b:386) claims that a "hierarchical based theory [of evolution] would not be Darwinism as traditionally conceived." This view restricts the term "Darwinism" to the most reductionist concept of the mathematical population geneticists. I agree with the nonreductionist Darwinians who believe in a hierarchical order of the evolutionary process and who have never seen species selection as conflicting in any way with Darwinism.

The recent arguments indicate that the answer to two important questions about species selection remain controversial: (1) Is species selection in conflict with (and/or independent of) individual selection? (2) Are there different kinds of species selection, with different authors defining species selection differently, or even in a contradictory manner?

Several authors have rejected species selection by considering it a strict alternative to individual selection (e.g., Hoffman and Hecht, 1986). Indeed, if one asks the uncompromising question, Is the individual or the species the target of selection?, one is forced to reject species selection. One can express this even more unequivocally by asking whether there are properties of species that are not properties of the individuals of which the species is composed. Most authors have conceded that such situations indeed occur, even Maynard Smith (1983:280), who on the whole is hostile to the concept of species selection. By contrast, I must state I do not know of a single species character that is not also part of the genotype of every individual. Furthermore, it had become a species character because it had been favored by individual selection.

In the classical situation, one species is superior to another because its individuals are better adapted and better able to utilize the resources of the environment. Even though competition between two species is seemingly involved, analysis establishes that the superiority is due to the greater success of the individuals of the "superior" species and that individual selection is involved. Whenever two species are competing with each other, the individuals of both species are, so to speak, merged into a single ecological population, and selection deals with the total of these individuals. The individuals of the superior species have a higher expectation of survival and successful reproduction than those of the inferior species, so that eventually only individuals of the superior species will survive. The traditional process of individual selection has thus led to the extinction of one species, that is, to species selection. In such a case there is no conflict between individual selection and species selection. This will be the case whenever the "characters in common" of the individuals of the one species provide competitive superiority over the individuals of another species.

It has, however, also been argued that one species may be superior to another not because of any superiority of the composing individuals but because the species as a whole has characteristics that give superiority to its members. For instance, it has been argued that forms of reproduction or of dispersal and colonization that favor speciation, being species-specific, would give such species greater competitive superiority and thus constitute species selection. I am not persuaded by this argument. All the stated characteristics are also the characteristics of individuals and, when first appearing in evolution, entered populations in a polymorphic condition. It was the selective advantage of the individuals with the new characteristics (and of their offspring) that led to the spread and eventual universal incorporation of these new characteristics. In other words, these species characters were established by individual selection. I have failed to find in the literature a single good example of a species characteristic that is not also a selectable characteristic of individuals.

It has become evident in recent years that the term "species selection" has been applied to diverse phenomena. I have distinguished three kinds of species selection (Mayr, 1986), while Maynard Smith (1983) recognizes four kinds. It would seem irrelevant how many kinds one recognizes as long as one accepts the conclusion that species selection is not in opposition to individual selection but is an additional process at a higher hierarchical level.

The term "species selection" is somewhat unfortunate for several reasons. The first is the stated misconception that there is a conflict between individual selection and species selection, a conflict that I have shown does not exist. A second reason is that in the case of individual selection one can rightly say that a property or behavior is for the good of the individual. There is, however, no property that is, strictly speaking, for the good of the species. A favored property is always for the good of the individuals of which the species is composed. It is for this reason that I prefer the term "species turnover" for what others call species selection. We can conclude this discussion by stating that there is no conflict between individual selection and species turnover. Individual selection is usually involved in species turnover, but the loss of species from a biota may also be caused by cosmic and climatic catastrophes or other stochastic processes. Yet the effects of any restructuring of the biota relate to a different hierarchical level from individual selection.

Pluralism and Simultaneity

My 1954 proposal of genetic revolutions during peripatric speciation was pluralistic. I emphasized that no major evolutionary innovations occurred in the origins of most new species. By contrast, the first statements of the punctuationists sounded quite categorical, and this led to polarization and resistance. Their more recent statements have been formulated in a more pluralistic manner: "gradual phyletic transformation can and does occur" (Gould, 1982a:84) and "the relative frequency of punctuated equilibrium differs across taxa and environments" (Gould 1986:439). This renunciation of absolute, all-or-nothing statements has greatly contributed to reducing the opposition.

A second source of opposition to punctuationism is more difficult to overcome. It is the tendency of many authors to present two simultaneously occurring processes as a choice between one or the other. This is what Maynard Smith (1983) does when he asks whether the increased rate of evolutionary change in small populations is due to Mayr's genetic revolutions or to natural selection. This is curious considering how strongly I stressed the role of natural selection in the reorganization of the genotype of founder populations. Maynard Smith apparently failed to realize that genetic revolutions deal with the first step of the selection process while selection proper is the second step. He does not accept the possibility that a thawing out of the congealed part of the genotype, possible in founder populations, might make it easier for natural selection to achieve a stable new balance. Too often (Mayr, 1982a), evolutionary factors or processes are presented as alternatives when in reality they occur simultaneously.

Speciation and Macroevolution

The great importance of the speciation event is that it links macroevolution with microevolution. The facts that the individual is the target of selection and that the population and the species are the locale of evolutionary change automatically reduce all macroevolutionary processes to the microevolutionary level. The actors in this process, however, are not genes but genotypes and gene pools, entire cohesive systems of genes. The important insight is that whatever happens either in microevolution or in macroevolution, and whatever genetic phenomena are involved, proceeds through the selection of individuals: "All the processes and phenomena of macroevolution and of the origin of higher categories can be traced back to intraspecific variation, even though the first steps of such processes are usually very minute" (Mayr, 1942:298). Admittedly, genetics has so far been unable to analyze that part of the genotype that does not ordinarily vary in a local population but is so tightly integrated that it gives the genus, the family, the order, or the phylum its particular character (Carson, 1975). But even this part of the genotype, when it varies, varies in individuals and is subject to the recombination-selection cycle of ordinary allelic variation.

For the extreme reductionists among the geneticists who look at evolution, even macroevolution, as a process of changing frequencies of genes, there is a complete continuity among all phenomena of evolution. But those who think of evolution also as a change of species and higher taxa, including Darwin, have always considered evolution as hierarchical in structure. No other component of Darwin's thinking was as readily and widely adopted as his theory of common descent, a strictly hierarchical theory. And most of the paleontological literature, largely devoted to an elucidation of common descent, was hierarchical in approach. The term "hierarchical approach" introduces perhaps a new terminology but not a new concept. I agree with Grant (1983) that "adherents of the synthetic theory have in fact [consistently] employed a hierarchical approach to problems of macroevolution" (p. 153).

Remaining Problems

The controversies about punctuationism have clearly revealed that there remain major gaps in our knowledge. Several of these have been mentioned above, but I will single out two of them for more detailed discussion.

Population Size and Rate of Evolution

For the past sixty years, biologists have debated whether evolution advances more rapidly in large populations (species) or in small ones. Beginning with R. A. Fisher and Sewall Wright, geneticists favored the large population as the locale of the most rapid evolutionary change. Under the assumptions that the gene is the target of selection and that one can calculate the rate of evolution of a population or species by combining the rates of change of various independent genes, they concluded that the larger the population, the faster evolution will occur. As recently as 1977, Wright made this claim for widespread species consisting of many populations, a claim not based on observed facts but on theoretical considerations. This theory includes the acceptance of group selection among local demes, a type selection questioned by nearly all those who have thoroughly studied the problem (Sober, 1984).

Wright's scenario also assumed that the amount of gene flow among the postulated semi-isolated populations is very low, an assumption which has fared poorly in recent research. To be sure, there are species with low gene flow, but these are species with spotty, isolated distributions. In all widespread populous species there seems to be far more unobserved gene flow than is usually believed (Slatkin, 1985). Here the gene flow is often restricted to certain years in which the species has been particularly successful and the population has reached high densities. There is little evidence in most species for the semi-isolated demes postulated by Wright.

The evolutionary stability of large, widespread species is supported by the observation that most of the species in the fossil record that display stasis are large, widespread species with the samples taken from central populations. The claim of rapid evolutionary advance in large, widespread, populous species was questioned, on a theoretical basis, as far back as 1957 by Haldane and more recently by some mathematical population geneticists (Newman, Cohen, and Kipnis, 1985; Lande, 1985).

By contrast, I (1942, 1954) and my followers were impressed by the striking difference between the parental population and many peripherally isolated founder populations and species. This evidence was enriched by the findings of Carson and associates on incipient species and neospecies of *Drosophila* in Hawaii. The relative rapidity with which new and often drastically different species of *Drosophila* originate through founder events suggests an enormously accelerated rate of evolution, as compared to the widespread and relatively uniform sibling species complexes of *Drosophila* with continental distributions (e.g., *D. affinis, D. obscura*). The theory that evolution in small founder populations is more rapid than in large continental populations is based on solid observational evidence in contrast to the theorizing of reductionist genetics. We have yet to find a situation in which the isolated peripheral allospecies essentially retain their ancestral condition, while the large central population has seemingly greatly diverged.

It is important to distinguish between the observational evidence and the genetic interpretation. I believe that the observational evidence for the greatly increased rate of evolution in small populations is far better established than the reverse. Whether this increased rate is due to the breaking up of previously existing balances and the formation of new ones, as postulated by Mayr (1954), or simply to stochastic processes (or that both are involved), is still an open question. By merely invoking stochastic processes, however, Newman and his associates (1985) and

Lande (1985) demonstrated that passage from one to another adaptive peak had a much better chance of success in a small population. By contrast, the chance of success can be vanishingly small in large populations. Hence, say Newman and his colleagues, their model "predicts stasis and punctuation for a small to moderate population but only stasis for a large population" (p. 401). This is consistent with the observations of the naturalists. The breaking up of epistatic balances in founder populations has now also been demonstrated experimentally (Goodnight, 1987; Bryant et al., 1986).

There is one argument, seemingly based on observational evidence, that would seem to refute the thesis that large, widespread species are uniform at a given time level, a uniformity continuing as stasis. It is the argument that geographic variation is prevalent in all widespread species, a notion that conflicts with the postulated temporal uniformity of such species. If this claim were substantiated, it would be a valid objection to punctuationism. The factual evidence is limited, however. On the basis of my own analysis of the bird fauna of North America (Mayr and Short, 1970) and of the New Guinea area (Mayr and Diamond, 1992) three counterarguments can be made against this thesis.

First, most widespread species are remarkably uniform. This is particularly true for planktotrophic marine invertebrates.

Second, most species displaying geographic variation vary clinically only in ecotypic characters affecting primarily size, proportions, and coloration, but show no significant evolutionary departures. This is true for virtually all the illustrations of phyletic gradualism discovered by P. D. Gingerich (1977) in early tertiary mammals. Nor is the thesis of the evolutionary uniformity of widespread species refuted by the discovery of the localized selection of a few genes controlling mimicry in the genus *Heliconius* (Turner, 1984). Such selection for local mimicry is the exact equivalent to substrate protective coloration in rodents and other animals selected for crypsis. Species with oligogenic geographic variation in crypsis may be exceedingly widespread, but the genes involved in this adaptation do not seem a suitable basis for speciation. Interestingly, such genes are rarely combined into domains. The fourteen or so variable characters of *Heliconius melpomene* that permit the mimicry of coexisting species are scattered over nine chromosomes, and in the case of *Heliconius erato* some seventeen or eighteen variable characters are dispersed on ten chromosomes (Turner, 1984). Such independence of the gene loci greatly facilitates a shift to new patterns, but it does not encourage the evolution of a new stable evolutionary type.

Finally, polytypic species with striking variation are invariably secondarily fused mosaics of former founder populations. Their variation cannot be used to argue against the evolutionary role of founder populations.

Structure of the Genotype

Another unresolved argument concerns the structure of the genotype. Mathematical geneticists base their calculations on the assumptions that all genes are more or less independent of each other and that recombination following crossing over can produce a virtually unlimited assortment of genotypes. Other evolutionists, including such geneticists as Mather, Dobzhansky, Lerner, Wallace, and Carson, believe that there are cohesive domains in the genotype maintained by epistatic balances and that many evolutionary phenomena are best explained by such an assumption. Reductionist geneticists believe that they can explain all phenomena of seeming cohesion (domains of the genotype) in terms of single genes. Their opponents are yet unconvinced. Considering the rapid progress of molecular genetics and the ongoing discovery of new molecular structures and interactions of the genotype, there is every reason to believe that this argument will soon be resolved.

The Contributions of Punctuationism to Evolutionary Theory

Even some of its opponents admit that punctuationism has had an enormously stimulating effect on evolutionary biology (Rhodes, 1984; Maynard Smith, 1984b; Gould, 1985). The controversy has brought to light numerous equivocations and has helped to clarify distinctions between alternatives, such as between phyletic and allopatric speciation, between phenotypic and taxic saltations, between various types of group selection, between the evolutionary potential of small and large populations, between an uncompromisingly reductionist and a more holistic concept of the genotype, between various concepts of species selection, and still others. To eliminate these equivocations it was not only necessary to clarify concepts but also to show that we needed a broader factual foundation. As Gould has correctly emphasized, one of the most important contributions of punctuationism has been its stimulation of fruitful empirical research, much of it still ongoing.

To be sure, the claims of some punctuationists, such as the prevalence of total stasis and the impossibility of evolutionary change without speciation, are clearly invalid. Furthermore, it has been shown that "speciational evolution" (perhaps a better term than "punctuationism") is fully consistent with Darwinism; and finally, that seeming evolutionary saltations, as indicated by the fossil record, can be explained without invoking systemic mutations or other mechanisms in conflict with molecular genetics. It is irrelevant for the theory of speciational evolution how relatively frequent evolutionary stasis is or how frequent the occasional occurrence of drastic reorganization during peripatric speciation.

Most of all, punctuationism has shown how one-sided has been the myopic focusing of paleontologists and population geneticists on the one-dimensional, transformational, upward movement of evolution. It finally brought general recognition to the insight of those who had come from taxonomy (E. Poulton, Rensch, Mayr) and had consistently stressed that the lavish production of diversity is the most important component of evolution.

What had not been realized before is how truly Darwinian speciational evolution is. It was generally recognized that regular variational evolution in the Darwinian sense takes place at the level of the individual and population, but that a similar variational evolution occurs at the level of species was generally ignored. Transformational evolution of species (phyletic gradualism) is not nearly as important in evolution as the production of a rich diversity of species and the establishment of evolutionary advance by selection among these species. In other words, speciational evolution is Darwinian evolution at a higher hierarchical level. The importance of this insight can hardly be exaggerated.

The replacement of transformational evolution (including Lamarck) in 1859 by Darwin's variational evolution necessitated a complete rethinking of the old question of gradual versus saltational. The adoption of any kind of saltational evolution before 1859 documented essentialism. Now we can have quasi-saltational evolution (on the geological time scale) by gradual population change. Consequently, all the arguments against saltational evolution made against the pre-1859 paradigm are irrelevant when applied to rapid variational evolution. I suspect that some of the nonbiologists contributing to this book have failed to see that after 1859 we have been in a different ballpark and must use different arguments.

After this paper had originally been submitted to the editors, a somewhat revised version was published in a volume of my recent essays (Mayr, 1988).

References

Ayala, F. J. 1983. Beyond Darwinism? The challenge of macroevolution to the synthetic theory of evolution. *PSA*. 2(1982):275–291 [Philosophy of Science Association].

Bernard, C. 1895. *Elements de Paleontologie*.

Bryant, E. H., et al. 1986. The effect of an experimental bottleneck upon quantitative genetic variation in the housefly. *Genetics* 114:1191.

Buch, L. v. 1825. *Physicalische Beschreibung der Canarischen Inseln*. Berlin: Kgl. Akad. Wiss., pp. 132–133.

Buzas, M. A., and S. J. Culver. 1984. Species duration and evolution: benthic Foraminifera on the Atlantic continental margin. *Science* 225:829–830.

Carson, H. L. 1975. The genetics of speciation at the diploid level. *Am. Nat.* 109:83–92.

Carson, H. L., and A. R. Templeton. 1984. Genetic revolutions in relation to speciation phenomena: the founding of new populations. *Annu. Rev. Ecol. Syst.* 15:97–131.

Chaline, J., and B. Laurin. 1986. Phyletic gradualism in a European Plio-Pleistocene *Mimomyx* lineage (Arvicolidae, Rodentia). *Paleobiology* 12:203–216.

Cheetham, A. H. 1986. Tempo of evolution in a Neogene bryozoan: rates of morphological change within and across species boundaries. *Paleobiology* 12:190–202.

Chetverikov, S. S. 1926. On certain aspects of the evolutionary process from the standpoint of modern genetics. *J. Exp. Biol.* A2:3–54 (Russian).

Darwin, C. 1859. *On the Origin of Species by Means of Natural Selection or the Preservation of Favored Races in the Struggle for Life*. London: Murray.

——. 1868. *The Variation of Animals and Plants under Domestication*, vols. 1 and 2. London: Murray.

de Ricqles, A., ed. 1983. Formes panchroniques et 'fossiles vivants'. *Bull. Soc. Zool. France* 108 (4): 529–673.

Dobzhansky, T. 1937. *Genetics and the Origin of Species*. 1st ed. New York: Columbia University Press.

——. 1951. *Genetics and the Origin of Species*. 3d ed. New York: Columbia University Press.

Eldredge, N. 1971. The allopatric model and phylogeny in Paleozoic invertebrates. *Evolution* 25:156–167.

Eldredge, N., and S. J. Gould. 1972. Punctuated equilibria: an alternative to phyletic gradualism. In T. J. M. Schopf, ed., *Models in Paleobiology*. San Francisco: Freeman, Cooper, pp. 305–332.

Eldredge, N., and S. M. Stanley, eds. 1984. *Living Fossils*. New York: Springer.

Fisher, R. A. 1930. *The Genetical Theory of Natural Selection*. Oxford: Clarendon Press.

Gilinsky, N. L. 1986. Species selection as a causal process. *Evol. Biol.* 20:249–281.

Gingerich, P. D. 1977. Patterns of evolution in the mammalian fossil record. In A. Hallam, ed., *Patterns of Evolution as Illustrated by the Fossil Record*. Amsterdam and New York: Elsevier, pp. 469–500.

——. 1984. Punctuated equilibria—where is the evidence? *Syst. Zool.* 33:335–338.

Goldschmidt, R. 1940. *The Material Basis of Evolution*. New Haven: Yale University Press.

——. 1952. Evolution, as viewed by one geneticist. *Am. Sci.* 40:84–98.

Goodnight, C. J. 1987. On the effect of founder events on epistatic genetic variance. *Evolution* 41:80–91.

Gould, S. J. 1977. The return of hopeful monsters. *Nat. Hist.* 86:22–30.

——. 1980. Is a new and general theory of evolution emerging? *Paleobiology* 6:119–130.

——. 1982a. The meaning of punctuated equilibrium and its role in validating a hierarchical approach to macroevolution. In R. Milkman, ed., *Perspectives in Evolution*. Sunderland, Mass.: Sinauer, pp. 83–104.

——. 1982b. Darwinism and the expansion of evolutionnary theory. *Science* 216:380–387.

——. 1983. [Answer to Schopf and Hoffman]. *Science* 219:439–440.

——. 1985. The paradox of the first tier: an agenda for paleobiology. *Paleobiology* 11:2–12.

——. 1986. Punctuated equilibrium: empirical response. *Science* 232:439.

Gould, S. J., and C. B. Calloway. 1980. Clams versus brachiopods. *Paleobiology* 6:394.

Gould, S. J., and N. Eldredge. 1977. Punctuated equilibria: the tempo and mode of evolution reconsidered. *Paleobiology* 3:115–151.

——. 1986. Punctuated equilibrium at the third stage. *Syst. Zool.* 35:143–148.

Grant, V. 1982. Punctuated equilibria: a critique. *Biol. Zentralbl.* 101:175–184.

——. 1983. The synthetic theory strikes back. *Biol. Zentralbl.* 102:149–158.

Haffer, J. 1974. *Avian Speciation in Tropical South America*. Cambridge, Mass.: Nuttall Ornithological Club, no. 14.

Haldane, J. B. S. 1957. The cost of natural selection. *J. Genet.* 55:511–524.

Hoffman, A. 1982. Punctuated versus gradual mode of evolution: a reconsideration. *Evol. Biol.* 15:411–436.

Hoffman, A., and M. K. Hecht. 1986. Species selection as a causal process: a reply. *Evol. Biol.* 20:275–281.
Huxley, A. 1982. Address of the president. *Proc. R. Soc. Lond.* A 379:ix–xvii.
Huxley, J. 1942. *Evolution: The Modern Synthesis*. London: Allen & Unwin.
Lande, R. 1985. Expected time for random genetic drift of a population between stable phenotypic states. *Proc. Natl. Acad. Sci. USA* 82:7641.
Lerner, I. M. 1954. *Genetic Homeostasis*. New York: Wiley.
Levinton, J. S. 1983. Stasis in progress: the empirical basis of macroevolution. *Annu. Rev. Ecol. Syst.* 14:103–137.
Levinton, J. S., and C. M. Simon. 1980. A critique of the punctuated equilibria model and implications for the detection of speciation in the fossil record. *Syst. Zool.* 29:130–142.
Lewontin, R. 1983. Kinds of evolution. *Scientia* 118:65–82.
Malmgren, B. A., W. A. Berggren, and G. P. Lohmann. 1984. Species formation through punctuated gradualism in planctonic Foraminifera. *Science* 225:317–319.
Mather, K. 1953. The genetical structure of populations. *Symp. Soc. Exp. Biol.* 7 (Evolution): 66–95.
Maynard Smith, J. 1983. Current controversies in evolutionary biology. In M. Grene, ed., *Dimensions of Darwinism*. Cambridge: Cambridge University Press, pp. 273–286.
——. 1984a. Palaeontology at the high table. *Nature* 309:401–402.
——. 1984b. The genetics of stasis and punctuation. *Annu. Rev. Genetics* 17:11–25.
Mayr, E. 1942. *Systematics and the Origin of Species*. New York: Columbia University Press.
——. 1954. Change of genetic environment and evolution. In J. Huxley, ed., *Evolution as a Process*. London: Allen & Unwin, pp. 157–180.
——. 1960. The emergence of evolutionary novelties. In S. Tax, ed., *Evolution after Darwin*, Vol. 1. Chicago: University of Chicago Press, pp. 349–380.
——. 1963. *Animal Species and Evolution*. Cambridge, Mass.: Harvard University Press.
——. 1970. *Populations, Species, and Evolution*. Cambridge, Mass.: Harvard University Press.
——. 1975. The unity of the genotype. *Biol. Zentralb.* 94:377–388.
——. 1981. Adaptation and selection. *Biol. Zentralb.* 101:161–174.
——. 1982a. *The Growth of Biological Thought*. Cambridge, Mass.: Harvard University Press.
——. 1982b. Speciation and macroevolution. *Evolution* 36:1119–1132.
——. 1982c. Processes of speciation in animals. In *Mechanisms of Speciation* (Rome symposium 1982). New York: Alan R. Liss, pp. 1–19.
——. 1984a. The concept of finality in Darwin and after Darwin. *Scientia* 118:97–117 (English); 119–134 (Italian).

——. 1984b. The triumph of evolutionary synthesis. *Times Literary Supplement* no. 4, 257, November 2, pp. 1261–1262.

——. 1984c. Evolution of fish species flocks: a commentary, introduction. In A. A. Echelle and I. Kornfield, eds., *Evolution of Fish Species Flocks*. Orono: University of Maine at Orono Press, pp. 3–11.

——. 1986. Review of Sober, Elliott. 1984. *The Nature of Selection: Evolutionary Theory in Philosophical Focus*. Cambridge, Mass.: MIT Press. *Paleobiology* 12: 233–239.

——. 1988. *Toward a New Philosophy of Biology*. Cambridge, Mass.: Harvard University Press.

Mayr, E., and J. Diamond. 1992. Birds of Northern Melanesia. Unpublished.

Mayr, E., and L. Short. 1970. *Species Taxa of North American Birds: A Contribution to Comparative Systematics*. Cambridge, Mass.: Nuttall Ornithological Club, no. 9.

Newman, C. M., J. E. Cohen, and C. Kipnis. 1985. Neo-Darwinian evolution implies punctuated equilibria. *Nature* 315:400–401.

Osborn, H. F. 1894. *From the Greeks to Darwin*. New York: Columbia University Press.

Penny, D. 1983. Charles Darwin, gradualism and punctuated equilibria. *Syst. Zool.* 32:72–74.

Remane, J. 1983. Selektion und Evolutionstheorie. *Palaeontol. Z.* 57:205–212.

Rensch, B. 1947. *Neuere Probleme der Abstammungslehre*. Stuttgart: Enke.

Rhodes, F. H. T. 1983. Gradualism, punctuated equilibrium and the *Origin of Species*. *Nature* 305:269–272.

——. 1984. [Reply to Gingerich]. *Nature* 315:401–402.

Rose, K. D., and T. M. Brown. 1984. Gradual phyletic evolution at the generic level in early Eocene Omomyid Primates. *Nature* 309:250–252.

Schopf, T. J. M. 1982. A critical assessment of punctuated equilibria. I. Duration of taxa. *Evolution* 36:1144–1157.

Schopf, T. J. M., and A. Hoffman. 1983. Punctuated equilibrium and the fossil record. *Science* 219:438–439.

Simpson, G. G. 1944. *Tempo and Mode*. New York: Columbia University Press.

——. 1953. *The Major Features of Evolution*. New York: Columbia University Press.

——. 1961. *Principles of Animal Taxonomy*. New York: Columbia University Press.

Slatkin, M. 1985. Gene flow in natural populations. *Annu. Rev. Ecol. Syst.* 16:393–430.

Sober, E. 1984. *The Nature of Selection*. Cambridge, Mass.: MIT Press.

Stanley, S. M. 1975. A theory of evolution above the species level. *Proc. Natl. Acad. Sci. USA* 72:646–650.

——. 1979. *Macroevolution: Pattern and Process*. San Francisco: Freeman.

Stebbins, G. L., and F. J. Ayala. 1981. Is a new evolutionary synthesis necessary? *Science* 213:967–971.

Turner, J. R. G. 1983. Mimetic butterflies and punctuated equilibria: some old light on a new paradigm. *Biol. J. Linn. Soc.* 20:277–300.

——. 1984. Darwin's coffin and Doctor Pangloss—do adaptationist models explain mimicry? In B. Shorrocks, ed., *Evolutionary Ecology*. Oxford: Blackwell, pp. 313–361.

Van Valen, L. 1982. Integration of species: stasis and biogeography. *Evol. Theory* 6:99–112.

Wallace, B. 1985. Reflections on the still-"hopeful monster." *Q. Rev. Biol.* 60:31–42.

Willis, J. C. 1922. *Age and Area*. Cambridge: Cambridge University Press.

Willmann, R. 1985. *Die Art in Raum und Zeit*. Berlin and Hamburg: Parey.

Wright, S. 1931. Evolution in Mendelian populations. *Genetics* 6:97–159.

——. 1932. The roles of mutation, inbreeding, cross-breeding, and selection in evolution. *Proc. Sixth Int. Congr. Genet.* 1:356–366.

——. 1977. *Evolution and the Genetics of Populations*. Vol. 3. Chicago: University of Chicago Press.

2

Punctuated Equilibrium in Fact and Theory

STEPHEN JAY GOULD

THE GENESIS AND HISTORY OF PUNCTUATED EQUILIBRIUM

Sigmund Freud once stated that psychic ontogeny had all the complexities of an imaginary (and impossible) Rome with every building, from Romulus to Pio Nono and Victor Emmanuel, existing simultaneously, several to each spot. For this reason, and therefore not so paradoxically, the worst possible person to ask about the genesis of a theory is the generator himself. How can anyone sort out the germ of original insight from the complex history of present convictions—for the current edifice, following Freud's metaphor, dwells inextricably with the original structure. Charles Darwin's autobiography, for example, may be the most misleading personal document ever published by a scientist.

I feel confident, however, about some broad outlines in the genesis of punctuated equilibrium, and I feel that these guidelines might disperse some persistent confusion, and also explicate the current role of this theory within evolutionary biology. Above all, Niles Eldridge and I followed the most important methodological principle of science by restricting the scope of punctuated equilibrium to a single and definable phenomenon that could yield testable data. We did not construe the theory as a cosmic, but vague and intractable, statement about *the* pace of evolution at all scales, or about *the* nature of change.

Niles Eldredge and I were graduate students in paleontology together under Norman D. Newell at the American Museum of Natural History.

At that time (early to mid-1960s) a mini-revolution was brewing within American paleontology—a restructuring of its principal focus from biostratigraphic correlation in the service of geology (and the petroleum industry) to a concern with biological problems of evolution. This movement, then a fledgling, has since flourished, numerically and intellectually (consider the journal *Paleobiology* since its founding in 1975, not to mention the outstanding success of paleobiology in such areas as mass extinction, the early history of diversification of life on earth, and patterns of speciation).

Most nonpaleontologists are surprised to learn how little biological work once went forward in American paleontology and how little biological training or knowledge most paleontologists possessed. The reasons are complex and woven into the curiously contingent histories of several disciplines—commerce, education, and government service among them. I cannot pursue this issue here (though it would provide a good Ph.D. topic for an aspiring sociologist of science). Norman Newell was the most biologically oriented of "old-guard" paleontologists and had personally fomented this restructuring by urging students to tackle evolutionary problems.

Niles and I went to study with Newell because we were primarily interested in evolution—a direction that was, at the time, still a rarity in paleontology. We began our professional lives with a commitment to engage in the empirical study of evolution as illustrated by the fossil record. We were both interested in small-scale, quantitative research on species and lineages, a concern fostered by our other adviser, John Imbrie, who taught us multivariate statistical analysis.

Now imagine the frustration of two hyperenthusiastic, idealistic, noncynical, ambitious young men captivated with evolution, committed to its study in the detailed fossil record of lineages, and faced with the following situation. The traditional wisdom of the profession held (correctly) that the fossil record of most species showed stability (often for millions of years) following geologically unresolvable origin. "Evolution," however, had long been restrictively defined as "insensibly graded fossil sequences"—and such hardly existed. A few old classics graced the pages of every textbook—precious examples of insensibly graded sequences viewed as exemplars of "evolution"—but most turned out to be wrong upon close scrutiny (I restudied two myself; see Gould, 1972, 1974). In other words, we were told that our primary data base contained virtually no examples of the phenomenon we wished to study.

This odd situation created no cognitive dissonance within the field, for paleontologists did have an explanation for why the phenomenon that regulated their record left no empirical trace at the small-scale level of lineages—and most paleontologists weren't much interested in evolution anyway. The record, we were told, was so imperfect that truly insensible fossil transitions left no trace in the rocks: if but two of a thousand steps are preserved, gradual sequences appear as abrupt breaks. Darwin had advanced this explanation to resolve his embarrassment with the fossil record (1859: chap. 9), and his argument had remained canonical within the profession ever since.

Niles and I had one advantage in combating this frustration. We had been well trained in the details of modern evolutionary theory. (I also had a fascination for history and philosophy of science. I had been strongly influenced by the then radical views of T. S. Kuhn, N. R. Hanson, and P. Feyerabend on the hold that theory imposes upon perception. I had studied Charles Lyell's works in detail and had recognized the central character of the textbook myth that Lyell's principle of uniformity had forged modern geology by discrediting biblically trained catastrophists: see my first paper, Gould, 1965, and my book, Gould, 1987). I had, in short, and by these influences, recognized the powerful hold that gradualism, as ideology, had exerted within British and American natural science. We had long discussions about whether insights from evolutionary theory might break the impasse that traditional explanations for the fossil record had placed before our practical hopes—for why would one enter a field where intrinsic limitations on evidence had wiped out nearly all traces of the phenomenon one wished to study? Traditional explanations for stasis and abrupt appearance had paid an awful price in sacrificing the possibility of empirics for the satisfaction of harmony.

Eventually we (primarily Niles) recognized that the standard theory of speciation—Ernst Mayr's allopatric or peripatric scheme (1954, 1963)—would not, in fact, yield insensibly graded fossil sequences when extrapolated into geological time, but would produce just what we see: geologically unresolvable appearance followed by stasis. For if species almost always arise in small populations isolated at the periphery of parental ranges, and in a period of time slow by the scale of our lives but effectively instantaneous in the geological world of millions, then the workings of speciation should be recorded in the fossil record as stasis and abrupt appearance. The literal record was not a hopelessly imperfect fraction of truly insensible gradation within large populations but an accurate reflection of the actual process identified by evolutionists as the chief motor of biological change. The theory of punctuated equilibrium was, in its initial formulation (Eldredge and Gould, 1972), little more than this insight adumbrated.

Punctuated equilibrium was not a grandiose theory of the nature of change, a Marxist plot, a cladistical cabal, an attempt to sneak the hopeful monster back into evolution, or a tortuous assault on the concept of adaptation. It was at first, and has always been: (1) A well-defined, testable theory about the origin of species and their geological deployment (not a general rubric for any old idea about rapidity at any scale, and specifically not a notion about saltation, the origin of new *Baupläne*, or mass extinction). (2) A theory based on the recognition that events judged as glacially slow in ecological time might appear instantaneous in geological resolution (the conversion of the peripheral isolate into a species, in particular). Some neontologists, misconstruing punctuated equilibrium as a theory of saltation, have charged that we made a disabling error in not recognizing the difference in scale between ecology and geology and in thinking that geological abruptness demanded some notion of true discontinuity. Quite the reverse: punctuated equilibrium is not a theory of saltation, and its anchor lies in this appreciation of scaling—particularly, in recognizing how an ecologically "slow" event like allopatric speciation must translate into the geological record. (3) An idea resolvable within the rubric of known mechanisms and causes (though requiring—see below—an expanded notion of hierarchy to understand how these causes operate at higher levels), not a proposal about new kinds of genetic changes. As a geological expression of the orthodox view about speciation, punctuated equilibrium could not have been born in genetic novelty (and never has entered this important arena).

Neither Niles nor I wears a halo of modesty, so we are not being disingenuously self-effacing in stating that we had no inkling about the extent of discussion and controversy that punctuated equilibrium would generate (at most, we hoped to stir up some self-examination among paleontologists). We had no hidden agenda (unless it was buried so deeply in our psyches that even we weren't aware). Moreover, our reaction to the influence of our theory is a mixture of profound gratification and simmering distress—for we know, in this world of good news–bad news, that the reasons are a curious mixture of the infuriating and the sublime.

The distressful centers on punctuated equilibrium's mode of passage into general culture through media reports. As always, some accounts were excellent and insightful (Lewin, 1980, 1986; Rensberger, 1980, 1981, 1984; Bakker, 1985; Lessem, 1987), the great majority were adequate, but some prominent comments were egregiously bad and harmful. The reason for all this attention centers on an unfortunate accident of history: the coincidence of professional interest in punctuated equilibrium with the high point of creationist resurgence, now largely defeated.

The Chicago macroevolution meeting of October 1980—the event that launched the public career of punctuated equilibrium—would normally have attracted no press coverage (beyond, perhaps, the news and comment sections of *Science* and *Nature*), but became a media event because a vulgar reading of "Darwin in death throes" coincided with the height of creationist lobbying for equal-time bills before state legislature. (Two were passed. Judge William Overton struck down the Arkansas law in early 1982, and the Supreme Court ended this disgraceful episode in American history by striking down the entire genre in June 1987.)

The cause of our profession was injured by a few prominent press reports that almost assimilated all questioning of strict neo-Darwinism to the creationist cause. Worst was the following lead from *Discover*, October 1980 (the editor later apologized to me personally for this lamentable lapse of fact and judgment): "Among paleontologists, scientists who study the fossil record, there is growing dissent from the prevailing view of Darwinism. Partly as a result of the disagreement among scientists, the fundamentalists are successfully reintroducing creationism into textbooks and schoolrooms across the U.S."

The conference itself then spawned an equally misleading report in *Newsweek* (November 3, 1980). I have a thick file of accounts in creationist publications as well but shall leave this aspect unmentioned except to say that one can fight back against misrepresentation—and that we did so effectively (my testimony in the Arkansas creationism trial centered on creationist dishonesty in the reporting of punctuated equilibrium). Distorted reporting cuts both ways. Since then, the vast majority of reports have been responsible and generally accurate, but one still finds the occasional hype and emptiness ("Science contra Darwin" in *Newsweek*, April 8, 1985; "Life after Darwin," *TWA Ambassador*, August 1986). I take particular pride in the international coverage that punctuated equilibrium has generated: in England (editorial in the *Guardian*, November 21, 1978; lead article by J. Davy in the *Observer*,

August 16, 1981); France (Blanc, 1982; Thuilier, 1981; exhibit at the Musée archéologique, Dijon, 1982; *Le Monde*, May 26, 1982); Italy (Salvatori, 1984); Spain (Sequeiros, 1981; Garcia-Valdecasas, 1982); New Zealand (Paterson, 1986); Japan (in *Kagaku Asahi*, February 1987); India (editorial, *The Times of India*, March 20, 1984); Venezuela (Mondolfi, 1986); and China (Beijing, *Ren Min Ri Bao* [People's daily], article by Lu Jizhuan, March 21, 1983).

The good side is more easily, and calmly stated. We are gratified beyond measure by the serious and thoughtful attention, often quite critical, that many of our most respected colleagues have accorded to punctuated equilibrium because they recognize it as a distinct, testable hypothesis about some of the most important issues in macroevolutionary theory (Maynard Smith, 1983a, b, 1984; Rhodes, 1983; Turner, 1986; Mahé and Devillers, 1981; Huxley, 1982; Newman, Cohen, and Kipnis, 1985; Milligan, 1986). The main reason for "all the fuss," contra R. Dawkins (1985), is that punctuated equilibrium has something new, interesting, and workable to say about the history of life.

Punctuated equilibrium also plugged into something bigger than we realized when we first wrote. We had developed a paleontological hypothesis about the two most prominent features of pattern in the geological life of most fossil species—their abrupt appearance (punctuation) and their subsequent stasis (equilibrium). (The name "punctuated equilibrium" may be elliptical, but it is not oxymoronic, for it collates these two separate and central characteristics of species.) The initial debate occurred among paleontologists and centered upon these empirical aspects of the fossil record (Gingerich, 1976; see Gould and Eldredge, 1977).

Subsequently, however, developing trends in what we regard as the most important reformulation now occurring within evolutionary theory—the elaboration of a hierarchical theory of selection with full understanding of transfers and interactions across levels—enlarged the context for both aspects of punctuated equilibrium and made our paleontological hypothesis a source of concern within the general theory. To some extent, punctuated equilibrium played a role in instigating the investigation of hierarchy theory; it certainly set an overt and friendly context. But mostly, it was swept along, fortunate to be in the right place at the right time. The next two sections discuss how the concept of punctuation instantiates, or at least makes necessary, the description (perhaps also the causation) of evolutionary trends as a sorting among species, and not as a simple extrapolation of processes occurring within single populations. The empirics of stasis also implicate higher-level sorting in the explanation of trends, and speak as well to questions of constraint in evolution and to the general critique of adaptation (Gould and Lewontin, 1979), an issue that also requires hierarchy to transcend the sterility of the adaptive-random dichotomy (Gould and Vrba, 1982).

In rereading our initial paper (Eldredge and Gould, 1972), I was struck by how groping and tentative our incipient perceptions of these wider contexts were when we first developed punctuated equilibrium. We spoke vaguely of homeostasis as a cause for equilibrium in our conclusion (p. 114) but could not, as committed adaptationists, sense the broader theme. More important, I distinctly remember my feeling that our section on trends (written by Niles) and on the phenomenon that we did not name but would later be called species selection (pp. 111–112) was the most exciting and potentially significant in the paper—but I could not, for the life of me, quite figure out why. We did have inklings of the shape of things to come, but I am now almost embarrassed to discover how little we initially exploited the aspects of punctuated equilibrium that subsequently became most interesting.

Punctuation and Hierarchy

Evolutionary trends—sustained and sequential changes of average character states within lineages—are the bread and butter of paleontology, the primary phenomenon worth noticing about macroevolution. (The most ludicrous of creationist distortions about punctuated equilibrium claims that our theory is the last-gasp effort of evolutionists to explain the empirics of a trend-free fossil record without admitting that the huge and bridgeless gaps among higher taxa are signs of creation *ex nihilo*. Punctuated equilibrium, needless to say, is primarily an unconventional theory about the genesis of trends viewed as the fundamental feature of life's history.)

Trends had traditionally been depicted and explained, on the extrapolative model, as products of directional selection upon a population simply extended through time—the anagenetic march of frequency distributions up a geological column. But punctuated equilibrium clearly implies that this model cannot account for trends if most species arise with geological abruptness and then remain stable throughout their duration. Under punctuated equilibrium, a trend must be described as a "higher-order" sorting among species treated as stable individuals—as an analogue of conventional variation and selection within populations, with speciation as the source of variation and differential births and deaths of species taking the part of natural selection. In this perspective, macroevolution would have to be viewed as "taxic" (Eldredge, 1979), or the story of the differential success of species treated as entities, not as "transformational," or the traditional tale of adaptively changing character states within continuously evolving populations. This was the insight that had excited me in our initial section on trends.

In interpreting this higher-level analogue of natural selection, we made our biggest conceptual mistake and unfortunately sowed much subsequent confusion. We did not name this analogue in our original work (Eldredge and Gould, 1972). Steven M. Stanley (1975, 1979) dubbed it "species selection"—and we responded enthusiastically, extending both the term and the concept (Gould and Eldredge, 1977).

The conceptual error is crucial to an understanding of evolutionary theory, and we are grateful to Elisabeth Vrba for working out the proper logic with us (Vrba, 1980; Vrba and Eldredge, 1984; Gould and Vrba, 1982; Vrba and Gould, 1986). Selection is a claim about causality—differential births or deaths based on characters of the objects being sorted. "Sorting" is the descriptive claim that such a differential exists, leading to the accumulation of some trait or other within a population or lineage. Since this distinction is so basic in logic and crucial in practice, we must wonder why it has not been clearly stated throughout the history of Darwinism. The answer seems to be that classical Darwinism is a causally nonhierarchical (however descriptively hierarchical) theory, committed (whenever possible) to locating the cause of evolutionary change at a single level of objects—natural selection working through the differential births and deaths of *organisms*. In such a one-level world, *sorting of* organisms will always be caused by *selection upon* organisms, and we need not distinguish sorting from selection in practice (however separate they may be in logic). In a hierarchical world, however, where selection may occur on "individuals" at a variety of levels (genes and species, as well as organisms), with effects propagating up and down the hierarchy to objects at other levels, sorting of individuals at any level may be caused either by selection at that level or by the ramifying effects of selection at another level (Vrba and Gould, 1986). It is therefore essential to recognize the difference between sorting (a description of differential birth and death) and selection (a causal claim about the basis of sorting).

Our misnamed "species selection" (cf. Stanley, 1975, and Gould and Eldredge, 1977) was really a claim about species sorting. This sorting may be caused by true species selection (see next paragraph), but it may also arise as an effect of conventional selection upon organisms. For example, the common trend to increasing brain size within mammalian lineages (Jerison, 1973) might be reducible to selection upon organisms if the trend arises by the greater longevity of big-brained species and if this enhanced duration reflects only the success of big-brained organisms in conventional competition.

We should speak of species selection only if the sorting among species has, as its causal basis, selection upon emergent and irreducible characters of species treated as entities. (Emergence is not a mystical notion of absolute novelty in higher-level traits. Emergent characters of species depend on states of organisms, just as the parts of an organismal phenotype depend on the properties of genes and cells. Emergent characters arise by nonadditivity and interaction among lower-level traits. In this sense, emergent features do not exist at the lower level, just as the purchasing power of a quarter is not inherent in the proportions of metals that make up its alloy.) Trends produced by differential death of species are more easily reducible to Darwinian selection on organisms, but trends due to differential birth (Gilinsky, 1986) more often fall into the domain of true species selection—for speciation is a property of populations (organisms do not speciate), whereas extinction is often a simple concatenation of deaths among organisms.

(Species selection is a difficult idea still beset with problems of concept and definition. The literature, unfortunately, contains two quite different traditions for its identification, and I am not sure that they coincide: the notion of emergent characters [see above] and the influence of group characters on the fitness of organisms [see Mayo and Gilinsky, 1987; Sober, 1984; Lloyd, forthcoming]).

In any case, species selection must be restricted and defined in the proper causal sense described above. Stanley (1979) and Mayr (this volume and 1982) have unfortunately continued to use "species selection" in the purely descriptive sense of sorting due to differential birth and death of species. The reason for concern is simple and concept-bound: true species selection is a challenge to the basis of Darwinian theory; species sorting, when reducible to selection on organisms, is thoroughly Darwinian. If hierarchy theory is, as I believe (Gould, 1982a, b), an important revision and extension of Darwinism, then true species selection probably has a high relative frequency in nature. If nearly all species sorting may be causally reduced to selection on organisms, then conventional Darwinism gets a big boost.

Punctuated equilibrium, by stressing the ineluctability of species sorting in the explanation of trends, has provided new insights into the nature of macroevolution in either case. If sorting is reducible to selection on organisms, trends must still be described at the species level—and such a description revises the old mode of portrayal as anagenesis within lineages on the extrapolative model. If sorting is often due to species selection—or, rather, has an important component of species selection, for all levels of selection may work simultaneously and in the same or opposite directions—then the new description of trends, stimulated by punctuated equilibrium, will have prompted a revision in evolutionary theory.

Stasis and Constraint

Of the two phenomena named in punctuated equilibrium, stasis has been more noted for two reasons. First, as several strong proponents of the modern Darwinian synthesis have fairly acknowledged (see Mayr, this volume; Maynard Smith, 1983a, b; Stebbins and Ayala, 1982), stasis as the history of most species was an unexpected and surprising result, even if resolvable, ex post facto, with conventional theory. (Punctuation is the geological translation of allopatric speciation, but most neontologists had assumed that fossil species would show prominent anagenetic change during their subsequent history—including frequent transformation into new species. They assumed, in other words, that geology would record a world of flux, rather than a world dominated by stability.)

Second, stasis is the more testable of the two phenomena named in punctuated equilibrium. Most punctuations are lost in imperfections of the fossil record (small populations, isolated at peripheries of parental ranges, and speciating in geologically unresolvable amounts of time). Unusually complete data in regions with high rates of sedimentation may preserve punctuations (Williamson, 1981; and subsequent discussion in Fryer et al., 1983; Williamson, 1985a, b), but stasis can be empirically tested in a majority of lineages. The 5-million-year history of a species may still include more gaps than evidence, but even a few percent of a temporally well-distributed record provides enough samples for testing stasis. (Imperfection does distort an idealized uniform sampling scheme, but we needn't space our ten samples exactly five hundred thousand years apart in order to affirm stasis. If a species looks at its end as it did at its beginning, and if a goodly number of well-distributed samples from the intervening times show no substantial excursions, then stasis has prevailed.)

More debate has centered on reasons for this unanticipated stasis. Committed selectionists have argued that stasis entails nothing surprising or at all indicative of incompleteness in our current accounts of evolutionary causation—for stasis can be maintained by the well-known process of stabilizing selection (Charlesworth, Lande, and Slatkin, 1982; Levinton, 1983; Lande, 1985). I have always found this argument curious in form, or at least excessively a prioristic in asserting that because one can, in principle, frame an explanation in conventional terms, somehow the issue becomes uninteresting or even closed without a test of alternatives that may fit the data better.

I continue to have one cardinal difficulty with the attribution of stasis to stabilizing selection—the scales are all wrong. Stabilizing selection can act powerfully within ecological balances of the here and now (Darwin gave us much data on overproduction of offspring and on the mortal struggle that occurs behind the face of nature "bright with gladness"). But geological stasis lasts for millions of years and, more relevant than the magnitude of time alone, persists through the usual pronounced vicissitudes of environment that invariably occur over such durations. (I am most impressed by the stability of so many species through the fluctuations of Pleistocene glacial-interglacial cycles—see Cronin, 1985, 1987, on ostracodes; White and Harris, 1977, on African pigs.) One can, of course, argue that these fluctuations did not form part of the species' effective environment and that relevant ecologies were stable, but how often will this form of special pleading work if stasis of species through ordinary geological shifts of environment be general?

I fully admit that my alternative also relies on gut feelings and, in part, on negative evidence—but I find the logic compelling enough to win a hearing for this view, though not (of course) to establish it. I would rather suppose that the profoundness and temporal depth of stasis are trying to tell us that change is *actively prevented*, rather than always potential but merely suppressed because no adaptive advantage would accrue. I strongly suspect, in other words, that stasis is primarily an active feature of organisms and populations, maintained by evolved genetic and developmental coherences, based largely on complex epistasis in genetic programs and the resilient and limited geometries of developmental sequences.

This is not a particularly radical or non-Darwinian proposal for several reasons. Many of these coherences are built by natural selection (but not all—see Gould and Lewontin, 1979; Goodwin, 1980; Kauffman, 1983). Stabilizing selection (see Williamson, 1986) has an internal component based on elimination of phenodeviants for their developmental abnormalities; it is not only a process of external policing by an unforgiving environment.

Constraints exerted by inherited genetic and developmental architecture now represent a major theme in evolutionary theory (see Maynard Smith, 1983b; Alberch, 1985, for example) and a welcome corrective to a previously exaggerated reliance on immediate externally driven selection. This corrective is reasserting the importance of history and structure upon a discipline that had veered too far toward immediate adaptation, thereby almost losing its anchor as a fundamental science of history (Gould, 1986).

The enormous frustration of this subject lies in our woeful ignorance of the mechanics of development, both at the molecular and tissue levels. Homeoboxes may provide a good start (Gould, 1985b; Gehring, 1985), but the excitement that they have generated (Gehring, 1987; Marx, 1986), combined with their limited implications for development so far, illustrate the rudimentary character of our understanding. Much shall be learned in the next decade or two, and though molecular studies will predominate, I hope that the stabilities demonstrated by punctuated equilibrium may aid our understanding or constraint, or at least set a climate that views its study as central.

The Wider Implications of Punctuated Equilibrium

Its Role in a Revised Evolutionary Theory

Beyond the empirical issue of recognition and occurrence (see next section), no aspect of punctuated equilibrium has been more contentious than its relationship with evolutionary theory, particularly with the Darwinian framework of the so-called Modern Synthesis (Mayr and Provine, 1980).

The confusion arises from failure to distinguish the basic geometry of the model from the wider implication of its frequency for the interpretation of macroevolution. The basic geometry itself—geologically instantaneous origin and later stasis of species—contains nothing outside the logic of modern Darwinism, as we and others have continually emphasized (Gould and Eldredge, 1977:117; Simpson, 1976). Punctuations are viewed as the geological translation of ordinary allopatric speciation; the constraints of stasis are seen as largely internal but potentially resolvable within the theory so long as perspectives extend beyond the restrictive domain of stabilizing selection. (Nonetheless, such pronounced stasis was unanticipated by most Darwinians, and its widespread recognition, leading to a reemphasis on internal factors of constraint, has contributed to a major shift of interest away from the optimality of immediate adaptation toward the study of how past history and internal architecture, both genetic and developmental, interact with current selection to mediate both stability and change. For the role of punctuated equilibrium in this welcome reemphasis on evolution as a science of history, we take great pride.)

Nevertheless, the potentially high relative frequency of punctuated equilibrium as a geometry of macroevolution does have broader implications for a more fundamental revision of evolutionary theory. Some critics have charged that we waffle on this issue, claiming either conformity with orthodoxy in order to ingratiate or novelty in order to inspire attention—but our position is as simple and consistent as the charge is anti-intellectual in that primal sense of substituting remarks ad hominem for analysis. The geometry of punctuated equilibrium may fall within conventional theory; the high relative frequency of the geometry, in its implications for the hierarchical perspective, may require revisions of the theory. We had not worked out this wider theme when we first wrote in 1972, largely because hierarchy itself was then a notion foreign to our understandings. Thus we now make broader claims for punctuated equilibrium within evolutionary theory than we did in 1972, but these are additions in a realm that we had not grasped when we first wrote, and we stand by our initial interpretation for the basic geometry of punctuated equilibrium.

I have argued (Gould, 1982b, 1986) that Darwin's theory of natural selection has a central logic focused on two propositions: (1) that natural selection plays a creative role in accumulating favorable changes through long sequences of successive steps, not merely a negative role in eliminating the unfit, and (2) that selection works by a struggle of organisms (not other entities like species, populations, or parts of organisms) for reproductive success. Allegiance to these statements forms a minimal criterion for conventional Darwinism. Darwinian naturalists may (indeed usually do) maintain a descriptively hierarchical view of nature acting simultaneously at several levels, but Darwinian causal explanations reduce this phenomenology to one locus in framing explanations based on selection—the struggle among organisms. Mayr's views (1982 and this volume) illustrate this point well. Mayr supports hierarchy in the description of nature but insists that all selection relevant to evolution occurs on organisms. He uses "species selection" in the purely descriptive sense that we call "sorting" and strongly maintains that all causal selection centers on organisms. In our terminology, his "species selection" is not selection on species at all, but a higher-level description of how ordinary selection on organisms appears from the perspective of species.

The restriction of selection to "struggle" among organisms was central to Darwin's research program and to his entire revolution in thought—for this radical idea cut through the Paleyan notion that a description of higher-order harmony required direct causation at these levels; Darwin argued that all phenomena traditionally ascribed to God's wisdom could arise as side consequences of nature's only causal process—the struggle among organisms for personal reproductive success. (Natural selection is the analogue of Adam Smith's struggle for profit. Natural selection yields ecological order incidentally, just as unrestrained personal competition leads to economic harmony, through the nonaction of the "invisible hand," in the theory of laissez-faire. Smith's vision provided Darwin's immediate inspiration for the fully articulated theory of natural selection—see Schweber, 1977; Gruber and Barrett, 1974.)

In breaking through this central logic, the hierarchical theory is non-Darwinian in its antireductionist claim that evolution proceeds by the simultaneous operation of selection at several levels (genes, organisms, and species among them), with complex transferences and interactions across levels. But hierarchy theory is not anti-Darwinian in the fundamental sense of rejecting selection as the primary agent of evolutionary causality. I have, rather, viewed hierarchy theory as a "higher" Darwinism because it takes the puzzles within conventional one-level selection theory and tries to resolve them by an enlarged view of selection acting at other levels as well. This theory is an expansion, not a refutation of

Darwinism—but it functions, both in practice and in logic, as a theory quite different in form and implication from the conventional one-level account.

Several critics of punctuated equilibrium have made the following logical mistake, thereby contributing to the acrimony of this discussion. Noting our claim that punctuated equilibrium represents something beyond Darwinism, but not grasping the notion of hierarchy and therefore confusing *beyond* with *against*, they assume that punctuated equilibrium is anti-Darwinian and therefore try to ally it with traditional forms of this genre (saltationism, in particular, a subject quite separate from punctuated equilibrium).

In asserting that punctuated equilibrium suggests an expansion with a big difference, not a refutation, of Darwinism, we are neither being diplomatic about its unconventionalities nor trying to occupy some middle ground between boredom and apostasy; our view arises directly from the logic of hierarchy theory and from the role of punctuated equilibrium in its development.

I believe that hierarchy theory in causal perspective will provide the most fundamental reconstruction of evolutionary theory since the Modern Synthesis of the 1930s and 1940s. I am not so vain as to think that punctuated equilibrium either constructed the theory or stands as its primary defense. Punctuated equilibrium is but one pathway to the elaboration of hierarchy, and probably not the best or most persuasive; that role will probably fall to our new understanding of the genome and the need for gene-level selection embodied in such ideas as "selfish DNA." But punctuated equilibrium was, historically, the main pathway by which macroevolution encountered this developing theory—and no causal theory of hierarchy can be complete without a firm understanding of the species level. (We must never forget that evolution is fundamentally about these larger entities—lions and tigers and things of C. H. Waddington's famous remark—however much the reductionistic traditions of science dictate that most research focus on lower-level phenomena.)

We may at least say that punctuated equilibrium directed the attention of macroevolutionists to the causally distinct status of species and that this insight helped to lead the larger discipline of evolutionary biology toward the expanded Darwinism of hierarchy theory. I think that I speak for Niles Eldredge as well in stating that we view the elaboration of punctuated equilibrium into a macroevolutionary basis for hierarchy theory as its major potential contribution, though we certainly take pride in the achievement of our original goal—the removal of the gradualist filter from paleontological perspectives and the recognition of punctuated equilibrium as a real and basic empirical pattern in life's history, not a feature of the record's imperfection, worthy only of our inattention or embarrassment.

A Note on Philosophies of Change in General

Uniformitarianism is a powerful vision of nature. In accord with the greatest contribution of late eighteenth-century political and intellectual turmoil to our view *de rerum natura*, it substituted the concept of change and flux as the essence of reality for our older comfort in stability and order based on fixed categories separated by unbridgeable gaps. It also introduced the primary gradualistic postulate that time equals effort—or that the magnitude of geological or biological alterations should be measured, as a primary correlation, by elapsed time available for the change. A fundamental corollary of this equation holds that objects undergoing change do not, by constraints of their structure, impede or "push back" upon transforming forces linked with time. Objects of change must be the structural equivalent of billiard balls, capable of responding linearly and isotropically to whatever temporal forces impinge upon them. It is, of course, our opposition to this vision that unites rapidity of change with stability of structure in the theory of punctuated equilibrium. Punctuation and stasis are opposite sides of the same coin in a critique of the uniformitarian notion of time and the nature of change. We hold that stability is a norm actively maintained and that change is a rare imposition—a quick transition between islands of stability in the sea of possible structures.

Punctuated equilibrium is but one specific manifestation, at one well-defined level, of this general alternative vision of change, or punctuational thinking in the large. Punctuated equilibrium is a theory about the origin and deployment of species in geological time—and nothing more. It is a theory about the second tier of time in my ranking of scales (Gould, 1985a). Punctuational theories for other scales have recently become popular—catastrophic notions of mass extinction at the third tier (Alvarez et al., 1980; Raup and Sepkoski, 1984) and new notions about the rapid restructuring of populations at the first (Carson and Templeton, 1984). In addition, punctuational ideas have become important in other disciplines (Kuhn, 1962, for example) and in general theories of natural change, particularly R. Thom's catastrophe theory

(1975) and I. Prigogine's notion of bifurcations (Prigogine and Stengers, 1984). This general movement in thought has several bases, including the popularity of structuralist thinking in philosophy (Foucault, 1972), the widespread recognition of uniformity and gradualism as a historical bias (see Gould, 1987), and (we would like to think) the empirical adequacy of punctuational change in a pluralistic world of many styles.

Punctuated equilibrium is not, most emphatically, either the origin or the main item in this general reassessment. Punctuated equilibrium is more a wave in this current—a theory for one level and one kind of phenomenon. We are proud to be part of this wider movement, and we have felt the excitement of being able to place our work into this broader context (see Gould, 1984); but punctuated equilibrium is not the context, only an item.

We have labored, almost to the point of pedantry, to keep clear this distinction between the specific theory of punctuated equilibrium and the wider context of punctuational thought. Nonetheless, colleagues have often criticized punctuated equilibrium by refuting another style and level of rapid change and then even asserting that its weaknesses are either our fault or carelessness and that punctuated equilibrium falls thereby.

Most egregious in this regard has been the long-standing and lamentable confusion of punctuated equilibrium with Richard Goldschmidt's saltationism—though we have continually, painstakingly, and clearly distinguished the two (Gould, 1982a). I do take an interest in Goldschmidt's ideas (Gould, 1982c), if only to redress a balance upset by previous derision among the orthodox (see Frazzetta, 1975:85). But some critics seem to think that because I co-authored the initial paper on punctuated equilibrium, everything I say subsequently about any subject is part and parcel of that research program. My interest in Goldschmidt springs largely from another source in my career—that leading to *Ontogeny and Phylogeny* (1977)—and I do assert an intellectual's birthright to multiple subjects of concern!

I have even heard the ad hominem charge that we purposely blur the specificity of punctuated equilibrium with the generality of punctuational change and that if we wish to further punctuated equilibrium, we should shut up about the generality and get on with it. I can only respond to such anti-intellectualism by stating that it is a scholar's duty to explore general contexts—but also, if not more so, to keep them distinct from specific and testable claims.

Criticisms and Proper Testing

Punctuated equilibrium has produced a good deal of debate, but this, in itself, pales to insignificance before the empirical work generated in its test upon scores of fossil lineages (symposium of Cope and Skelton, 1985, for example). Fruitfulness in research, with results both pro and contra, has provided our greatest satisfaction with punctuated equilibrium.

In this context of immediate utility, criticism of punctuated equilibrium has rested on three distinct claims, summarized below with an obviously partisan refutation.

1. Punctuated equilibrium is false or vacuous in principle or logic. I regard this attempt to dismiss the theory as fatuous, or at least in obvious discord with dozens of empirical studies that purport to test the theory in definite cases. (I may accuse my colleagues of many things, but not crass stupidity, and I doubt that so many would have spent so much time gathering data to test a theory that is, prima facie, devoid of meaning or false a priori.) P. D. Gingerich (1984), for example, has tried to win his argument by definition in proclaiming that gradualism is simply the equivalent of empiricism in paleontology and that stasis should be viewed "as gradualism at zero rate."

These comments puzzled and frustrated me until I realized, and confirmed in conversation with Gingerich, what he meant. He has recast gradualism not as an empirical statement about rate (with alternatives that can be tested), but as a claim for material continuity between succeeding stages in a genealogical series (punctuated equilibrium is then miscast as a claim for abrupt and absolute discontinuity). By these vacuous definitions, Gingerich's two statements are correct, of course—empiricism demands material continuity, and stasis is continuity without change (zero rate). But clearly, unless we are all monumental fools, these cannot be the definitions that have fueled this debate. Punctuation and gradualism are testable alternative hypotheses about rates in continuous genealogical sequences, not statements about material continuity versus an almost mystical notion of abruptness. Gingerich was correct in his earlier statement, and he did not waste (as his later views seem to hold) years of his life in an important series or papers dedicated to testing punctuated equilibrium (Gingerich, 1976; 1978). "Their [Eldredge's and my] view of speciation differs considerably from the traditional paleontological view of dynamic species with gradual evolutionary transitions, but it can be tested by study of the fossil record" (1978:454).

2. Punctuated equilibrium is a claim about the world, but it is untestable given the nature of the fossil record. A series of statements are folded into this common position (Levinton, 1983; Turner, 1986), but they form one coherent critique about paleontological limits to the recognition of biological species: the punctuational pattern of geologically abrupt appearance and subsequent long-term stasis of morphologically defined entities may be true, but if these entities cannot be equated with species, then the causal and explanatory postulates of punctuated equilibrium fail. Prominent among the charges are that since many events of speciation include no major shift of morphology, too many episodes of cladogenesis are unresolvable by fossils; segments of paleontological stasis may inextricably include not one species but several "cryptic" or "sibling" taxa undistinguishable in form but fully distinct as biological species; morphological differences may arise for so many reasons other than speciation (polymorphism within populations, ecophenotypic effects) that we cannot equate paleontological change with biological cause.

John Turner (1986), for example, in the most cogent critique of punctuated equilibrium yet published, invokes a gastronomical simile to compare our theory with a fork made of three prongs. The first—the punctuational pattern itself—he accepts as supported in a sufficient number of cases by the fossil record. The third—extension of the theory to explanation of trends as a higher-level sorting of species, perhaps importantly influenced by species selection—he regards as "an important extension of evolutionary theory into a hitherto little explored territory" (p. 206). The second—explanation of the punctuational pattern as a result of speciation—he regards as both central and untestable, "a naive misinterpretation of the fossil record" (p. 206) for reasons stated above.

Although we are grateful for this much from a formerly harsh critic (Turner, 1983, 1984)—the third prong, after all, is our most important theoretical claim, while our primary aim has always been to win recognition for the first prong as an observational fact of nature—I do acknowledge the centrality of the second prong and would make the following comments about species and the fossil record:

(a) The classical nature and long duration of this issue must be recognized; these concerns about species did not arise with punctuated equilibrium, and the critics of punctuated equilibrium did not invent the arguments. No subject has been more widely discussed among paleontologists (Weller, 1961; McAlester, 1962; entire volume and symposium of Sylvester-Bradley, 1956); it even bears a standard name as "*the* species problem in paleontology." Long discussion, of course, does not betoken resolution, but the pedigree has some relevance against those who imagine that they have discovered a gaping hole in paleontological logic, one that practitioners had never acknowledged before.

(b) I have never grasped the relevance of the charge (true of course) that many events of speciation entail little or no overt morphological change. Our claim is not that all speciation brings paleontologically visible change, but the very different notion that, *when* such change does occur, it accrues in concert with episodes of speciation. Similarly (for the same illogic has often been advanced in this case), the fact that most peripheral isolates do not form species is no argument against the claim that speciation, *when* it occurs, happens in peripherally isolated populations. If change at speciation is the "only game in town," then the evolution of phenotypes (including long-term trends) depends on speciation, even if most events of speciation yield little or no change.

(c) Similarly, I do not comprehend the argument about cryptic or sibling species, perhaps the most common point advanced against us. To be sure, we acknowledge that morphologically defined species may "hide" siblings; the situation is no different, of course, for modern species, the vast majority of which are defined morphologically and resident, so far as our knowledge goes, largely in museum drawers. If many periods of stasis include multiple species, then our argument is only strengthened—for the constraints upon change then affect several related forms, thereby providing affirmation in replication, whereas single incidents are only episodes.

(d) If identification of fossil species was always, and could only be in principle, grounded on morphology alone, critics would have a point. But we use an array of criteria beyond morphology in favorable cases—and these criteria provide both the standard resolution for "the species problem in paleontology" and the basis for our assertion that the biospecies has relevance as a paleontological subject. These additional criteria include geographical distribution and, especially, sympatric occurrence of species as a sign of biological distinctness, not mere morphological separation.

These criteria provide the major, and conceptually simple, test for cladogenesis (prong 2) as the basis of empirical patterns for punctuated equilibrium (prong 1): do ancestors continue to live after descendants branch off? If ancestors never appear in the record after descendants originate, then we do not know whether the punctuation has been caused by cladogenesis as punctuated equilibrium proposes, or by anagenesis at punctuational tempos (as a variety of mechanisms, including Wright's shifting balance—see Wright, 1982—would envision under certain environmental circumstances). But when ancestors survive, as they often do (Williamson, 1981; Vrba, 1980; White and Harris, 1977), then we have evidence for cladogenetic origin.

3. Punctuated equilibrium is testable but unsupported. This is the proper arena of debate, and an open issue at the moment. The question cannot be resolved, either pro or contra, by single cases, however well defined or documented. Natural history is not like a stereotype of physics, with crucial experiments disproving "laws" of nature that must be universally applicable. Virtually all questions in natural history boil down to issues of relative frequencies. Relative frequencies are notoriously difficult to establish in a multifarious world not only so full of cases per se but replete with cases of distinct individuality. This property of natural history explains why its great questions—including issues so fundamental as the power of adaptation and natural selection—are so difficult to answer and so long (and disputatiously) resident in the literature. Punctuated equilibrium is no different from other classical debates in natural history in this regard; it is neither easier nor more difficult to resolve than most conflicts within its genre.

As trained natural historians, we never made (as often charged) a claim for the exclusivity of punctuated equilibrium. Our assertion has always been for a relative frequency sufficiently high that the predominant amount of morphological change accumulated in evolutionary trends must be generated in punctuational events of cladogenesis. Our first paper (1972:97) simply tried to set up a model contrary to orthodox gradualism; while stating our hopes, we explicitly disclaimed any knowledge of actual frequencies: "How pervasive, then, is gradualism in these quasi-continuous sequences? A number of authors . . . have claimed that most species show little or no change throughout their stratigraphic range. But though it is tempting to conclude that gradual, progressive morphological change is an illusion, we recognize that there is little hard evidence to support either view." In our second paper (1977:119), we were even more explicit:

> We never claimed either that gradualism could not occur in theory, or did not occur in fact. Nature is far too varied and complex for such absolutes;

> Captain Corcoran's "hardly ever" is the strongest statement that a natural historian can hope to make. Issues like this are decided by relative frequency. . . . The fundamental question is not "whether at all" but "how often." Our unhappiness with gradualism arose from *its* status as restrictive dogma. For it has the unhappy property of excluding *a priori* the very data that might refute it. Stasis is ignored as "no data," while breaks are treated as imperfect data. . . . The explicit formulation of punctuated equilibria should lead to the casting of a wider net for data to test the relative frequency of evolutionary tempos. . . . Of course, we do not champion punctuated equilibria as liberal pluralists with no suspicion about the final outcome. We do regard punctuated equilibrium as by far the most common tempo of evolution—and we do assert that gradualism is both rare and unable in any case—given its characteristic rate—to serve as a source for major evolutionary events.

The first fifteen years of punctuated equilibrium have produced a wealth of elegant and decisive studies about individual cases, but we still do not have a proper sense of relative frequencies (and, more important, their differences across taxa, times, ecologies, and environments) as the consequence of a methodological impediment that is only now being rectified in the best modern studies.

Tabulation of cases from the literature tells us little about overall frequencies because the great majority were selected from much larger contemporary faunas on the strength of an a priori impression that the chosen species had undergone some change. Our motto and rallying cry—stasis is data—has still to take hold fully. For example, I have the impression (and this claim will be important if verified) that species of planktonic forams show unusually high frequencies of gradualistic change. But I also know that essentially all studies were done on lineages suspected or known to show change a priori. (Students desiring to undertake such a project, for example, often go to my good friend Bill Berggren, who, with his encyclopedic knowledge of Tertiary microfaunas, advises them to study this or that species because it has an "interesting" record of change.) Scores of other species, suspected of not changing or at least with undocumented change, have never been studied. The situation is even worse for data tabulated from literature before 1972 or from studies carried out without the context of this debate—for here the tyranny of reporting only change and treating stability as "nothing happening," and therefore not a phenomenon at all, reigns supreme. I am not even complaining of bias against punctuated equilibrium in the selection of cases—for some of our most elegant affirmations involve refuted cases of classical gradualism, as for coiling in the *Gryphaea* sequence (Hallam, 1959; Gould, 1972)—but of a skewed methodology that cannot, in principle, resolve the issue in either direction.

An instructive analogy may be made with electrophoretic studies of genetic variability. For more than half a century, geneticists could not answer the fundamental question, How different, on average, are two organisms within a species? They were stymied because, in the classical Mendelian analysis of pedigrees (the only technique then available), genes had to vary before they could be identified at all (so that inheritances of alternate alleles could be traced). Clearly, a biased sample, containing only genes that vary, cannot resolve the question. Such a sample greatly overestimates the average difference because nonvarying genes, a large majority according to all major theories (and now factually affirmed), will remain unassessed, and these must greatly reduce the average difference properly measured across all genes. The question can be answered only by taking a random sample of genes without foreknowledge of whether or not they vary.

Electrophoresis resolved this problem by permitting such random samples (Lewontin, 1974), and some of the initial results were surprising. For example, average differences among human races turn out to be astonishingly small, despite major external variation in such genetically controlled characters as skin and eye color. Measurement of average differences based only on these overtly varying genes would have yielded a completely spurious and elevated value. Similarly, compilation of gradualism based only on study of species known or suspected to change yields meaningless estimates of relative frequency for average evolutionary tempos.

By analogy with electrophoresis, this problem can be resolved only by a random sample of lineages, selected without knowledge of their tempos beforehand. Several sampling schemes are available for selecting taxa unbiased with respect to their larger group. One might, for example, consider a well-determined monophyletic clade and study all species within it. Or one might study all species with sufficient record in a particular geological formation, not just those that "look promising," or all measurable lineages of a higher taxon within a specified geographical province over a particular stretch of time. The possibilities are extensive, but all include the requirement of no preselection based on suspicions about change and complete (or at least randomized) enumeration within a chosen group.

As an example of major differences attained between right and wrong ways of sampling, two contrasting studies were presented at the North

American Paleontological Convention, Boulder, Colorado, in 1986. A. D. Barnovksy calculated relative frequency of punctuated equilibrium versus anagenetic transformation for Pleistocene mammals based exclusively on previously published reports in the literature. The two modes were close to equally frequent. Donald Prothero then reported on a field study for *all* mammalian lineages found in Oligocene rocks of the Big Badlands of South Dakota. All lineages remained in stasis, and all new forms entered the record with geological abruptness. He found no cases of gradual anagenesis. Of course, the differences might be real; perhaps the Pleistocene did witness a much higher frequency of gradualism. But I suspect that Barnovsky's result records the bias of literature; people tend to publish only on "interesting" lineages that are changing, and published works therefore impose a powerful bias toward gradualism (just as classical Mendelian analysis identified only varying genes). But Prothero studied *all* lineages for a time and place, without preconception about their modes or tempos—and all his lineages matched the predictions of punctuated equilibrium.

I am delighted that this message about proper sampling schemes for study of relative frequencies seems to be getting through. Three recent articles are exemplars in this regard. In the most detailed and elegant study done since this debate began, A. H. Cheetham (1986, 1987) treated all upper Miocene to lower Pliocene species of the monophyletic bryozoan genus *Metrarhabdotos* from the Dominican Republic. In an equally well-designed study with a different sampling scheme, Steven M. Stanley and X. Yang (1987) measured all lineages in four families of bivalves with sufficient record from Pliocene to Recent times in the western Atlantic. Both analyses found an overwhelming frequency of punctuated equilibrium and no role for anagenesis in situ in the production of new species or new morphologies. These studies, done in a methodologically proper way, indicate that punctuated equilibrium may have, as we originally proposed, a dominant relative frequency in the histories of most lineages. These studies are also admirable in treating a comprehensive range of nonredundant characters by multivariate morphometric analysis. Too much previous work has centered indecisively on single characters (usually abstracted from a total pattern because they seem, by qualitative observation before statistical analysis, to show change).

The Status of Punctuated Equilibrium Today

The debate on characteristic evolutionary modes and tempos remains lively and unresolved. I can envision a graded array of possible outcomes ranging from greatest pleasure for my adversaries to greatest satisfaction for myself.

1. Perhaps punctuated equilibrium is not a phenomenon at all, and its literal appearance either marks the record's imperfection or represents an artifact of biased preservation for biological data in geological archives. We have come far enough to dismiss this minimal option as false. Punctuated equilibrium happens, and often.

2. Perhaps punctuated equilibrium maintains a relative frequency too low to matter much. This option represents my principal fear, but I doubt that it can be correct for two major reasons. First, the subjective impression, based on a lifetime of experience, of too many distinguished taxonomic experts on particular groups affirms the dominance of the punctuational pattern within important taxa (Ager, 1986, on Mesozoic brachiopods; Fortey, 1985, on trilobites; Rickards, forthcoming, on graptolites, for example). Second, initial studies based on proper randomized or unbiased enumerative sampling schemes all affirm punctuated equilibrium overwhelmingly (Cheetham, 1986, 1987; Stanley and Yang, 1987).

3. Perhaps punctuated equilibrium is frequent, but our interpretations are overblown, and the pattern only records an orthodox form of neo-Darwinism geologically expressed. This would require that stasis be due to stabilizing selection and that all trends necessarily described as species sorting be fully reducible to selection at the organismic level. I have already given my reasons for doubting these interpretations.

4. Perhaps our unorthodox interpretations are sound in logic and argument—as our harsh but fair critic Turner allows (1986)—but the relative frequency of punctuated equilibrium is too low to permit these new insights much scope in the actual run of evolution. Or (more favorably for us), perhaps the relative frequency of punctuated equilibrium is high, but too few of its cases require the invocation of unorthodox interpretations for stasis or higher-level selection. The uniting theme in these two versions of outcome 4 admits the logic of our interpretation but denies its empirical importance. This claim is akin to R. A. Fisher's intriguing argument (1958) that species selection must be sound in logic but rare in occurrence. Fisher's argument fails in my view because it requires that stasis does not occur for reasons of constraint, but I greatly appreciate that this most brilliant of all Darwinians clearly appreciated the logical necessity for such a category.

From my own biased point of view, I think that we have come far enough to dismiss options 1 and 2. Options 3 and 4 now represent the position that opponents of punctuated equilibrium must defend. I believe that the second version of option 4 (high frequency of punctuated equilibrium but low frequency of cases requiring unorthodox interpretation) might be correct and now represents the most cogent critique of my larger expectations for punctuated equilibrium as embodied in options 5 and 6.

5. The relative frequency of punctuated equilibrium is high, and a large percentage of its cases require unorthodox interpretations for stasis and higher-level selection. This is the position that I have tried to defend throughout this chapter.

6. Punctuational styles of change will also prove to be important at other levels of organization and tiers of time both above and below punctuated equilibrium. Punctuated equilibrium shall therefore become an item and partner in a more general reinterpretation for the structure of history and the nature of change.

Wherever punctuated equilibrium comes to rest, whether it be deemed false and mundane or empirically true and ramifying in implication, I have at least had the unexampled pleasure of developing, under its aegis, a personal way of thought that challenged the fundamentals of my formal training and led me to reassess the basic postulates and procedures of my profession. I have been forced, above all, to take what had been regarded as an absence of data—the stability of lineages—and to recast it as a phenomenon that may aid our understanding of evolution. Intellectual life is largely a struggle to find more adequate taxonomies that reorder the world's complexity in more tractable ways. Stasis is data, and nature transcends the limits of our scale by operating Darwin's great insight about selection at several levels of its hierarchy.

References

Ager, D. V. 1986. Evolutionary patterns in Mesozoic Brachiopoda. In P. R. Racheboeuf and C. C. Emig, eds., *Biostratigraphie du Paleozoique* 4:33–41.

Alberch, P. 1985. Problems with the interpretation of developmental sequences. *Syst. Zool.* 34:46–58.

Alvarez, L., W. Alvarez, F. Asaro, and H. V. Michel. 1980. Extraterrestrial cause for the Cretaceous-Tertiary extinction. *Science* 208:1095–1108.

Anonymous. 1978. Missing, believed non-existent. *Guardian*, November 21.

Anonymous. 1982a. Evolution biologique et humaine. Ville de Dijon, Musée archeologique.

Anonymous. 1982b. L'evolution est-elle brusque ou graduelle? *Le Monde*, May 26.

Anonymous. 1984. Revolution in evolution. *The Times of India*, March 20.

Bakker, R. T. 1985. Evolution by revolution. *Science 85*, November, pp. 72–81.

Blanc, M. 1982. Les théories de l'évolution aujourd'hui. *La Recherche*, January, pp. 26–40.

Carson, H. L., and A. R. Templeton. 1984. Genetic revolutions in relation to speciation phenomena: the founding of new populations. *Annu. Rev. Ecol. Syst.* 15:97–131.

Charlesworth, B., R. Lande, and M. Slatkin. 1982. A Neo-Darwinian commentary on macroevolution. *Evolution* 36:474–498.

Cheetham, A. H. 1986. Tempo of evolution in a Neogene bryozoan: rates of morphologic change within and across species boundaries. *Paleobiology* 12:190–202.

——. 1987. Trends of evolution in a Neogene bryozoan: are trends in single morphological charcters misleading? *Paleobiology* 13:286–296.

Cope, J. C. W., and P. W. Skelton, eds. 1985. Evolutionary case histories from the fossil record. *Palaeontol. Assoc. Special Papers* no. 33.

Cronin, T. M. 1985. Speciation and stasis in marine Ostracoda: climatic modulation of evolution. *Science* 227:60–63.

——. 1987. Evolution, biogeography, and systematics of *Puriana*: evolution and speciation in Ostracoda, III. *J. Paleontol.*, Suppl. to no. 3 (Memoir 21), vol. 61.

Darwin, C. 1859. *On the Origin of Species*. London: Murray.

Davy, J. 1981. Once upon a time. . . . *Observer*, August 16.

Dawkins, R. 1985. What was all the fuss about? *Nature* 316:683–684.

Eldredge, N. 1979. Alternative approaches to evolutionary theory. In J. H. Schwartz and H. B. Rollins, eds., *Models and Methodologies in Evolutionary Theory*. Bull. Carnegie Mus. Nat. Hist. 13:7–19.

Eldredge, N., and S. J. Gould. 1972. Punctuated equilibria: an alternative to phyletic gradualism. In T. J. M. Schopf, ed., *Models in Paleobiology*. San Francisco: Freeman, Cooper, pp. 305–332.

Eisher, R. A. 1958. *The Genetical Theory of Natural Selection*. New York: Dover.

Fortey, R. A. 1985. Gradualism and punctuated equilibria as competing and complementary theories. In J. C. W. Cope and P. W. Skelton, eds. *Paleontol. Assoc. Special Papers* no. 33. pp. 17–28.

Foucault, M. 1972. *The Archaelogy of Knowledge*. New York: Pantheon Books.

Frazzetta, T. H. 1975. *Complex Adaptations in Evolving Populations*. Sunderland, Mass.: Sinauer.

Fryer, G., P. H. Greenwood, and J. F. Peake. 1983. Punctuated equilibrium, morphological stasis, and the palaeontological documentation of stasis: a biological appraisal of a case history in an African lake. *Biol. J. Linn. Soc.* 20:195–205.

Garcia-Valdecasas, A. 1982. La teoria de la evolucion: los terminos de una controversia. *Rev. Univ. Complutense de Madrid*, pp. 153–158.

Gehring, W. J. 1985. The homeobox: a key to the understanding of development. *Cell* 40:3–5.

——. 1987. Homeotic genes, the homeobox, and the spatial organization of the embryo. *Harvey Lect.* 81:153–172.

Gilinsky, N. L. 1986. Species selection as a causal process. *Evol. Biol.* 20:249–273.

Gingerich, P. D. 1976. Paleontology and phylogeny: patterns of evolution at the species level in early Tertiary mammals. *Am. J. Sci.* 276:1–28.

——. 1978. Evolutionary transition from ammonite *Subprionocyclus* to *Reedsites*—punctuated or gradual? *Evolution* 32:454–456.

——. 1984. Punctuated equilibria—where is the evidence? *Syst. Zool.* 33:335–338.

Goodwin, B. C. 1980. Pattern formation and its regeneration in the Protozoa. In G. W. Gooday, ed., *The Eukaryotic Microbial Cell.* Soc. General Microbial Symp. No. 30. Cambridge: Cambridge University Press, pp. 377–404.

Gould, S. J. 1965. Is uniformitarianism necessary? *Am. J. Sci.* 263:223–228.

——. 1972. Allometric fallacies and the evolution of *Gryphaea*: a new interpretaion based on White's criterion of geometric similarity. In T. Dobzhansky et al., eds., *Evolutionary Biology* 6:91–118.

——. 1974. The evolutionary significance of "bizarre" structures: antler size and skull size in the "Irish Elk," *Megaloceros giganteus. Evolution* 28:191–220.

——. 1977. *Ontogeny and Phylogeny*. Cambridge, Mass.: Belknap Press of Harvard University Press.

——. 1982a. The meaning of punctuated equilibrium and its role in validating a hierarchical approach to macroevolution. In R. Milkman, ed., *Perspectives on Evolution*. Sunderland, Mass.: Sinauer, pp. 83–104.

——. 1982b. Darwinism and the expansion of evolutionary theory. *Science* 216:380–387.

——. 1982c. The uses of heresy: an introduction to Richard Goldschmidt's *The Material Basis of Evolution*. New Haven: Yale University Press, pp. xiii–xlii.

——. 1984. Toward the vindication of punctuational change. In W. A. Berggren and J. A. Van Couvering, eds., *Catastrophes and Earth History*. Princeton: Princeton University Press, pp. 9–34.

——. 1985a. The paradox of the first tier: an agenda for paleobiology. *Paleobiology* 11:2–12.

——. 1985b. Geoffrey and the homeobox. *Nat. Hist.* 94 (November): 12–23.

——. 1986. Evolution and the triumph of homology, or why history matters. *Am. Sci.* (January–February): 60–69.

——. 1987. *Time's Arrow, Time's Cycle*. Cambridge, Mass.: Harvard University Press.

Gould, S. J., and N. Eldredge. 1977. Punctuated equilibria: the tempo and mode of evolution reconsidered. *Paleobiology* 3:115–151.

Gould, S. J., and R. C. Lewontin. 1979. The spandrels of San Marco and the Panglossian paradigm: a critique of the adaptationist programme. *Proc. R. Soc. Lond.* B 205:581–598.

Gould, S. J., and E. S. Vrba. 1982. Exaptation—a missing term in the science of form. *Paleobiology* 8:4–15.

Gruber, H. E., and P. H. Barrett. 1974. *Darwin on Man.* New York: E. P. Dutton.

Hallam, A. 1959. On the supposed evolution of *Gryphaea* in the Lias. *Geol. Mag.* 96:99–108.

Huxley, A. 1982. Address of the president. *Proc. R. Soc. Lond.* 214:137–152.

Jerison, H. J. 1973. *The Evolution of the Brain and Intelligence.* New York: Academic Press.

Kauffman, S. A. 1983. Developmental constraints: Internal factors in evolution. In B. C. Goodwin, ed., *Development and Evolution.* Cambridge: Cambridge University Press, pp. 195–225.

Kuhn, T. S. 1962. *The Structure of Scientific Revolutions.* Chicago: University of Chicago Press.

Lande, R. 1985. Expected time for random genetic drift of a population between stable phenotypic states. *Proc. Natl. Acad. Sci. USA* 82:7641–7645.

Lessem, D. 1987. Punctuated equilibrium. *Boston Globe*, April 6.

Levinton, J. S. 1983. Stasis in progress: the empirical basis of macroevolution. *Annu. Rev. Ecol. Syst.* 14:103–137.

Lewin, R. 1980. Evolutionary theory under fire. *Science* 210:883–887.

——. 1986. Punctuated equilibrium is now old hat. *Science* 231:672–673.

Lewontin, R. C. 1974. *The Genetic Bases of Evolutionary Change.* New York: Columbia University Press.

Lloyd, Lisa. Forthcoming. Book on hierarchy and units of selection. (Department of Philosophy, University of California, San Diego.)

Lu, J. 1983. Is the Darwinian theory of evolution really wrong—a highlighted debate. (In Chinese) *Ren Min Ri Bao* [People's daily], Beijing, March 21.

Mahé, J. and C. Devillers. 1981. Stabilité de l'espèce et évolution: la théorie de l'equilibre intermittent ("punctuated equilibrium"). *Geobios* 14:477–491.

Marx, J. L. 1986. The continuing saga of "homeo-madness." *Science* 232:158–159.

Maynard Smith, J. 1983a. The genetics of stasis and punctuation. *Annu. Rev. Genet.* 17:11–25.

——. 1983b. Evolution and development. In B. C. Goodwin, ed., *Development and Evolution.* Cambridge: Cambridge University Press, pp. 33–45.

——. 1984. Palaeontology at the high table. *Nature* 309:401–402.

Mayo, D. G., and N. L. Gilinsky. 1987. Models of group selection. *Philos. Sci.*.

Mayr, E. 1954. Change of genetic environment and evolution. In J. Huxley, A. C. Hardy, and E. B. Ford, eds., *Evolution as a Process.* London: Allen & Unwin, pp. 157–180.

——. 1963. *Animal Species and Evolution*. Cambridge, Mass.: Harvard University Press.
——. 1982. Speciation and macroevolution. *Evolution* 36:1119–1132.
Mayr, E., and W. Provine. 1980. *The Evolutionary Synthesis*. Cambridge, Mass.: Harvard University Press.
McAlester, A. L. 1962. Some comments on the species problem. *J. Paleontol.* 36:1377–1381.
Milligan, B. G. 1986. Punctuated evolution induced by ecological change. *Am. Nat.* 127:522–532.
Mondolfi, J. C. 1986. Y el eslabon perdido, que? Caracas, *Espada Rota*, January–February.
Newman, C. M., J. E. Cohen, and C. Kipnis. 1985. Neo-Darwinian evolution implies punctuated equilibria. *Nature* 315:400–401.
Paterson, M. 1986. Story of evolution still unfolding. *New Zealand Herald, Auckland, Weekend Magazine*, July 5.
Prigogine, I., and I. Stengers. 1984. *Order Out of Chaos*. New York: Bantam Books.
Raup, D. M., and J. J. Sepkoski, Jr. 1984. Periodicity of extinctions in the geologic past. *Proc. Natl. Acad. Sci. USA* 81:801–805.
Rensberger, B. 1980. Recent studies spark revolution in interpretation of evolution. *New York Times*, Tuesday, November 4.
——. 1981. The evolution of evolution. *Mosaic*, September–October, pp. 14–22.
——. 1984. New challenges to Darwin. *This World*, December 23, pp. 15–16.
Rhodes, F. H. T. 1983. Gradualism, punctuated equilibrium and the *Origin of the Species*. *Nature* 305:269–272.
Rickards, R. B. Forthcoming. Anachronistic, heraldic and echoic evolution: new patterns revealed by extinct, planktonic hemichordates.
Salvatori, N. 1984. Paleontologi a confronto sui tempi e i modi dell'evolutione. *Airone*, December, p. 34.
Schweber, S. S. 1977. The origin of the *Origin* revisted. *J. Hist. Biol.* 10:229–316.
Sequeiros, L. 1981. La evolucion biologica, teoria en crisis. *Razon y Fe*, December, pp. 586–593.
Simpson, G. G. 1976. The compleat paleontologist. *Annu. Rev. Earth Planetary Sci.* 4:1–14.
Sober, E. 1984. *The Nature of Selection*. Cambridge, Mass.: MIT Press.
Stanley, S. M. 1975. A theory of evolution above the species level. *Proc. Natl. Acad. Sci. USA* 72:646–650.
——. 1979. *Macroevolution: Patterns and Process*. San Francisco: W. H. Freeman.
Stanley, S. M., and X. Yang. 1987. Approximate evolutionary stasis for bivalve morphology over millions of years: a multivariate, multilineage study. *Paleobiology* 13:113–139.

Stebbins, G. L., and F. J. Ayala. 1982. Is a new evolutionary synthesis necessary? *Science* 23:967–971.

Sylvester-Bradley, P. C., ed., 1956. *The Species Concept in Palaeontology*. London: Systematics Assoc.

Thom, R. 1975. *Structural Stability and Morphogenesis*. Reading, Mass.: W. A. Benjamin.

Thuillier, P. 1981. Darwin et C°. Brussels: Editions complexe.

Tice, D. J. 1986. Life after Darwin. *TWA Ambassador*, August, pp. 43–50.

Turner, J. 1983. "The hypothesis that explains mimetic resemblance explains evolution"—The gradualist-saltationist schism. In M. Greene, ed., *Dimensions of Darwinism*. New York: Cambridge University Press, pp. 129–169.

——. 1984. Why we need evolution by jerks. *New Sci.*, February 9, pp. 34–35.

——. 1986. The genetics of adaptive radiation: a Neo-Darwinian theory of punctuational change. In D. M. Raup and D. Jablonski, eds., *Patterns and Processes in the History of Life*. Berlin: Springer Verlag, pp. 183–207.

Vrba, E. S. 1980. Evolution, species and fossils: how does life evolve? *S. Afr. J. Sci.* 76:61–84.

Vrba, E. S., and N. Eldredge. 1984. Individuals, hierarchies and processes: towards a more complete evolutionary theory. *Paleobiology* 10:146–171.

Vrba, E. S., and S. J. Gould. 1986. The hierarchical expansion of sorting and selection: sorting and selection cannot be equated. *Paleobiology* 12:217–228.

Weller, J. M. 1961. The species problem. *J. Paleontol.* 35:1181–1192.

White, T. D., and J. M. Harris. 1977. Suid evolution and correlation of African hominid localities. *Science* 198:13–21.

Williamson, P. G. 1981. Palaeontological documentation of speciation of Cenozoic molluscs from Turkana Basin. *Nature* 293:437–443.

——. 1985a. Punctuated equilibrium, morphological stasis and the palaeontological documentation of speciation. *Biol. J. Linn. Soc. Lond.* 26:307–324.

——. 1985b. In reply to Fryer, Greenwood, and Peake. *Biol. J. Linn. Soc. Lond.* 26:337–340.

——. 1986. Selection or constraint? A proposal on the mechanisms of stasis. In K. S. W. Campbell and M. F. Day, eds., *Rates of Evolution*. London: Allen & Unwin.

Wright, S. 1982. Character change, speciation and the higher taxa. *Evolution* 36:427–443.

3

The Empirical Case for the Punctuational Model of Evolution

STEVEN M. STANLEY

There is no question that well-established species evolve. The punctuational hypothesis (Eldredge, 1971; Eldredge and Gould, 1972) asserts that these species usually evolve very sluggishly, however, and that the failure of the fossil record to document the gradual unfolding of major evolutionary trends reflects this relative stasis, rather than a general deficiency of the record of well-established species. According to this hypothesis, the inadequacy of the fossil record is on a much finer scale: the record fails to document events of rapid evolutionary transformation. The second aspect of this hypothesis is that these rapid events account for most evolution in the history of life. To test this assertion, we can formulate two mutually exclusive hypotheses (Stanley, 1975, 1979). The first holds that most evolutionary change occurs by the gradual transformation of established species, which is to say by long-term phyletic evolution (the gradualistic model); the second asserts, alternatively, that most evolutionary change is associated with rapid branching events, or speciation (the punctuational model). If the phyletic and speciational components of evolution are nearly equal, we cannot hope to employ fossil data to distinguish between them. If one component dominates, however, there are ways in which the fossil record will allow us to test one possibility against the other. There is still a third pattern of change to be considered—a stepwise (or staircase) pattern in which well-established species occasionally undergo rapid change in between long intervals of approximate stasis or slow phyletic change. In this staircase pattern, most change takes place during brief intervals without branching.

The preceding formulation allows us to avoid couching the issue in extreme positions, as unfortunately happened in the immediate aftermath of Eldredge and Gould's 1972 paper, when some workers assumed that the punctuational model associated virtually all evolutionary change with speciation. Even ten years later, a conspicuous review of the controversy stated that "the model of punctuated equilibria, as meant here, claims only that apart from very short periods of very rapid, but not necessarily discontinuous, evolution in very small populations, the rate of morphologic evolution is close to zero" (Hoffman, 1982). I know of no punctuationist who harbors so extreme a view.

Many major evolutionary transformatioms have been concentrated in episodes known as adaptive radiations—rapid origins of many taxa from relatively small ancestral groups. The gradualistic model can be tested by investigating what changes have occurred within well-established species that have played a role in adaptive radiation. Hypothetically, if we were to find that all of the varied species constituting an adaptive radiation underwent little evolutionary change once they were well established, we would have to invoke the punctuational alternative, the staircase alternative, or a combination of the two: most of the changes accounting for the diversity of species would have to have been concentrated in rapid events—even though the fossil record might never document the details of these steps. I have labeled this test of the gradualistic model the Test of Adaptive Radiation and applied it chiefly to the adaptive radiation of the mammals that commenced at the start of the Cenozoic Era, following the demise of the dinosaurs, which left the adaptive zone of terrestrial tetrapods largely vacant.

Even as late as the early Eocene, which spanned approximately 3 to 5 million years, the mammalian adaptive radiation continued: numerous new genera and families were arising. The excellent early Eocene fossil record of mammals in the Big Horn Basin of Wyoming reveals that, though many mammals were evolving, mammalian species were typically very stable. D. M. Schankler (1980) plotted the stratigraphic ranges of about seventy species of this region and found that the large majority existed for at least a million years and many survived 2 million years or more. At this time, Wyoming supported a subtropical jungle occupied by mammal species that must have made up a large fraction of the total North American mammalian fauna. Many of these species were terminated by true extinction, but to give gradualism the benefit of the doubt, we can postulate that species disappeared by pseudo-extinction—by evolving enough that at some arbitrary point they were recognized as new species. Even this charitable posture allows us to array at most three or four species end-to-end to span the entire early Eocene. The key point is that the set of populations assigned to any single species in the fossil record encompasses only a very small range of morphologies. An evolutionary lineage consisting of only three or four species, each grading into the next, will not display enough change that the earliest and latest populations will be assigned to different families. Seldom will even a generic transition be recognized. Thus the gradualistic model fails this test. Higher taxa must have been arising primarily via brief events of rapid transformation. These events were the loci of most evolutionary change.

This approach to the issue at hand has been criticized with the suggestion that paleontological species are merely figments of taxonomists' imaginations (Maynard Smith, 1981)—that many of the entities designed as species may actually embody substantial evolutionary change. Two recent studies have blunted this challenge. One is a thorough study of the pattern of evolution within *Metrarabdotos*, a genus of marine bryozoans (Cheetham, 1986). The other is a similarly comprehensive study of a large group of marine bivalve mollusks (Stanley and Yang, 1987).

Stasis in a Clade of Marine Bryozoans

Bryozoans, or moss animals, encrust hard substrata in seas throughout the world. Alan Cheetham's study focused on Caribbean species of the bryozoan genus *Metrarabdotos*, which appear to form an isolated, monophyletic clade. A small fragment of a *Metrarabdotos* colony serves to characterize the colony morphometrically, and to study the pattern of evolution of the genus, Cheetham took advantage of the great abundance of such fragments in a remarkably continuous stratigraphic section of marine sediments in the Dominican Republic. His study was based on a segment of this section representing the time interval extending from about 8 to 3.5 million years before the present. The mean interval between successsive samples was 0.16 million years, although no species was found in every sample between its lowermost and uppermost occurrences. Cheetham's morphometric analysis was based on the measurement of forty-six characters for each colony. That such a set of characters adequately depicts overall morphology is illustrated by the strong correlation between bryozoan skeletal form and soft anatomy

(Winston, 1981). Employing the Mahalanobis multivariate statistical distance, Cheetham assigned fossil colonies to species by cluster analysis, and he then reconstructed the phylogeny for the thirteen late Neogene species by connecting nearest-neighbor clusters using morphological distances and stratigraphic positions.

Cheetham's primary question was whether rates of evolutionary change within species (phyletic evolution) could account for transitions between species that formed rapidly and evolved slowly thereafter. To answer this question he adapted a technique devised by Brian Charlesworth (1985). This technique determines whether the minimum rate of evolution required to form one species from another is significantly greater than the mean rate of evolution within the ancestral species. Sampling proved adequate to compare degrees of morphological change within nine species to degrees of morphological change between ancestral and descendant species. In every case, the rate across the species boundary exceeded that within the ancestral species, and in eight of these cases, this difference was statistically significant. The strength of the punctuational pattern was heavily underscored by the observation that in every case the mean rate of evolution for an established species was not significantly different from zero.

For two reasons, Cheetham's study not only ruled out phyletic evolution as the dominant source of evolution within the *Metrarabdotos* clade but also ruled out staircase evolution. First, there is a numerical problem: the clade was diversifying so that there were no potential ancestors for the origin of many of the new species by phyletic transition. Second, many ancestral and descendant species overlapped in time.

Stasis in a Set of Marine Bivalve Lineages

The other comprehensive analysis, which also revealed a prevalence of approximate stasis, focused on bivalve mollusks that burrow in the sea floor (Stanley and Yang, 1987). Our starting point for this study was an arbitrarily assembled group of species of early Pliocene age (about 4 million years old). For inclusion, an early Pliocene species had to belong to one of four particular higher taxa, have an adult size larger than a specified limit (for easy measurement), be characterized by a weakly convex shell (to minimize distortion when characters were projected onto a plane for measurement), be represented by a fossil population of about twenty well-preserved specimens falling within a narrow size range, and have potential descendant populations in the modern world—congeneric populations more similar to it than to any other early Pliocene species. These rules for qualification yielded nineteen early Pliocene species, each of which was compared to the most similar living species. Morphological variables for this comparison were distances between features of the shell interior—muscle scars, hinge teeth, and the like—and lengths of lines connecting points on the shell margin and thus defining the shell outline. There were twenty-four of these variables, and all measurements were normalized for size by expressing them as a fraction of the square root of shell area. The twenty-four variables provided information not only on shell form but also on a wide array of traits of soft anatomy and behavior. There is no reason to assume that the collective rate of evolutionary change for these traits should have differed from that of other adaptive features.

The question that we asked in comparing the fossil and living forms was, How much evolution has occurred in the course of 4 million years? The choice of the most similar living species as the descendant form was a matter of parsimony: it was assumed to be the most likely phyletic descendant. In twelve cases, the most similar living species, by conventionl taxonomy, belonged to the same species as the fossil population. In four cases, the living species has traditionally been assigned to a different species than the fossil population but was considered possibly or probably to be the actual direct phyletic descendant. In the three remaining cases, the living species was a Pacific form that was almost certainly not the true phyletic descendant, meaning that it was separated from the early Pliocene form by at least one speciation event, which possibly occurred before 4 million years ago. To be conservative, however (to risk attributing more change to phyletic evolution than it actually produced), we assumed that the Pacific species were the direct phyletic descendants of the early Pliocene forms.

We compared fossil and living bivalve species by means of the Mahalanobis multivariate statistic. Recognizing that no two populatious are identical, we sought a way to give meaning to values of the statistic. The chosen yardstick was the equivalent set of values obtained in comparing geographically separated populations belonging to individual recent species that were employed in the study. The result was striking evidence for approximate stasis. Nearly all comparisons of 4-million-year-old and recent populations yielded multivariate distances within the range of intraspecific geographic variability. We also examined mean percentage differences between populations for each of the twenty-four parameters. The set of differences in all parameters for all comparisons between fossils and recent specimens turned out to be remarkably similar to the comparable set of differences between conspecific recent populations.

For four lineages, we were able to extend the temporal interval of our analysis. Three of the early Pliocene lineages could be traced back to a time 17 million years before the present, and one lineage, back to a time 13 million years before the present. Even over these great spans of time, evolutionary differences were virtually all within the range measured for conspecific recent populations. For the *Linga pensylvanica* lineage, discriminant analysis actually misassigned three of twenty-four 17-million-year-old shells to the recent populations. Populations approximately 1 and 2 million years old were available for three of these ancient lineages, and the entire set of population for each of these revealed a weak zigzag pattern. Such a pattern may represent actual zigzag evolution, within narrow limits, or simply haphazard sampling of geographic variability at various points. In any event, it represents approximate evolutionary stasis over a long interval of geologic time.

Is Staircase Evolution an Alternative?

Even if approximate stasis has characterized lineages throughout nearly all of their existence, the possibility remains that they have undergone episodic phyletic change during brief intervals so that the overall pattern has been stepwise change without branching. B. A. Malmgren et al. (1983) have identified such a pattern for the *Globorotalia conoidea–G. inflata* lineage of planktonic foraminifera, in which the evolutionary step occurred close to the Miocene-Pliocene boundary. The change in shape achieved by this evolutionary step was not enormous, however, and its functional significance remains unknown. In addition, it is not evident that comparable patterns have typified multicellular organisms.

I pointed out earlier that the study by Cheetham (1986) rules out transition from species to species by abrupt phyletic steps as the dominant mode of change. The frequent occurrence of staircase evolution is more generally opposed by the history of long, narrow clades, that is, ones that have survived for great intervals of time with little branching. By definition, all such clades have experienced few speciation events; for this reason, the punctuational model predicts that they should exhibit little evolutionary change. The gradualistic and staircase models make no such prediction; according to these models, long, narrow clades should experience a normal range of phyletic rates. As it turns out, all extant long, narrow clades are represented in the modern world by taxa (such as aardvarks, snapping turtles, and alligators) that closely resemble their ancient ancestors: the prediction of the punctuational model is upheld. This constitutes the Test of Living Fossils (Stanley, 1975, 1979). Despite these arguments, we do not yet have a clear picture of the relative importance of rapidly divergent speciation and sudden phyletic steps (the staircase pattern) in the history of life in general.

Less Comprehensive Studies

What is unique about the studies of Cheetham (1986) and Stanley and Yang (1987) is that they represent comprehensive and seemingly unbiased tests of the importance of gradual phyletic evolution. Each dealt with a large set of arbitrarily chosen, true evolutionary lineages and employed a large set of morphologic parameters that represent not merely a few skeletal features but a wide variety of anatomical and behavioral traits.

Numerous paleontological studies, in contrast to those described above, have purported to document substantial gradual phyletic evolution. A review of these reveals that most are rendered equivocal or diminished in importance by at least one of three deficiencies:

1. Studies of this type too numerous to list have failed to demonstrate that two populations or taxa compared as ancestor and descendant are, in fact, segments of a single evolutionary lineage. In other words, for each of these studies the possibility remains that one or more speciation events have been overlooked. A large proportion of such studies were undertaken in the era when gradualism dominated paleontological thinking so that there seemed no need for special concern about the evolutionary contribution of speciation. Even since the punctuational hypothesis was advanced, many studies have been published purporting to document phyletic evolution without accounting for major gaps in the fossil record. Some of these, especially those focusing on Pleistocene mammals, have failed even to provide convincing evidence of the precise ages of included populations (e.g., Chaline and Laurin, 1986).

2. Most studies claiming to measure gradual phyletic evolution have focused on lineages thought in advance to embody substantial evolu-

tion. In other words, no attempt has been made to assess an arbitrarily or randomly selected set of lineages so as to include examples of stasis wherever they may occur. Gould and Eldredge (1977:116), in condemning the widespread failure of paleontologists to acknowledge that "stasis is data," have accurately portrayed the history of research on phyletic evolution. One of the best-documented examples of substantial phyletic evolution is that of the previously mentioned *Globorotalia conoidea–G. inflata* lineage of planktonic foraminifera for the interval between about 8 million years ago and the present (Malmgren and Kennett, 1982). During this span of time, this lineage appears to have undergone changes in skeletal (test) shape, including decrease in number of chambers in the final whorl and rounding of the periphery (both of which were the net result of numerous short-term, back-and-forth trends). Even this study has been challenged with the allegation that its conclusions may not apply to all geographic areas; ancestral morphologies may have persisted outside the narrow geographic region of the investigation (Scott, 1983). In addition, we must address the more general issue of bias in the choice of a lineage to study. The ancestral populations of the *Globorotalia conoidea–G. inflata* lineage existed about 8 million years ago with more than thirty other species of planktonic foraminifera. Four million years later more than twenty of these species survived as the same nominal species; most of the rest had suffered extinction. This does not imply perfect stasis, but it does indicate very sluggish change. It is precisely this sluggish change that has seldom been quantified because it has seemed uninteresting.

3. Paleontologists have not only tended to assess phyletic evolution for lineages in which they have believed, in advance, that it accomplished a great deal, but they have also tended to measure change only for particular characters that, at the outset of their study, they believed had undergone substantial change. This bias has been especially pronounced in the preferential choice of metrics that reflect only body size. As it turns out, size is far more flexible in evolution than is shape. A summary by L. Van Valen (1974) of published rates of evolution illustrates the bias. Sixty-four of the seventy-four rates listed here represent alleged evolutionary changes in size. The relative ease with which body size can evolve is illustrated by the history of marine bivalves and ammonoids of the European Jurassic (Hallam, 1975). Adult size is difficult to assess for bivalves, in which growth is indeterminate and can be measured in only an approximate way. Even so, large fossil populations are available for many species of the intensively studied European Jurassic, and the substantial increases in body size reported by A. Hallam

must be generally valid. Some of these species appear to have doubled or trebled their adult size in the course of 10 or 20 million years. Hallam (1978) noted, however, that changes in morphologic traits other than size were rare in Jurassic bivalves, and rates of change that he calculated for size in some species are much higher than any rates that Stanley and Yang (1987) calculated for shape in Neogene bivalves. These and other data suggest that it is a general rule of evolution that size is far more labile than shape.

There is no question that many assessments of phyletic evolution remain equivocal, but the bald statement of Antoni Hoffman (1982) that "there is no way to decide which of the two ideal models is more commonly fitted by real evolutionary lineages" is unjustified. The comprehensive studies recently conducted for marine bryozoans (Cheetham, 1986) and bivalves (Stanley and Yang, 1987) illustrate how the issue can be resolved. I should emphasize, however, that the punctuational pattern, with which these studies accord, allows for some gradual phyletic evolution of shape and for substantial changes in body size.

The Meaning of Species Longevities

The comprehensive morphometric studies of Cheetham (1986) and Stanley and Yang (1987) also suggest that, even in the absence of morphometric data, assessments of species longevities of the sort summarized earlier can provide valid evidence for evolutionary stasis. Cheetham found that the morphologic characters contributing most to interspecific differences in his multispecies discriminant analysis for ascophoran bryozoans were the ones that taxonomists have traditionally employed in the recognition of species. In other words, his analysis included taxonomically important characters. In the study of Stanley and Yang (1987), every bivalve lineage that had previously been assigned to a single species was found to encompass virtually no net directional change over any time span considered (up to 17 million years). Furthermore, two lineages that previously had each been divided into ancestral and descendant chronospecies were found to encompass virtually no net directional change: they deserve placement in a single species.

These observations support the generalization that when a competent taxonomist assigns conspecific status to two well-preserved fossil populations that lived at different times, these populations ordinarily are not separated by any more than a trivial amount of evolution. Numerous examples of such comparisons also undercut the criticism that fossil remains fail to offer a good estimate of total phenotype. Admittedly, species-level taxonomy for fossil members of some groups is very poor, either because the skeleton provides relatively little information (frogs are an example) or because fossil skeletons are usually fragmentary (birds are an example). Taxa of this type are of little value in assessing rates of evolution. By contrast, the skeletons of the cheilostome bryozoans and large siphonate bivalves employed in the studies described above are complex entities whose attributes reflect a wide variety of anatomical, physiological, and behavioral traits. From a statistical standpoint, it would be absurd to claim that for some reason these traits were preferentially immune to evolution—that they constitute a biased sample of phenotypic change (Schopf, 1982).

The North American and European species of the sycamore tree (*Platanus*) provide an interesting example of how morphological stasis can prevail while reproductive compatibility, the determinant of conspecific status, is retained. The two groups of sycamores have been isolated for at least 30 million years, yet they are not only very similar in morphology but also yield fertile hybrids (Stebbins, 1950).

Criticism has also been leveled at the fossil record for its failure to reveal the identity of sibling species—reproductively isolated taxa that exhibit nearly identical morphologies (Schopf, 1982). Ironically, this failure actually strengthens the case that species longevities make for the punctuational model. There is no question that many single "lineages" recognized in the fossil record must actually encompass two or more very similar species that overlapped in time. Under such circumstances, when approximate stasis has been maintained for millions of years, the stable "lineage" represents not a single example of stasis but two or more examples.

That speciation events separating sibling species have yielded very little divergence has no bearing on the punctuational model, however numerous sibling species may be in phylogeny. The punctuational model says nothing of the mean degree of adaptive change per speciation event. Rather, it claims that however infrequent rapid and pronounced evolutionary transitions may be, these events account for most of the total evolutionary change in phylogeny.

Rapidly Divergent Speciation

It has been asserted that paleontologists cannot test the punctuational model without documenting rapidly divergent speciation (Maynard

Smith, 1981). This demand ignores the fact that comprehensive, unbiased analyses of rates of phyletic change test the punctuational model against gradualism by following the conventional hypothetico-deductive method that pervades modern science. Cheetham (1986) in effect pitted three mutually exclusive hypotheses against one another: one holds that most evolutionary change in phylogeny is gradual, the second that most change is achieved by rapidly divergent speciation events, and the third that most change occurs in rapid steps not associated with branching. If two or all three of the potential components of evolution are subequal in importance, test results could be equivocal—and, perhaps, of limited interest. If, however, as Cheetham found, approximate stasis prevailed and most ancestral and descendant species overlapped in time, the only alternative is to account for most change by invoking rapidly divergent speciation events. In the judgment in favor of the punctuational model, as in many areas of science, a hypothesis is favored as valid not because it can be directly observed to be true but because the alternatives are disproven.

Although the conclusions I describe do not depend on our understanding the precise conditions under which rapidly divergent speciation occurs or the mechanisms that are involved, it is obviously desirable to comprehend these things. To date, the strongest claim for the paleontological documentation of rapidly divergent speciation is by P. G. Williamson (1981) in his controversial study of apparent adaptive radiation in a late Cenozoic African lake. Here relatively small, isolated populations underwent enormous morphological shifts during an interval estimated to have spanned between five thousand and fifty thousand years. At issue is whether these shifts constituted speciation or only ecophenotypic change. Neontological studies, often conducted within the framework of historical geology, have documented a number of examples of very rapid speciation.

Many of the examples of rapid evolutionary divergence of geographically isolated populations have yielded degrees of morphological separation judged to be only subspecific in rank (see Mayr, 1963:578–580). Nonetheless, these changes have generally occurred during intervals of time measured in hundreds or thousands of years, which are very brief in relation to the millions of years that have characterized species longevities. In other cases, species-level divergence has occurred during similarly brief intervals. For example, G. L. Bush (1969) concluded that a species of fly belonging to the genus *Rhagoletis* and feeding on apple evolved within the past one hundred years from a species feeding on hawthorne. Similarly, E. C. Zimmerman (1960) deduced that banana-feeding species of the genus *Hedylepta* in Hawaii evolved from their palm-feeding ancestors within the past one thousand years; here the key piece of evidence is that Polynesians introduced the banana plant to Hawaii only about one thousand years ago. The planktonic copepod *Cyclops dimorphus* has apparently evolved within the past thirty years; it is restricted to the Salton Sea, a body of water that formed in California between 1905 and 1907 as a result of human activities (Johnson, 1953).

Some examples of speciation in the recent past have constituted adaptive radiations. Fourteen species of cyprinid fishes endemic to Lake Lanao display features of the teeth and jaws that differ markedly from those of species in other lakes; Lake Lanao apparently came into being just ten thousand years ago (Myers, 1960).

Other examples of rapid speciation have produced new taxa that could easily be recognized as new genera. *Cyprinodon diabolis* is a small species of pupfishes confined to a tiny thermal spring in Death Valley, California, where its population size may never have exceeded five hundred individuals. It is so different in form from other closely related species as to deserve status in a separate genus, and it has been retained in *Cyprinodon* only because its ancestry is so well understood (Miller, 1950). J. H. Brown (1971) concluded that "*Cyprinodon*" *diabolis* must have evolved during the past twenty thousand or thirty thousand years, after the most recent glacial interval ceased to affect the region.

In one interesting botanical case, we have evidence not only of speciation in the recent past but of the genetic changes that occurred. J. Lewis (1962) concluded that at least half of the twenty-eight species of the plant genus *Clarkia* in California have had sudden origins. The best-studied example is of *C. lingulata*, which is known from only two colonies peripheral to the geographic range of its obvious ancestor, *C. biloba*. The descendant species has one more chromosome than its ancestor and is also unique in the presence of a large translocation and at least two paracentric inversions. The two species also differ in petal shape and produce sterile hybrids. It appears that changes in the pattern of gene arrangement have effected the morphological and reproductive divergence.

The Genetics of Rapidly Divergent Speciation

Ernst Mayr (1954) introduced the notion that rapid speciation might explain the sudden appearance of evolutionary novelties in the fossil record; in effect, he set the stage for the punctuational model of evolution. At the same time, Mayr introduced the idea that rapid speciation might entail the founder effect—the situation wherein an atypical individual or small group of individuals becomes geographically isolated from other members of its species and gives rise to a new species. Mayr was invoking the theoretical conclusions of Sewall Wright (1931, 1932) that small, isolated populations should display small degrees of genetic variability and should experience inbreeding and random changes in genetic constitution (genetic drift). Mayr pioneered in associating these events with speciation, suggesting that the change of genetic environment experienced by small, isolated (founder) populations should change the selective value of individual genes. He coined the term "genetic revolution" to describe the wholesale genetic changes that, under these circumstances, could lead to major evolutionary transitions.

Building on this idea and relying on his studies of the genetics of adaptive radiation in Hawaiian *Drosophila*, Hampton L. Carson (1968, 1975) introduced the concept of the population flush. He suggested that the genetic architecture of a species could be divided into an open and a closed system. The closed system was seen as constituting a coadapted gene complex that, though not easily disrupted, could be altered by genetic drift during a speciation event. As the first step in the population flush model, a founding population or individual invades a new habitat. Carson's chromosomal studies suggest that for Hawaiian drosophilids the founder has often been a single gravid female. The newly formed population then undergoes a flush—an interval of rapid population growth characterized by relaxed selection pressure under unusual ecological circumstances. In Carson's model, recombination during the population flush introduces new genetic variations that entail altered pleiotropic balances. As the flush nears its end and the population approaches the carrying capacity of the environment, selection is resumed. With the coadapted gene complex having been disrupted, however, and with selection operating in a new environment, the population moves to a new adaptive peak, evolving into a new species.

Speciation by way of very small populations may be more common than has generally been believed. Not to be overlooked here is the new evidence from studies of mitochondrial DNA that all modern humans have descended from a single female (Cann, Stoneking, and Wilson, 1987).

A. R. Templeton (1980) has invoked founder populations in a different model of rapidly divergent speciation: genetic transilience. Here the ancestral population is envisioned as encompassing a substantial amount of genetic variability. Inbreeding, however, places alleles in new genetic backgrounds, altering pleiotropic relationships between major genes. Selection then operates on alleles in new genetic contexts, yielding dramatic evolutionary divergence. A key aspect of the transilience model is that it relies on the effects of only a small number of key genes, which regulate other genes. More generally, regulatory genes now play a prominent role in most considerations of rapidly divergent speciation.

Templeton's observations lend support to the idea (Stanley, 1976) that sexual modes of reproduction prevail among eukaryotes primarily because of the importance of recombination to speciation. Asexual clades are simply clones, and they seldom diversify rapidly enough to offset normal rates of extinction. Asexual taxa crop up here and there in phylogeny but seldom survive for long.

Templeton's transilience model is more than an untested hypothesis. He conducted a remarkable experiment that duplicated many of the particulars of the model and yielded many of the predicted results. Employing asexual strains of *Drosophila mercatorum* that produce homozygous diploid offspring, he established clones, thereby mimicking the ultimate founder effect—the origin of a population from a single individual. By creating a unique condition of total homozygosity, the experiment also severely altered the genetic environment for important genes in a manner analogous to genetic drift in nature. The experiment also situated the population in a strange laboratory environment. One rare homozygous genome that was extremely well adapted to parthenogenesis was found to be incompatible with the coadapted genome of the sexual parent stock. This homozygous genome had passed through a selective bottleneck. Another homozygous genome turned out to be genetically compatible with the sexual parent stock. In other words, the two homozygous genomes possessed different forms of coadaptation. Even so, these two genomes were not found to differ from each other in any structural enzyme locus that was investigated. Templeton concluded that a small number of regulatory genes were primarily responsible for the transition to the asexual forms. One of these forms, in addition to possessing a new reproductive system, also differed from its sexual progenitors in being larger, darker, slightly different in wing shape, and more reluctant to initiate flight—and also in having a shorter generation time and greater egg-laying capacity. The genetic material responsible for these features was not introduced by mutation in the course of the experiments but was already present. The implication here is that potentially useful alleles may remain sequestered for some time before becoming situated in a genetic environment in which they can be fixed by drift or selection.

This artificial evolutionary transition produced a new species in just a few generations. Similarly, J. E. Powell (1978) produced reproductive isolation between populations of *Drosophila pseudoobscura* in the laboratory by engineering repeated population flushes, and J. N. Ahearn (1980) accomplished the same result for *Drosophila silvestria* by subjecting a laboratory stock to severe population reductions so as to induce random genetic drift. None of these experiments yielded a new species that was strikingly unusual—a form that might, for example, approach distinctiveness at the generic level. We must recognize, however, that nature operates a very much larger laboratory over a vastly longer interval of time. Furthermore, markedly divergent speciation events, which might seem unlikely, must be extremely rare. Only a small fraction of speciation events in nature are markedly divergent, and simple calculations reveal that successful speciation events of any type are uncommon. Five million years approximates mean species duration for many animal taxa, and the diversity of life on earth does not ordinarily change greatly in that course of time. This means that, on the average, a species gives rise to only about one new species during its 5-million-year lifetime. These considerations relate to species that become well established. Clearly, many small isolates that are technically new species die out almost instantaneously on a geological scale, never playing a significant role in the ecosystem or giving rise to other species. Inasmuch as many species generate myriads of isolates over millions of years, there is no basis for doubting the possibility of divergent speciation as a major source of evolutionary change.

Summary

I began this review by describing how general patterns of evolutionary stasis have been uncovered by two morphometric studies that have employed fossil data in a uniquely comprehensive and unbiased manner, in both morphological and taxonomic terms. These two studies also focused on real lineages. Many other studies have treated only body size, which has been much more flexible in evolution than has shape, or they have compared ancestral and descendant populations that were not known with certainty to be segments of a single phyletic lineage or were chosen for study because they were believed to be separated by substantial evolutionary change. Proponents of the gradualistic model of evolution now face the challenge of undertaking comprehensive and unbiased studies revealing that phyletic evolution predominates in other segments of phylogeny.

The two comprehensive and unbiased studies that have been completed to date lend credence to previously conducted evaluations of species longevity that have indicated that established species typically undergo relatively little evolution in the course of 10^6–10^7 million years. Paleontological data suggest that stepwise changes in established species may be less common than rapidly divergent branching (speciation) events, but this issue is not fully resolved.

I have also reviewed biological studies in the field and laboratory which indicate that, in calling for rapidly divergent speciation, the punctuational model is not demanding the impossible. Distinctive species have formed in nature during recent intervals in the order of 10^2–10^4 years. Artificial speciation in the laboratory has been achieved even more rapidly, with founder effects, genetic drift, and shifting influence of regulatory genes all playing significant roles. Events of markedly divergent speciation are very rare. The relatively smaller number of these occurrences required by the punctuational model cannot be viewed as improbable, given the great number of small isolates that are generated within higher taxa in the course of geological time.

References

Ahearn, J. N. 1980. Evolution of behavioral reproductive isolation in a laboratory stock of *Drosophila silevstria*. *Experimentia* 36:63–64.

Brown, J. H. 1971. The desert peepfish. *Sci. Am.* 225 (November):104–110.

Bush, G. L. 1969. Sympatric host race formation and speciation in frugivorous flies of the genus *Rhagoletis* (*Diptera tephritidae*). *Evolution* 23:237–251.

Cann, R. L., M. Stoneking, and A. C. Wilson. 1987. Mitochondrial DNA and human evolution. *Nature* 325:31–36.

Carson, H. L. 1968. The population flush and its genetic consequences. In R. C. Lewontin, ed., *Population Biology and Evolution*. Syracuse, N.Y.: Syracuse University Press, pp. 123–137.

——. 1975. The genetics of speciation at the diploid level. *Am. Nat.* 109:73–92.

Chaline, J., and B. Laurin. 1986. Phyletic gradualism in a European Plio-Pleistocene *Mimomys* lineage (Arvicolidae, Rodentia). *Paleobiology* 12:203–216.

Charlesworth, B. 1985. Some quantitative methods for studying evolutionary patterns in single characters. *Paleobiology* 10:308–318.

Cheetham, A. H. 1986. Tempo of evolution in a Neogene bryozoan: rates of morphologic change within and across species boundaries. *Paleobiology* 12:190–202.

Eldredge, N. 1971. The allopatric model and phylogeny in Paleozoic invertebrates. *Evolution* 25:156–167.

Eldredge, N., and S. J. Gould. 1972. Punctuated equilibria: an alternative to phyletic gradualism. In T. J. M. Schopf, ed., *Models in Paleobiology*. San Francisco: Freeman, Cooper, pp. 82–115.

Gould, S. J., and N. Eldredge. 1977. Punctuated equilibria: the tempo and mode of evolution reconsidered. *Paleobiology* 3:115–151.

Hallam, A. 1975. Evolutionary size increase and longevity in Jurassic bivalves and ammonites. *Nature* 258:439–446.

——. 1978. How rare is phyletic gradualism and what is its evolutionary significance: evidence from Jurassic bivalves. *Paleobiology* 4:16–25.

Hoffman, A. 1982. Punctuated versus gradual mode of evolution: a reconstruction. *Evol. Biol.* 15:411–437.

Johnson, M. W. 1953. The copepod *Cyclops dimorphys* Kiefer from the Salton Sea. *Am. Midl. Nat.* 49:188–192.

Lewis, H. 1962. Catastrophic selection as a factor in speciation. *Evolution* 16:257–271.

Malmgren, B. A., W. A. Berggren, and G. P. Lohmann. 1983. Evidence for punctuated gradualism in the late Neogene *Goborotalia tumida* lineage of planktonic foraminifera. *Paleobiology* 9:377–389.

Malmgren, B. A., and J. P. Kennett. 1982. The potential of morphometrically based phylo-zonation. *Mar. Micropaleontol.* 7:285–296.

Maynard Smith, J. 1981. Macroevolution. *Nature* 289:13–14.

Mayr, E. 1954. Change of genetic environment and evolution. In J. Huxley, A. C. Hardy, and E. B. Ford, eds., *Evolution as a Process*. London: Allen & Unwin, pp. 157–180.

——. 1963. *Animal Species and Evolution*. Cambridge, Mass.: Harvard University Press.

Miller, R. R. 1950. Speciation in fishes of the genera *Cyprinodon* and *Empertrichthys* inhabiting the Death Valley region. *Evolution* 4:155–163.

Myers, G. S. 1960. The endemic fish fauna of Lake Lanao, and the evolution of higher taxonomic categories. *Evolution* 14:323–333.

Powell, J. E. 1978. The founder-flush speciation theory: an experimental approach. *Evolution* 32:465–474.

Schankler, D. 1980. Faunal zonation of the Wellwood Formation in the central Big Horn Basin, Wyoming. *Univ. Mich. Pap. Paleontol.* 24:99–114.

Schopf, T. J. M. 1982. A critical assessment of punctuated equilibria. 1. Duration of taxa. *Evolution* 36:1144–1157.

Scott, G. H. 1983. Biostratigraphy and histories of upper Miocene-Pliocene

Globorotalia, South Atlantic and Southwest Pacific. *Mar. Micropaleontol.* 7:369–384.
Stanley, S. M. 1975. A theory of evolution above the species level. *Proc. Natl. Acac. Sci. USA* 72:646–650.
——. 1976. Clades versus clones in evolution: why we have sex. *Science* 190:382–383.
——. 1979. *Macroevolution: Pattern and Process*. San Francisco: W. H. Freeman.
Stanley, S. M., and X. Yang. 1987. Approximate evolutionary stasis for bivalve morphology over millions of years: a multivariate, multilineage study. *Paleobiology* 13:113–139.
Stebbins, G. L. 1950. *Variation and Evolution in Plants*. New York: Columbia University Press.
Templeton, A. R. 1980. Modes of speciation and inferences based on genetic distances. *Genetics* 94:1011–1038.
Van Valen, L. 1974. Two modes of evolution. *Nature* 252:290–300.
Williamson, P. G. 1981. Paleontological documentation of speciation in Cenozoic molluscs from Turkana Basin. *Nature* 293:437–443.
Winston, J. E. 1981. Feeding behavior of modern bryozoans. In T. W. Broadhead, ed., *Lophophorates: Notes for a Short Course*. University of Tennessee, Department of Geological Sciences. *Stud. Geol.* 5:1–21.
Wright, S. 1931. Evolution in Mendelian populations. *Genetics* 16:97–159.
——. 1932. The roles of mutation, inbreeding, crossbreeding and selection in evolution. *Proc. Sixth Int. Congr. Genet.*, pp. 356–366.
Zimmerman, E. C. 1960. Possible evidence of rapid evolution in Hawaiian moths. *Evolution* 14:137–138.

4

Punctuated Equilibria, Rates of Change, and Large-Scale Entities in Evolutionary Systems

NILES ELDREDGE

Archaeology and paleontology, often confused in the public's mind, share more than a common source of data in the earth beneath our feet. Both are charged with documenting the actual results of history, be it the history of the evolutionary process as preserved in the sequence of faunas and floras in the fossil record or the history of human sociocultural evolution as preserved in the archaeological record. In this chapter I seek a still deeper analogy between the two disciplines. As I see it, both archaeology and paleontology are poised and ready to enter the ranks of what, for want of a better term, we might call "functional" science, conferring an alternate status to the traditional perception of them as purely "historical" sciences.

Asked what is unique and potentially powerful about their data, both archaeologists and paleontologists generally respond with the simple word "time." As skimpy and scantly incomplete as their data generally are, both disciplines are unique within their larger fields of evolutionary and sociocultural inquiry in providing the exact temporal backdrop against which the actual course of historical events has been played out. Yet, until recently, relatively few contributions to an understanding of evolutionary mechanisms (or simply processes of history) have been forthcoming from these disciplines. This is because the actual processes of change (or of that vastly more common pattern, stasis, meaning stability or virtual lack of change) are commonly held to be understood solely in the short-term phenomena of currently functioning systems.

This epistemological assertion—that we can understand processes of stasis and change in both biological and sociocultural systems only by studying the structure and function of currently existing systems (see, e.g., Dobzhansky, 1937, and Carson, 1981, for evolutionary biology)—is usually not distinguished from the ontological claim that those processes and phenomena observed in the short term are both necessary and entirely sufficient to account for all historically observed phenomena, including those of the paleontological and archaeological records. Thus the role of archaeology as well as of paleontology has been simply to document the actual events of history, leaving to evolutionary biologists (mainly, in the past fifty years, geneticists) and sociocultural theorists working with various aspects of human behavior and sociocultural organization the task of elucidating the mechanisms of stasis and change in those systems.

Science seeks to explain phenomena (classes of events, in particular) in terms of causal interactions among entities. Thus in evolutionary biology, physical attributes of organisms are held to undergo adaptive modification according to the differential reproductive success of organisms that vary with respect to that trait within a breeding population—the principle of natural selection. It is my central contention here that in restricting analysis to those entities and interactive processes that are easily accessible to human observers over a few minutes, days, months, or years (a procedure easily defended on epistemological grounds) automatically and arbitrarily limits the range of entities, hence of interactive processes, that are admitted to causal theories of stasis and change. And such excluded entities and processes may well prove to be important to a full theory of history in both biological and sociocultural systems.

In particular, I argue that it is the main legacy of the evolutionary biological theory of punctuated equilibria, which came out of paleontology, that large-scale, interactive biotic entities of which organisms are parts do in fact exist; that their natures can be specified; and that their causal interactions can thus be elucidated. In setting out the argument, I focus first and most intensely on such large-scale biological entities simply because those are the systems with which I am familiar. I then argue that much the same approach can be taken with large-scale sociocultural entities, thus bringing archaeology more directly into the business of elaborating genuine causal scientific theory. My goal is to show that these two "historical sciences" have not been taking full advantage of their gift of time in not recognizing the nature, or even the existence, of large-scale systems. I am emphatically not seeking to forge any unified theory of biological and cultural evolution because I believe that humans are organisms and have evolved biologically; but I am also convinced (for the usual reasons, especially differences in modes of hereditary transmission of traits) that cultural and biological aspects of human existence are largely decoupled (see, for example, Bock, 1980, on the incommensurability of biological and cultural patterns of variation within *Homo sapiens*).

Punctuated Equilibria: Themes and Content

The notion of punctuated equilibria arose in an attempt to bring evolutionary theory more closely in line with general patterns of evolutionary history that are manifest in the fossil record (Eldredge, 1971; Eldredge and Gould, 1972—in which the term was coined; see Eldredge, 1985a, for a general account of the history and content of the concept). The paleontologists who reviewed the first edition of Darwin's (1859) *On the Origin of Species* (see Hull, 1973, for a compendium of reviews) all expressed some degree of surprise that Darwin failed to mention the marked degree of nonchange (or stasis, as it is now generally termed; Eldredge and Gould, 1972) that is characteristic of most species after their earliest appearance in the fossil record. Thus it was known in Darwin's day that the bulk of a species' history was spent in relative stability, a fact that Darwin interwove into his narrative by the sixth edition (Darwin, 1872). Yet evolutionary biology since Darwin's time has unquestionably seen evolution as a matter of slow, steady, and progressive transformation of the adaptive features of organisms. Punctuated equilibria, as its central empirical facet, sought an explanation for the overlooked phenomenon of marked stability, responding to a pattern in which adaptive evolutionary change seems to be concentrated in (relatively) brief episodes, "punctuating" vastly longer intervals when little or no change occurs.

Darwin's central task had been to demonstrate that life has in fact had a history—that all living forms are descended from a remote common ancestor, that there has been "descent with modification," as he put it, in the intervening eons. His success derived in no small part from his description of a powerful engine of evolutionary change—natural selection. Natural selection, Darwin's direct, natural analogue of the artificial selection of breeders, stood as the materialistic substitute for a supernatural Creator as the molder of design in the organic world. And Darwin's depiction of that organic world as a scene of constant struggle and competition over both resources and mates made the corollary seem overwhelmingly inevitable: those variations that give an edge will tend to be disproportionately passed along to succeeding generations. Even if environments do not change, variants will always appear that play the game of life better than ever, and thus improvement can be expected. But environments do change, and with great regularity. Thus it seemed (to Darwin and his successors, e.g. Simpson, 1944; see Eldredge, 1985b) wholly inevitable that the history of life would be one of constant modification of organismic properties shifting to keep pace with an ever-fickle environment. Adaptive change has seemed ineluctable and inevitable in the Darwinian tradition.

Such a view, handed down intact to the present time as the central picture of both evolutionary change and evolutionary history, has infused all branches of evolutionary biology from population genetics to paleontology. To be sure, dampening effects—constraints—are known in all fields; opposing forces (e.g., of selection and mutation) as well as simple genetic recombination have been seen as conservative factors in population genetics, and the intricate delicacy of developmental systems ensures that reduced viability is the common result of substantial alteration of developmental pathways.

In paleontology, exceptions to the expectation that evolutionary history should ordinarily be recorded as progressive series of gradually transmogrified organisms have generally been explained in another fashion altogether: as the natural outcome of faulty data. Darwin himself helped found the modern discipline of taphonomy (the study of the formation of the fossil record) in partial answer to the riddle of the disturbing lack of many examples of progressive change through the rock record. The fossil record does indeed have gaps, and, for the most part, paleontologists have ever been willing to explain the discrepancy between expected and observed evolutionary patterns as the simple effect of the haphazard way organisms die, become preserved, survive in fossilized form for as many as hundreds of millions of years, are discovered, taken to the laboratory, and analyzed.

Yet stasis stands out as a bald fact of the fossil record; most species that have been recorded over a significant interval of time (for marine invertebrate organisms, typically in the few, but up to 10 or 15, millions of years; for terrestrial vertebrates, more on the order of hundreds of thousands up to a few millions of years), little change is observed. More precisely, the differences observed between (interpreted) ancestral and descendant species generally involve characters that are not observed to vary, let alone change in the direction of the descendant, within the ancestral species before the appearance of its presumed descendant (see Eldredge and Cracraft, 1980:285ff.). The same pattern has been noted in the modern-day biota as well: patterns of variation within species often do not correspond to the differences observed between closely related species. Thus, though a species may display geographic variation to some degree, and some (generally oscillatory) change might occur throughout the history of a species, the differences observed between species in an ancestral-descendant lineage do not usually arise as a simple outgrowth of variation and evolutionary transformation within species.

It is the central goal of punctuated equilibria to explain these observations by bringing to bear a subset of evolutionary theory quite distinct from the general expectation of patterns of gradual adaptive change through time. For though the differences between species in the main involve functional characteristics of organisms and appear to have arisen in the manner envisioned by Darwin—through natural selection—the first question to be asked is, Why is adaptive change seemingly concentrated into brief episodes? The answer still preferred by many evolutionary biologists is simply that everyone knows that evolutionary rates vary and that peculiar concatenations of ecological and historical factors will conspire at some times to foster change and at others to induce stability (see, e.g., Hecht and Hoffmann, 1986).

Yet a more general theory that can be applied to the common patterns of stasis and punctuation in evolutionary history is the theory of speciation: species are understood to be (sexual) reproductive communities, with reproduction going on within the community and with little (and, usually, no) reproduction going on with members of other such reproductive communities. Speciation, in this sense, is the origin of one or more descendant reproductive communities from an ancestral reproductive community. This is in stark contrast to the alternative, and still very general, conceptualization of species as collections of anatomically similar organisms. Thus, to Darwin, the evolution of species is a matter of the transformation of anatomical properties of organisms to the point at which an identifiably new group of similar organisms has emerged; hence the title of his book, which dealt with "species" strictly as ephemeral collections of similar organisms—arbitrarily designated subdivisions of lineages that were undergoing regular and progressive, if not utterly constant, transformation through time. This is the sense in which most paleontologists to this day use the term "species."

In the late 1930s and early 1940s, when the Modern Synthesis was in the throes of formulation, the geneticist Theodosius Dobzhansky and the systematist Ernst Mayr each wrote books on evolutionary theory. Both were concerned to add one item to the general Darwinian picture of intergradational transformation of the adaptive properties of organisms. For both men realized that *discontinuity* ranks equally with adaptive *diversity* as a phenomenon to be explained by evolutionary theory. There is a panoply of organic form in biotic nature, and it seemed to them (as it does to me and to virtually every other evolutionary biologist) that natural selection, working on a groundmass of variation whose ultimate source is genetic mutation, is the effective agent underlying the production of such diversity. Yet that diversity is not smoothly continuous in the living world (though both men expected that it would be so were the fossil record complete); rather, diversity seems to come in packages—in "species." Thus species, both as clusters of discontinuity in the spectrum of adaptive diversity and as communities of reproductive interaction, required explanation.

Because natural selection produces adaptive change gradationally, as a continuum of change, Mayr and Dobzhansky both argued that discontinuity must arise through disruption of reproduction. That is why both men defined species as reproductive communities, eschewing any reference to general organismic similarity in form. Dobzhansky (1937) in particular was concerned to develop the notion of isolating mechanisms, a theme elaborated on by Mayr (1942); both zoologists argued that the most common initial source of reproductive isolation derived from physical (i.e., geographic) isolation, with subsequent divergence. Their views continue to serve as the basis of the modern understanding of the process of speciation. Recently H. E. H. Paterson (1985 and references therein) has further clarified the matter, pointing out that all sexually reproducing organisms have species-specific-mate recognition systems that allow conspecifics to recognize one another for mating purposes (either simple chemical recognition of sperm and egg in seawater or complex bits of behavior that more literally involve recognition).

The point of Paterson's contribution is that isolation may lead to divergence in the specific mate-recognition system (SMRS), which is the minimal requirement for speciation to occur. Such a change may be accompanied by no, little, or a great deal of change in the vastly larger set of organismic adaptations—those pertaining to the soma, the economic adaptations of organisms. But though no economic change may accompany SMRS change, the pattern to be explained by punctuated equilibria says that economic change in general is not to be expected unless as an accompaniment to SMRS change. Thus the original statement of punctuated equilibria took the theory of allopatric (geographic) speciation and, as Mayr and Dobzhansky argue, saw the onset of "reproductive isolation" (now best understood as Paterson's SMRS disruption) as causally implicated, not only in the disruption of the spectrum of adaptively generated diversity but in the instigation and preservation of adaptive change. Precisely how and why economic change, when it occurs at all, generally accompanies SMRS disruption remains an unresolved and somewhat contentious problem.

Implications of Punctuated Equilibria: Large-Scale Biotic Systems

Allopatric speciation is a theory of species birth—how new species arise from old. Fragmentation of an ancestral species into two descendants may amount to a roughly 50-50 division; perhaps more commonly, new species arise as relatively small, peripheral populations become geographically isolated and reproductively and (generally) economically disjunct from the parental population. Both extreme possibilities have analogues in the asexual reproductive activities of various organisms: single-celled organisms commonly reproduce by simple division, whereas the polyps of the aquatic coelenterate *Hydra* reproduce by budding tiny descendants that grow and eventually detach. In either case, there are two individual organisms where once there had been one—precisely the case in speciation. Ancestral species rarely become extinct in the process of giving rise to descendants.

Likewise, it is clear that all species eventually suffer extinction—just as all organisms eventually die. Thus species are spatiotemporally bounded entities; like organisms, they have births, histories, and deaths. Organisms are parts of species, and because organismic traits tend to remain conservative during the vast bulk of a species' history, species tend to remain recognizably discrete from one another throughout their entire histories. In other words, there are some intriguing parallels between organisms and species. In particular, specific instances of both categories—organisms and species—deserve to be called *individuals.*

Just such a suggestion was made by Michael Ghiselin (1974), who argued that species, which have names, are defined ostensively. Thus they are more analogous to a corporation, such as General Motors, than to a substance such as gold, which is, at least theoretically, a spatiotemporally unrestricted class. Though the class "gold" is defined with reference to certain atomic properties (i.e., any atom with atomic weight 79 is ipso facto an atom of gold), each particular atom of gold is a spatiotemporally restricted individual, having had a beginning, a history, and, eventually, a demise. We do not name individual atoms of gold because no one has as yet conceived of a useful purpose or practicable way for doing so.

As with gold, so with species. The Linnaean category "species" is indeed a class: any community of reproductively interacting (sexual) organisms, with reproductive adaptations (SMRS) unique to that community, belongs to the class "species." Each particular species, each with its own SMRS, is an individual instance, a taxon of species rank that is named, viz. *Homo sapiens*.

Particular species have long been considered to be classes (like gold): collections of adaptively similar organisms, for example, any extant hominid with a cranial capacity greater than, say, 1,200 cc would be an organism of the class *Homo sapiens*. The difference between the two ontological conceptualizations of species is critical: if we see particular species as groups of similar organisms, species have no particular ontological status; we are free to accept an evolutionary theory that sees the entire history of life as no more than a matter of stochastic and (especially) deterministic (i.e., natural selection) processes that form, maintain, and modify the phenotypic and underlying genotypic properties of organisms.

If, however, species are spatiotemporally bounded entities, with births, histories, and deaths, factors influencing their births and deaths would obviously have to be taken into account in any evolutionary theory. Moreover, through specifying the nature of such entities—species and other large-scale biotic entities that might be recognized—the roles played by such entities, plus any interactive processes in which they might be engaged, could also be specified. When species are viewed as classes, evolutionary dynamics are seen only in terms of the interactions of smaller-scale entities (individuals), such as organisms and genes—and disciplines that deal with species and monophyletic taxa of higher Linnaean rank (i.e., systematics and paleontology) are perforce entirely "historical," constrained as they are merely to record the results of the evolutionary process, the events that are the outcomes of interactions among lower-level entities.

The original formulation of punctuated equilibria, elaborated independently of Ghiselin's formal argument on the individuality of species, offers a case in point. For if species characteristically remain stable through the bulk of their histories, how do we explain directional change throughout a lineage of ancestral-descendant species? Traditionally, such directionality was seen as a simple accumulation of directional natural selection: small amounts of change accrue in relatively brief intervals; over the vaster stretches of geological time, truly large-scale, directional change can accrue, the reflection of so-called ortho-selection. Yet the prevalence of stasis means that such a simple model cannot be an accurate description of a process underlying the production of admitted examples of long-term directional change in evolution—as in the increase in hominid brain size over the past 4 or 5 million years.

Originally, we (Eldredge and Gould, 1972) simply suggested that trends likely reflect a pattern of differential species survival superimposed over a pattern of speciation in which adaptive changes occur under the aegis of natural selection. Steven M. Stanley (1975) and many subsequent authors since generalized this notion of species selection to embrace a host of theoretical models that would systematically bias births or deaths—speciations or extinctions—of component species within a lineage. Given the diversity of mechanisms available, Elisabeth Vrba and I (1984) suggested that the preferable, neutral term for the general phenomenon would be "species sorting." For present purposes, the important point is that long-term directional changes within lineages through evolutionary time are now almost universally considered by theorists to involve differential comings and goings of species interpreted as entire entities, as well as selection-mediated adaptive change of the phenotypic properties of organisms.

Thus, in barest outlines, is the general nature of a change in paleontology—from a position of chronicler of the major events in life's evolution, to one of active contributor, to a consideration of causal processes effective during that evolutionary history. The job of writing the theory that is implicit in this newly perceived structure has barely begun. The question for the rest of this chapter is, To what extent are these changes in paleontological thinking relevant to a consideration of the histories and evolution of social systems in general and to the study of human history and archaeology in particular?

What Are Social Systems?

My consideration of the nature of social systems stems explicitly from my background as an evolutionary biologist and paleontologist. Sociobiology, of course, has recently been widely touted in biological circles as an achievement of a general scientific—and explicitly evolutionary—approach to social behavior and social systems. In general, sociobiology represents a relatively straightforward application of contemporary versions of neo-Darwinism: social behavior is generally interpreted in adaptive terms, and specifiable traits are thought to be developed, retained, and modified according to their contribution to the probabilities of reproductive success of the organisms (or their close kin), in accord with the standard form of ultra-Darwinism to be found, for example, in the works of Richard Dawkins (e.g., 1976, 1982, 1987).

The view that behavioral traits, like any phenotypic trait, can be understood fundamentally as devices for reproductive success amounts to an elision in an evolutionary theory of causal processes underlying the origin, maintenance, and modification of adaptations. We observe, for example, that sharp teeth, a fusiform body, and a capacity (including behavioral repertoire) for remaining motionless, then darting rapidly from ambush to seize prey, is characteristic of the general feeding adaptations of the redfin pickerel, *Esox americanus americanus*. We can assume, in this instance, that the behaviors and morphologies functioning in the observed manner were evolved for the purpose we see them fulfill—in other words, they are adaptations for feeding. To claim that such feeding adaptations are actually devices for spreading the genes underlying such morphologies and behaviors is consistent with Dawkins's notion of "selfish genes" but is not the best possible description of nature. The sharp teeth, fusiform body, and so on are for efficient capture of prey; they are not for enhancing reproductive success. Yet it is, of course, true, as Darwin and Wallace taught us long ago, that fish that eat more efficiently will tend on average to leave more offspring—bearing the qualities of their parents—than will those less well endowed. Natural selection is a matter of differential economic success biasing reproductive success. To elide the statement and claim that (economic) adaptations arise for the enhancement of reproductive success—instead of seeing selection as a side effect of relative economic success—is to provide a distorted description of the origin and (worse) the present-moment functioning of a phenotypic trait complex. Not least among the difficulties arising from such a worldview is that the genetically based instructions for building such systems somehow emerge as of greater importance than the systems they build—a general difficulty with the ultra-Darwinian notion of selfish genes.

Sociobiologists, as I have pointed out elsewhere (Eldredge, 1987), tend to describe social systems as if they arise and are maintained by a competitive struggle among organisms to pass on their genetic information. G. F. Oster and E. O. Wilson (1978) provide a refreshing countertendency; in their monograph of hymenopteran caste evolution, these authors devote the greater amount of their attention to what they call "ergonomics," that is, the economic behavior(s) of castes within insect colonies. Such behavior (which amounts to the growth, differentiation, and maintenance of the soma of all organisms, plus additional aspects in social organisms) typically dominates the vast bulk of any organism's life. Relatively few organisms spend a disproportionate amount of their time (and economically based energy) on reproductive matters—queens in hymenopteran colonies being an obvious exception.

In nonsocial organisms, there is an increasing distinction between soma and germ line—between anatomical areas devoted to the two great classes of biological process involving matter-energy transfer and replication and transmission of genetic information (reproduction)—as the skein of organismic complexity, from prokaryotic bacteria through multicellular eukaryotic organisms (plants, fungi, and animals) is surveyed. As I and colleagues have argued elsewhere (Eldredge and Salthe, 1984; Eldredge, 1985b, 1986), the dichotomy between the two classes of biological process becomes even more pronounced when we look at larger-scale biological entities. Organisms are simultaneously parts of two very different hierarchically arrayed systems. On the one hand, organisms are parts of local populations (of conspecifics), which are in turn parts of local economic systems (ecosystems); local ecosystems are parts of larger, interconnected regional economic systems—and so on, until the entire biosphere of the earth is considered.

On the other hand, organisms are parts of species, which in turn are parts of monophyletic taxa. These are the entities of traditional interest to evolutionists. They are historical entities and, on a moment-by-moment functional basis, are best seen as packages of genetic information that stock the economic arena—the ecosystems in which the moment-by-moment business of life is conducted. I have argued (especially Eldredge, 1985b, 1986) that evolutionary theory in general has been based on a faulty appreciation of the nature of these large-scale entities. In particular, there has been a strong tendency to confound the two sets of processes and hierarchical systems. Cross-genealogical economic systems (ecosystems) are all but completely absent from standard evolutionary discourse, and, instead, species and monophyletic taxa are depicted as economic entities, occupying "niches" and "adaptive zones."

The point of all this biological ontology is simple. If there truly is a formal dichotomy between large-scale biotic entities, if economic and genealogical systems above the level of organisms are indeed separate, all social systems emerge as reintegrations of the reproductive and economic activities of organisms. From this perspective, social systems are very much hybrid affairs—stable entities arising from a complex commingling of reproductive and economic organismic activities. And as would perhaps be expected, the form assumed by such a reintegration varies greatly. In most hymenopteran colonies, division of labor remains strict, with particular castes assigned particular economic or reproductive roles. The blending of the twin themes arises because all bees of a colony share the same mother, and the economic activities of nonreproducing castes are extended in part in procuring energy sources for reproductive activities.

Mammalian social systems as a rule follow a strikingly different course of economic and reproductive integration. In human societies, family units are economic and reproductive cooperatives, with male(s) and female(s) performing purely economic tasks (however much these tasks may be the same or differ) as well as (different) reproductive functions and a variety of economic activities directly pertaining to reproduction, as in care of offspring.

Yet even within human society, there is a tendency to segregate reproductive activities from economic ones.[1] Thus complex and generally hierarchically internested entities are rife in the economic sector of any complex society. An individual *Homo sapiens* typically belongs to some form of family, but also, in complex societies, may be part of a department within a division of a company within a multinational corporation—the prime source of economic input to the individual's life. There are myriad other categories of "economic" organizations, most of which are also hierarchically structured, to which that same individual may belong—fraternal organizations, sports teams, religious and political groups, and others. To a hierarchically minded biologist, human social organization offers the clearest examples of hierarchically structured systems to which organisms can belong.

Thus, beyond the simple example of families, human social organiza-

1. "Economic" pertains to all processes of matter-energy transfer in organisms used for the differentiation, growth, and maintenance of the soma; used this way, "economics" is a more general concept than the usual meaning of the term in the social sciences.

tion in general clearly reveals the duality of organismic existence in general, reflecting the clear and obvious distinction between economic and reproductive activity to be drawn. Even the hierarchical associations of families reflect this dichotomy. Families are hierarchically structured in at least two separate ways: they are genealogically structured into hierarchies of relationships, and they are cross-genealogically economically structured, as when clusters of (generally) cross-genealogical (unrelated) families form villages, towns, and similar communities of varying degrees of geopolitical integrity. Such associations of families are obviously of economic, rather than reproductive, importance.

And yet, in human society, there is far greater interplay between the economic and reproductive activities of component organisms than in nonsocial organisms. A coral spends much of its time respiring and, when favorable conditions prevail, with tentacles fully extended, acting as a sort of microcarnivore, removing tiny (and, in some species, even larger) organisms from the seawater. Occasionally, provided there is enough energy beyond the needs simply to maintain body tissues, the coral will reproduce. (Organisms in general do not reproduce unless they have an excess of energy—reproduction being a physiological luxury not needed for an organism's survival.) In humans and other mammalian social organisms, the situation is far more complex; not only do normal Darwinian considerations—namely, that relative economic success impinges on probability of reproductive success—obtain as in all sexual organisms, but reproductive success may have an equally direct impact on economic success. For example (and as I have discussed in Eldredge, 1987), evolutionists and eugenicists have often decried the propensity for members of disadvantaged socioeconomic classes to produce more children than members of higher classes—an apparent contradiction in the eyes of many early biologists (see Kevles, 1985). The work of another Victorian—Karl Marx—supplied the answer to the puzzle: reproductive proclivity is a source of labor, thus of direct economic import to the family as a potential source of wealth.

Thus social systems introduce a complexity to the causal network of interaction between the economic and reproductive activities of organisms. Not only does economic success determine (stochastically) reproductive success, but reproductive success (measured in means other than pure numbers—as in "quality" as well) has many potential economic ramifications. One son playing shortstop for the New York Yankees may well mean more to the old-age economic welfare of parents than five whose own economic fates are far less grand. Such considerations are unimportant in a biological evolutionary context because older people are generally past the age of reproduction, but from the standpoint of the fabric and stability of a social system, they may be very relevant indeed.

Stability and Change in Social Systems

There is an epistemological bind in the recognition of long-term, large-scale systems as entities. Although it is by no means a logical, formal requirement, it is obviously helpful to their recognition if such entities remain stable, thus recognizably more or less the same for extended periods of time. Thus stasis greatly aids in the recognition of a species over intervals of millions of years in the fossil record. Such stability is also of importance in recognizing long-term social entities, such as the "Anasazi" culture of the American Southwest (Berry, 1982).

Yet stability is not prima facie evidence that what is involved is a large-scale entity. Stasis in biological systems involves the anatomical (or phenotypic in general) properties of organisms; evolutionary nonchange in such features no more ensures that species are "individuals" than does stasis in cultural traits imply that nation-states are spatiotemporally bounded entities. Yet if there is independent evidence and reason supporting interpretations that large-scale systems such as species and nation-states are spatiotemporally bounded entities, it becomes abundantly clear that stability in phenotypic and cultural traits (respectively) is of enormous practical utility in recognizing such systems.

Once the possibility that such systems are large-scale entities is entertained, however, their existence affects interpretations of stasis and change in biological or cultural traits. Just as we ask how speciation and species extinction affect patterns of stasis and change in organismic adaptations, we also must ask, What are the general patterns of stasis and change in the histories of cultural traits? What governs stability? What prompts change? What forms of interactions between large-scale sociocultural entities act to promote either stasis or change of cultural traits?

The work of historian F. J. Teggart (esp. 1925) provides some insight into how such a line of inquiry can be developed. Teggart noted parallel problems between biological and social theories of evolution.[2] Teggart claimed that there are three main patterns underlying stasis and change

2. For example, Teggart (1925) said he turned to Darwin to provide a model for

in historical cultural systems. First and foremost, is stability, governed by custom and other factors; the strongest signal of history is simply nonchange.

Second, there is a tendency for long-term change gradually to accrue within a sociocultural system. Teggart's example involves language—the drift in linguistic usage that happens so rapidly as to be detectable within the span of a single human lifetime. R. Lande (1976) has discussed seemingly comparable, albeit proportionately even slower, patterns of morphological drift through time within species in the fossil record.

But such gradual change (contrary to the basic models of both biological and cultural evolution) does not seem to be the stuff of "real" evolutionary (or historical) change. New species (even genera and families) may come and go, all the while some lineages are slowly accruing gradual change through time; and the differences between closely related (or ancestral-descendant) species often do not seem to arise as a smooth function of patterns of variation (of gradual, protracted, directional change) within the ancestral species. And so it is, Teggart writes, with cultural change: new traits often appear abruptly and generally are not the outgrowth of slow, steady change through time.

Teggart's view of how substantial cultural change is effected in history is particularly instructive, for Teggart saw most substantial change concentrated into historical events, which result from interactions between large-scale sociocultural systems. Egyptian society, for example, was profoundly altered with the introduction of horse-drawn chariots when the Hyksos dominated the Egyptians for a while during the interregnum between the Middle and New Kingdoms.

I cite Teggart's theories of patterns and process of historical change only to illustrate the potential value of recognizing the existence of large-scale entities and the downward causal influence that their presence and interactions may have on stasis and change of cultural traits. Teggart clearly sees sociocultural systems embedded hierarchically within larger (regionally organized) entities, and he sees interactions (largely but not exclusively portrayed as collisions) between like entities (e.g., nation-states) as the source of the events that account for patterns

---

evolutionary thought only to be dismayed that Darwin argued that adaptive change is perforce a matter of gradual and progressive change—in spite of abundant evidence to the contrary. Teggart recognized much the same rhetoric underlying what little cultural evolutionary theory existed.

of abrupt change that interrupt vastly longer sequences of internal stability and gradual accrual of generally negligible amounts of change.

M. S. Berry's (1982) work on the sequence of cultures in the North American Southwest offers a similar example. Rather than interpreting the pattern as a linear history, in which change sometimes occurred rapidly and at other times at a more leisurely pace, Berry argues that the patterns of stasis interrupted by brief spurts of rather profound cultural change do not represent linear evolution, but rather a sequence of habitation and replacement. The Anasazi are a historical whole, as regionally diverse and temporally modified as they were. They were replaced by another cultural system, not as a smooth evolutionary outgrowth but because the Anasazi were eventually (and rather abruptly) no longer able to occupy their territory. The point is not that all sociocultural change should now be seen to come in brief spurts interrupting vastly longer periods of stasis—as a direct analogue to patterns of stasis and change commonly encountered in the biological evolutionary realm. Nor, certainly, is my aim to call for some sort of megatheory of stasis and change that permeates both biological and sociocultural systems. Rather, the point is to explore the possibility that social systems may profitably be viewed as hierarchically arrayed stable entities—large-scale spatiotemporally bounded entities. If so, then it must be true that a theory of evolution, even one that focuses exclusively on biological or cultural traits of organisms, cannot be complete if it addresses only mechanisms of stasis and change of such traits on a generation-by-generation basis. Human sociocultural systems—because they are complex amalgams of the economic and reproductive behaviors of *Homo sapiens* organisms—automatically have the capacity for interaction (as do all biological economic systems); some anthropologists have already treated the capacity of social units to split—to give rise to other such entities. Thus the further intriguing possibility emerges that such social entities may themselves be susceptible to a form of selection directly analogous to large-scale natural selection—something of an irony because neither cultural selection (because the probability of reproductive success is unaffected by variance in the distribution of cultural traits) nor species selection (because species cannot be economic interactors) are themselves wholly legitimate analogues of true natural selection.

Events that stem from the interactions and behaviors of large-scale entities in both the biological and sociocultural realms constitute biological and sociocultural history—evolution. Births and deaths of such large-scale systems are important in themselves; the possibility that lower-level events—the invention, maintenance, modification, and eventual disappearance of both biological and cultural traits, which are the traditional objects of focus in both biological and cultural evolutionary analyses—may also be largely determined by interactions and behaviors of large-scale entities makes the recognition of the importance of such systems all the more critical in the further elaboration of both biological and cultural evolutionary theory.

References

Berry, M. S. 1982. *Time, Space and Transition in Anasazi Prehistory.* Salt Lake City: University of Utah Press.

Bock, K. 1980. *Human Nature and History: A Response to Sociobiology.* New York: Columbia University Press.

Carson, H. L. 1981. Untitled letter on Chicago macroevolution conference. *Science* 211:773.

Darwin, C. 1859. *On the Origin of Species.* Facsimile ed. New York: Atheneum, 1967.

——. 1872. *On the Origin of Species.* 6th ed. London: Murray.

Dawkins, R. 1976. *The Selfish Gene.* Oxford: Oxford University Press.

——. 1982. *The Extended Phenotype: The Gene as the Unit of Selection.* Oxford: W. H. Freeman.

——. 1987. *The Blind Watchmaker.* New York: Norton.

Dobzhansky, T. 1937. *Genetics and the Origin of Species.* Reprint. New York: Columbia University Press, 1982.

Eldredge, N. 1971. The allopatric model and phylogeny in Paleozoic invertebrates. *Evolution* 25:156–167.

——. 1985a. *Time Frames.* New York: Simon and Schuster.

——. 1985b. *Unfinished Synthesis.* New York: Oxford University Press.

——. 1986. Information, economics and evolution. *Annu. Rev. Ecol. Syst.* 17:351–369.

——. 1987. The evolutionary context of social behavior. In G. Greenberg and E. Tobach, eds., *Hierarchy and the Evolutionary Context of Social Behavior.* Hillsdale, N.J.: L. Erlbaum, 19–30.

Eldredge, N., and J. Cracraft. 1980. *Phylogenetic Patterns and the Evolutionary Process.* New York: Columbia University Press.

Eldredge, N., and S. J. Gould. 1972. Punctuated equilibria: an alternative to phyletic gradualism. In T. J. M. Schopf, ed., *Models in Paleobiology.* San Francisco: Freeman, Cooper, pp. 82–115.

Eldredge, N., and S. N. Salthe. 1984. Hierarchy and evolution. *Oxford Surv. Evol. Biol.* 1:182–206.

Ghiselin, M. J. 1974. A radical solution to the species problem. *Syst. Zool.* 23:536–544.

Hecht, M. K., and A. Hoffman. 1986. Why not neo-Darwinism? A critique of paleobiological challenges. *Oxford Surv. Evol. Biol.* 3:1–47.

Hull, D. L. 1973. *Darwin and His Critics.* Chicago: University of Chicago Press.

Kevles, D. J. 1985. *In the Name of Eugenics.* New York: Knopf.

Lande, R. 1976. Natural selection and random genetic drift in phenotypic evolution. *Evolution* 30:314–334.

Mayr, E. 1942. *Systematics and the Origin of Species.* Reprint. New York: Columbia University Press, 1982.

Oster, G. F., and E. O. Wilson. 1978. *Caste and Ecology in the Social Insects.* Monographs in Population Biology 12. Princeton: Princeton University Press.

Paterson, H. E. H. 1985. The recognition concept of species. In E. S. Vrba, ed., *Species and Speciation. Transvaal Mus. Monogr.* 4:21–29.

Simpson, G. G. 1944. *Tempo and Mode in Evolution.* Reprint. New York: Columbia University Press, 1984.

Stanley, S. M. 1975. A theory of evolution above the species level. *Proc. Natl. Acad. Sci. USA* 72:646–650.

Teggart, F. J. 1925. *Theory of History.* New Haven: Yale University Press. Reprint. 1977, Berkeley: University of California Press.

Vrba, E. S., and N. Eldredge. 1984. Individuals, hierarchies and processes: towards a more complete evolutionary theory. *Paleobiology* 10:146–171.

5

Twenty Years Later: Punctuated Equilibrium in Retrospect

Antoni Hoffman

One of the most hotly debated issues in modern evolutionary theory is the relationship between micro- and macroevolution. This controversy is at the core of many current, in particular paleobiological, challenges to the neo-Darwinian paradigm of evolutionary biology (Hoffman, 1989).

Microevolution encompasses phenotypic change within evolving biological species as well as taxonomic diversification of the biosphere by speciation. These phenomena are effects of microevolutionary processes, that is, the mechanisms that introduce change to the genetic pools of individual species. The microevolutionary processes always operate within a particular framework of extrinsic, ecological conditions and on particular species composed of more or less interconnected populations of organisms with their historically established genotypes and phenotypes. In other words, microevolutionary phenomena are caused by the action of various genetic evolutionary forces (natural selection, genetic drift, mutation pressure, various molecular processes) on a set of biological initial conditions, which determine the starting point for microevolution, and within the limits imposed, biological boundary conditions, which determine the range of biological states that microevolution can potentially bring about. The initial conditions are given by the evolutionary history of each evolving species. The boundary conditions, in turn, are set by the ecology of a species (including both the biotic and abiotic contexts) and its biological constitution.

Microevolution is thus induced by genetic evolutionary forces, which are universal. They operate in all biological species, though their interrelationships and relative significance vary substantially among particular cases and hence they have very different effects depending on the extrinsic, ecological context and the initial and boundary biological conditions. These forces are all genetic. Natural selection, however, is conceivable solely within an environmental context. It results from variation in genetic fitness, that is, a hereditary variation in the probability of reproductive success caused by hereditary phenotypic variation among individual organisms in a population. This variation is obviously dependent on the local environmental context. Microevolution is thus inextricably interwoven with environment.

This general description of microevolution appears to be widely accepted by both the advocates and the critics of the neo-Darwinian paradigm. The meaning of macroevolution, by contrast, is much more ambiguous. It is often variously understood even by participants in a single scientific discussion, and its meaning has changed considerably since the 1940s and 1950s, when it was widely employed by such classics of neo-Darwinism as George Gaylord Simpson and Bernhard Rensch.

Macroevolution is now most generally understood to encompass all the phenomena observed and described beyond the level of individual species in both space and time. Macroevolutionary phenomena include, for example, origins of new ground body plans, changes in frequency distribution of particular phenotypic characters in various organic groups or even in the entire biosphere, distribution of rates and modes of evolution in various organic groups, rates origination and extinction of species (or higher taxa) and their relationship to diversity in various groups, ecosystems, or even the biosphere. Individual biologists and paleontologists emphasize one or another category of these phenomena as the most important, or interesting, aspect of macroevolution, but all agree that investigation of all these phenomena is an indispensable component of evolutionary studies. The controversy about macroevolution focuses, however, on whether these macroevolutionary phenomena result solely from the operation of microevolutionary processes, constrained by their initial and boundary biological conditions and summed over the enormous diversity of biota and over the huge spans of ecological time, or whether they reflect the action of a separate class of biological, indeed macroevolutionary, processes.

It is one of the main tenets of the neo-Darwinian paradigm that macroevolution is nothing but microevolution extended over the vast number of species, ecological situations, and time dimensions. This claim is now often rejected. Many biologists and paleontologists postulate the existence of specifically macroevolutionary processes which operate above the level of species, independently of genetic evolutionary forces (e.g., Stanley, 1979; Eldredge, 1985; Gould, 1985; Salthe, 1985). Many others disagree (e.g., Dawkins, 1986; Hecht and Hoffman, 1986; Levinton, 1988; Hoffman, 1989). Historically, the contemporary macroevolutionary challenges to neo-Darwinism have largely stemmed from punctuated equilibrium.

Punctuated Equilibrium: Original Formulation

The concept of punctuated equilibrium has undergone a long and rather confusing evolution during the nearly twenty years since its first publication. In their original article, Niles Eldredge and Stephen Jay Gould (1972) proposed that the microevolutionary processes should, and indeed do, lead to a macroevolutonary phenomenon unexpected by evolutionary biologists and paleontologists. This phenomenon is a distinctly bimodal frequency distribution of genotypic, and consequently also phenotypic, rates of species evolution. Eldredge and Gould rejected the idea they called phyletic gradualism, which was regarded as orthodox within the neo-Darwinian paradigm—that the evolution of species is always a slow and gradual change of the entire species, driven by orthoselection and hence leading in the same adaptive direction over very long time intervals. In their view, phyletic gradualism maintains that evolution entails a continual adaptation of species to unidirectional change in at least one physical environmental parameter; this evolution, furthermore, proceeds at an approximately constant rate. Eldredge and Gould wrote that phyletic gradualism contradicts the neo-Darwinian paradigm, which predicts that the evolutionary history of particular phylogenetic lineages consists of very long time intervals of stasis—when essentially no changes take place—interrupted from time to time by rapid speciation events. Speciation occurs because small peripheral isolates are subject to extreme selective pressures in atypical environmental settings and therefore natural selection leads to their genetic, and consequently also phenotypic, separation from their formerly conspecific populations. According to Eldredge and Gould, the neo-Darwinian theory of microevolution predicts the macroevolutionary pattern of punctuated equilibrium. It implies that the norm in evolutionary history of any species is genetic and phenotypic homeostasis, which can break down solely by allopatric speciation in small marginal populations at the periphery of the species' range. (In a later article, Gould and Eldredge [1977] accepted also a possibility, and even plausibility, of sympatric speciation, but the precise mode of speciation has little significance for either the validity or the implications of punctuated equilibrium.)

Punctuated equilibrium thus postulates that almost all evolutionary change should be concentrated in speciation events, whereas the change taking place within particular species, between successive speciation events, is insignificant and virtually nonexistent. The frequency distribution of evolutionary rates should be bimodal—very high rates at speciation and zero rates at other times. According to Eldredge and Gould (1972), exactly such is the pattern of evolutionary rates revealed by the fossil record. Paleontologists document almost exclusively very long-ranging species separated from each other by large morphological gaps that reflect discontinuities produced during speciation: stasis occasionally interrupted by rapid change, hence the term "punctuated equilibrium."

Punctuated Equilibrium: Contrasting Versions and Their Evaluation

Clearly, punctuated equilibrium was formulated in opposition to phyletic gradualism, which was understood as unidirectional evolution at a constant rate caused by orthoselection in the entire species. At the same time, however, punctuated equilibrium portrayed speciation as a necessary condition for any significant evolutionary change of organisms. Consequently, it could, and indeed used to, be variously interpreted.

On the one hand, its original formulation could simply be understood as rejection of phyletic gradualism, as negation of the universality, or even merely commonness, of unidirectional evolution of species occurring at an approximately constant pace. Max Hecht and I have termed this interpretation the *weak version* of punctuated equilibrium (Hecht and Hoffman, 1986; Hoffman, 1989). This version holds that, generally, the evolution of species does not proceed for a long time in the same direction and at the same rate but varies in both direction and rate. That this interpretation of punctuated equilibrium is not a straw man but a real proposition is well evidenced by recent articles by Eldredge (1984) and Elisabeth Vrba (1985). They write explicitly that the concept of punctuated equilibrium tells only that the rate of evolution fluctuates along particular phyletic lineages and that speciation may, though does not have to, be associated with considerable phenotypic change.

The weak version of punctuated equilibrium does not introduce anything new to evolutionary biology and paleontology. Within the neo-Darwinian paradigm of evolutionary theory, it is entirely trivial. This assertion is abundantly documented by the classic works of Simpson (1944, 1953), and even a critical reading of Darwin's *Origin of Species* proves this point beyond any reasonable doubt (Penny 1983; Rhodes 1983). One might argue that regardless of what Darwin or Simpson did or did not actually write on the tempo and mode of evolution, the universality of phyletic gradualism was postulated by the version of neo-Darwinism that dominated American evolutionary biology and paleontology in the 1950s and 1960s, when Eldredge and Gould grew up and received their education. Punctuated equilibrium originated in that particular sociological and intellectual context, and it might constitute a rebellion against its authors' teachers and peers. In fact, however, the classic American textbooks of the time, even those explicitly cited by Eldredge and Gould (1972), do not advocate phyletic gradualism as defined by the proponents of punctuated equilibrium. For example, R. C. Moore, C. G. Lalicker, and A. G. Fischer (1952) wrote in their *Invertebrate Fossils*—the standard textbook of paleontology in the 1950s and 1960s—that periods of slow and gradual, or even arrested, evolution alternate in individual lineages with bursts of explosive evolutionary change; they also considered rapid allopatric speciation as the main mode of species origination. It seems therefore that the weak version of punctuated equilibrium does not differ at all from the standard version of neo-Darwinism.

On the other hand, the original formulation of punctuated equilibrium can be interpreted as the proposition that gradual phenotypic change is almost absent from the evolution of phyletic lineages and that periods of complete phenotypic stasis of species are interrupted solely by speciation events. This interpretation constitutes the *strong version* of punctuated equilibrium (Hecht and Hoffman, 1986; Hoffman, 1989). Several articles by Gould and Eldredge (1977; Gould, 1980, 1982) and Steven M. Stanley (1982a, b) clearly demonstrate that the strong version does not distort or exaggerate the meaning its proponents attributed to punctuated equilibrium.

This version of punctuated equilibrium has been repeatedly tested by paleontologists over the past fifteen years. Unfortunately, such empirical tests in the fossil record can never be decisive because the paleontological data do not allow for unequivocal identification of gradual and punctuated evolution of species (Hoffman and Reif, 1990). To demonstrate beyond a reasonable doubt the occurrence of gradual evolution in a geological setting, one cannot only show that an organic group underwent a gradual and significant phenotypic change over a certain area. One also has to show that the area was sufficiently large and ecologically heterogeneous to rule out an alternative explanation for the gradual phenotypic change such as migration of a set of conspecific populations, or closely related species, representative of clinal variation and tracking their wandering habitats. This is a very difficult, often impossible, task because it requires precise time correlation of biological events taking place in a variety of areas distant from one another. Nevertheless, and contrary to repeated assertions by the advocates of punctuated equilibrium, the paleontological literature does contain a considerable number of convincing instances of gradual phenotypic evolution (see bibliography in Levinton, 1988; Hoffman, 1989), and new examples continue to appear (e.g., Abe et al., 1988; Wei and Kennett, 1988; Fenster et al., 1989; Sorhannus, 1990).

It is much more difficult, if possible at all, to demonstrate or to refute cases of punctuated phenotypic evolution in the fossil record. Apart from the very rare case of direct paleontological documentation of branching of a phyletic lineage which would provide compelling evidence for cladogenetic speciation, event-speciation is identified in paleontology with profound morphological change. The association of such "speciation" with phenotypic change, then, becomes inevitable, although it may well be more apparent than real. In turn, all the instances of rapid appearance in a given area of a new population, clearly distinct from its potential predecessors, may simply represent its immigration from other areas. In other words, the morphological gaps between related species may indeed reflect the notorious incompleteness of the fossil record, as Darwin claimed long ago. Furthermore, a lot of genetic data prove beyond doubt the absence of any unequivocal, one-to-one relationship between phenotypic, and also genetic, evolution and speciation; speciation may be associated with very big as well as with very little change (e.g., Ayala, 1975; Avise and Ayala, 1976; Douglas and Avise, 1982).

Therefore, instances of punctuated phenotypic evolution are generally less convincing than those of gradual evolution, even if they are sometimes fairly strong (e.g., Ager, 1983; Kelley, 1983; Cronin, 1985; Cheetham, 1986; Stanley and Yang, 1987). Their credibility is further weakened because the fossil record is full of stratigraphic gaps, and the statistical correlation between punctuated phenotypic evolution and stratigraphic incompleteness is striking (McKinney 1985). Perhaps most interesting, though not surprising, is the paleontological evidence for phenotypic evolution intermediate in mode between gradual and punctuated evolution. In some cases, the rate of phenotypic evolution underwent a considerable acceleration without any detectable relation to branching of the phyletic lineages (e.g., Cisne et al., 1982; Malmgren, Berggren, and Lohmann, 1983); in other cases, speciation occurred relatively rapidly but certainly gradually on the evolutionary time scale (e.g., Lazarus, 1986; Sorhannus et al., 1988). Even the much touted case of punctuated evolution of freshwater mollusks in the Lake Turkana section in East Africa (Williamson, 1981) is now being described as speciation—that is, taking for granted this controversial interpretation (see, for example, Fryer, Greenwood, and Peake, 1983)—within perhaps fifty thousand generations (Williamson, 1985), which definitely is not rapid by biological standards.

This brief discussion of the empirical paleontological data does not lead to refutation of the proposition that punctuated phenotypic evolution occurs in nature, but it shows that the strong version of punctuated equilibrium—the claim that gradual phenotypic evolution occurs at best exceptionally, at orders of magnitude less commonly than punctuated evolution—is blatantly false. There are no empirical data to support it and considerable data to contradict it.

Between 1977 and 1982, however, the advocates of punctuated equilibrium took this strong version for granted and, to explain the postulated macroevolutionary phenomenon, proposed that there is a fundamental biological difference between the processes of microevolution and speciation (Gould and Eldredge, 1977; Gould, 1980, 1982; Stanley, 1979, 1982a, b). They invoked the theories of speciation put forth by Ernst Mayr, Hampton Carson, and Sewall Wright to support the claim for the possibility—even necessity—of speciation always being very rapid, almost instantaneous, because of genetic revolution, which they tended to interpret in a Goldschmidtian way as the occurrence of point mutations with very large phenotypic effects leading to immediate origination of new high-rank taxa. In this view, speciation is a macroevolutionary phenomenon caused by a macroevolutionary process, irreducible to the processes of microevolutionary change operating within the framework of a species' environment and biological constitution. This is why Gould (1980) wrote of the neo-Darwinian paradigm as being effectively dead. Gould (1980, 1984) went so far as to postulate that the process of speciation is one of "discontinuous evolution," thus proposing what I have termed the *radically strong version* of punctuated equilibrium (Hoffman, 1989).

The idea that paleontological data could, in spite of their proverbial crudeness and incompleteness, force biologists to undertake a far-reaching revision of their perspective on microevolution, macromutations, and speciation had to, and indeed did, provoke a strong reaction. For even if the paleontological data did indicate a strict bimodality of morphological evolutionary rates, and hence a very high frequency of punctuated phenotypic evolution, this macroevolutionary pattern would fit a wide variety of quantitative models of microevolutionary processes (e.g., Ginzburg, 1981; Kirkpatrick, 1982; Petry, 1982; Newman, Cohen, and Kipnis, 1985; Lande, 1986). The theorists invoked by Gould and Stanley to support the implications of punctuated equilibrium outright rejected their perspective on speciation as a macroevolutionary process (Carson, 1982; Mayr, 1982; Wright, 1982) because even if speciation were indeed always due to genetic revolution, it would still remain a gradual process of population change (Carson and Templeton, 1984). Yet even the assumption about universality, or even merely a considerable commonness, of genetic revolution is not self-evidently true (Barton and Charlesworth, 1984).

The concept of macromutations as the main mechanism of speciation appears today implausible, to say the least. Although there is no logical way to disprove such a mechanism's potential operation in nature, the genetic data suggest that point mutations with large phenotypic effects generally are adaptively disadvantageous, and whenever they are not, they may not bring about speciation (Turner, 1981, 1983; Larson, 1983).

The strong version of punctuated equilibrium and its macroevolutionary implications are thus indefensible; in fact, no one seems willing to advocate it any longer. In a 1985 article Gould explicitly rejected this interpretation of punctuated equilibrium. He wrote that punctuated equilibrium is about neither the distribution of evolutionary rates, the mechanisms of speciation, nor the problem of macromutations and their role; and he referred the origination of new species to microevolution. Within the framework of natural sciences, the radically strong version is even less defensible because it clearly represents a metaphysical option, or a worldview, rather than scientific theory (Rieppel, 1987). It is, moreover, a worldview that runs counter to the entire tradition of natural sciences and makes it impossible to explain causally a historical pattern (Hoffman, 1989). Therefore, it cannot be satisfying for evolutionary biology.

One more version of punctuated equilibrium asserts that the evolutionary history of the majority of species consists mainly of a very long period of evolutionary stasis, when the species remains in homeostasis with its environment and essentially does not evolve. This *moderate version* of punctuated equilibrium, as Max Hecht and I have termed it (Hecht and Hoffman, 1986; Hoffman, 1989), constitutes the basis for some of the modern attempts to expand the neo-Darwinian paradigm into a hierarchical theory of selection (Eldredge, 1982, 1985, 1988; Gould, 1982, 1985).

Long-term evolutionary stasis of species, however, cannot be tested in the fossil record. Palentological data consist solely of a small sample of phenotypic traits—little more than morphology of the skeletal parts—which does not allow an inference to be made about changes in a species' genetic pool or even about change versus stability of the frequency distribution of phenotypes in a phyletic lineage. The claim that long-term evolutionary stasis is the norm in the evolution of species cannot be either confirmed or refuted. It is untestable.

This moderate version of punctuated equilibrium, however, could also be interpreted even more modestly as proposing that many, perhaps even the majority of, species exhibit long-term phenotypic stasis in many respects. Under this interpretation, though, it again borders on triviality. It is widely accepted not only by the advocates of punctuated equilibrium but also by the theory's ardent opponents (e.g., Hoffman, 1982; Gingerich, 1985). There are several examples in its favor, derived from both paleontology (e.g., Estes, 1975; Winston and Cheetham, 1984) and neontology (e.g., Wake, Roth, and Wake, 1983; Williams, 1984). But there is no way to estimate quantitatively the relative frequencies of gradual phenotypic evolution and long-term phenotypic stasis because it is impossible to draw a random sample of phyletic lineages across the diverse taxonomy and across the environmental spectrum of the biosphere in which the mode of phenotypic evolution could be empirically determined. The examples provided by Alan Cheetham (1988) and Steven M. Stanley and X. Yang (1987) in support of punctuated equilibrium, although spectacular and promising as the first steps in the right direction, each focus on only a single environmental regime, a single geographic area, and a single organic group (long known to be particularly favorable for punctuated patterns of phenotypic change). And, of course, the phenotypic stasis documented by paleontologists may well be more apparent than real, as the evolutionary change may have occurred chiefly in components of the phenotype that are not preserved in the fossil record.

Moreover, even if this weak formulation of the moderate version of punctuated equilibrium is correct, it does not have any significant implications for the neo-Darwinian paradigm because there is a very wide variety of biological mechanisms which are fully concordant with this paradigm and yet can conceivably lead to phenotypic stasis. Stabilizing selection and coevoltionary homeostasis with environment can bring about phenotypic stasis in a relatively constant environment (Charlesworth, Lande, and Slatkin, 1982; Barnard, 1984). Ontogenetic canalization of the phenotype, or the capability of effecting the same adult phenotype despite genetic and environmental changes, can maintain phenotypic stasis even under changing environmental conditions (Hoffman, 1982). Finally, and most simply, populations and species are capable of migration and active tracking of their wandering habitats to which they have become adapted in the course of their evolution. As long as their habitats are not destroyed or changed fundamentally while they are located in a cul-de-sac of a kind, the environment may not pose any challenge to be met by evolution.

Thus the debate on punctuated equilibrium may well be finally closed (Levinton et al., 1986). Contrary to the triumphant proclamations of Gould and Eldredge (1986), punctuated equilibrium has not been established as a truth that now everyone claims to have always known. Punctuated equilibrium has several different versions, and depending on the version accepted it is either trivial—that is, known ever since Darwin—or plainly wrong, or untestable. It cannot therefore justify a causal distinction between micro- and macroevolution. It does not indicate operation of a separate class of macroevolutionary processes.

Punctuated equilibrium cannot serve as an argument in the current discussion on macroevolution. Even less can it be assigned the role of a hallmark of a fundamental change in the Western perspective on the world. Eldredge and Gould (1972) portrayed punctuated equilibrium as a rebellion against the Western paradigm of continuity and ascribed themselves to the Marxist tradition of thought with its emphasis on abrupt, qualitative change. Gould (1984) particularly pointedly claimed that punctuated equilibrium was a paradigm shift in natural sciences. A prominent literary critic took this interpretation even further and depicted the relationship of the prose of Jorge Luis Borges and his fellow postmodernists to the classic novel in a close analogy to punctuated equilibrium as a revolution against nineteenth-century gradualism (Levine, 1986). Both the weak and the moderate versions of punctuated equilibrium, however, contain no revolutionary ideas and do not speak for abruptness of change. The strong version, however, clearly is an aborted revolution, dismissed by the evidence and, consequently, by its authors themselves. This is not to say that the paradigm of continuity must win overall—either in science or on a broader cultural scale—but only that punctuated equilibrium is irrelevant to discussion of this issue.

Theoretical Implications of Punctuated Equilibrium

As a problem in evolutionary biology and paleontology, punctuated equilibrium has reached the point at which there is nothing left to debate. The discussion that has brought us to this point, however, should be greatly credited for two merits. On the one hand, it stimulated exciting empirical research that has resulted in some fascinating data, and it also enforced much higher standards of paleontological work on the tempo and mode of evolution in species-level lineages. On the other hand, it helped bring the concept of species as logical individual and the theory of species selection to the focus of evolutionary biology.

The latter two ideas are part and parcel of the current drive in evolutionary biology to establish macroevolution not only as a category of phenomena but also as a class of processes, different from the microevolutionary ones envisaged by the neo-Darwinian paradigm (Salthe, 1985; Eldredge, 1988). Historically, both these ideas are independent of punctuated equilibrium. Species *qua* individual was proposed by Michael T. Ghiselin (1974) and David L. Hull (1980). Species selection was discussed—though not under this heading—by Ronald Fisher, Sewall Wright, and George Gaylord Simpson. Nevertheless, both these concepts were taken up and widely popularized by the proponents of punctuated equilibrium (e.g., Stanley, 1979; Eldredge, 1982; Gould, 1982; Vrba and Eldredge, 1984). Moreover, Vrba (1980, 1983) proposed in the context of punctuated equilibrium and species selection still another macrevolutionary hypothesis—effect macroevolution or species drift (see Levinton, 1988; Hoffman, 1989)—which is also presented as a challenge to the neo-Darwinian paradigm. These ideas and their relations to punctuated equilibrium are therefore also worth some consideration here.

The concept of species *qua* individual represents a stance in the old philosophical debate on the logical status of species. Ghiselin and Hull argue that given that species evolve, they cannot be regarded as classes of individual organisms because they have no essential features—no essence of any given species which an individual would have to possess to belong to a particular species. The biological criteria employed to define species include reproductive isolation, fertility of offspring, and sharing in a mate-recognition system. They thus invoke mechanisms that reflect the cohesion of species as a whole but not particular characters that each individual organism would have to bear. Species, moreover, are historical entities, localized in both space and time. Each has its beginning, its history, and its ending; they cannot reappear once they have gone extinct. Ghiselin and Hull further argue that if species are not logical classes, they must be logical individuals, and indeed species seem to have the attributes prerequisite for such an interpretation of their logical status. Individuals, however, may be subject to selection. Individuality is a necessary condition for an entity to be a unit of selection (Hull, 1980). One might therefore speak of species selection and not only of natural selection at the level of individual organisms within populations and species.

The relationship of this argument to punctuated equilibrium has appeared obvious to Eldredge, Gould, and Stanley. For if the strong version of punctuated equilibrium were true, all species would indeed have their beginnings and endings unambiguously defined and recognizable; their historical identity would be absolutely unquestionable. And if the evolutionary stasis of species were a norm in the history of the biosphere, all macroevolutionary trends reconstructed by paleontologists would demand explanation by reference to a macroevolutionary process because microevolutionary processes would be unable to effect any change beyond the level of a single species. Such an explanation could be provided by species selection, or the process of selection for those features of phyletic lineages which either favor speciation or help resist extinction. It could also be provided by species drift, or the process of spread of a trait through its accidental association with groups that happen to proliferate relative to other groups.

Neither the concept of species as logical individuals nor the theories of species selection and species drift depend, however, on the validity of punctuated equilibrium. Even the most ardent and determined neo-Darwinians are ready to accept species selection and drift as processes that can potentially operate in nature, though they outright reject punctuated equilibrium (Dawkins, 1982; Maynard Smith, 1983; Hecht and Hoffman, 1986; Mayr, 1986; Turner, 1986). Species that are gradually and slowly transformed into their descendant species can also be subject to selection for some characters favoring speciation or hindering extinction. Their traits can also spread or decrease in frequency because of a variety of random effects. The question is not whether such processes can occur in nature but whether they do indeed occur, and if they do, then what is their significance in evolution? Species drift most certainly occurs in nature because every speciation event and every species extinction produces this phenomenon. This cannot be a reason to decouple macroevolution from microevolution. A thorough examination of the alleged instances of species selection however, shows that not even a single good example of species selection has thus far been described (Hecht and Hoffman, 1986; Hoffman, 1989).

The problem of the logical status of species belongs to philosophy rather than to evolutionary biology, and even there it remains highly controversial (e.g., Rosenberg, 1985; Rieppel, 1986). The opposition between eternal and immutable classes, on one hand, and historical and evolving individuals, on the other, is not exhaustive. Species may well be sets of individuals—perhaps fuzzy sets since it is often unclear whether an individual organism should be assigned to one species or another—because there is no reason to believe that every set of elements is eternal, invariant, and immutable (Kitcher, 1984). Most important in the present context, however, this issue has little relevance for evolutionary biology. The biologist does not draw any conclusions about the species being investigated from a philosophical consideration of their logical status but focuses on the biology of these species. The philosopher may on this basis discuss the general category "species" and attempt some generalizations about its logical status. But this issue has nothing to do with punctuated equilibrium. Species that evolve slowly and gradually also are historical entities, localized in both space and time, equipped with some cohesion. Such species can also be regarded as either logical individuals, or sets, or classes, depending on the philosopher's philosophical perspective.

Thus punctuated equilibrium is neither necessary nor sufficient to validate the new developments in evolutionary theory. Its theoretical implications are much less than breathtaking. Its heuristic utility, however, has been great. It deserves credit for that.

REFERENCES

Abe, K., R. A. Reyment, F. L. Bookstein, A. Honigstein, A. Almogi-Labin, A. Rosenfeld, and O. Hermelin. 1988. Microevolution in two species of ostracods from the Santonian (Cretaceous) of Israel. *Hist. Biol.* 1:303–322.

Ager, D. V. 1983. Allopatric speciation—an example from the Mesozoic Brachiopodo. *Palaeontology* 26:555–565.

Avise, J. C., and F. J. Ayala. 1976. Genetic differentiation in speciose versus depauperate phylads: evidence from the California minnows. *Evolution* 30:46–58.

Ayala, F. J. 1975. Genetic differentiation during the speciation process. *Evol. Biol.* 8:1–78.

Barnard, C. J. 1984. Stasis: a coevolutionary model. *J. Theor. Biol.* 110:27–34.

Barton, N. H., and B. Charlesworth. 1984. Genetic revolutions, founder effects, and speciation. *Annu. Rev. Ecol. Syst.* 15:133–164.

Carson, H. L. 1982. Speciation as a major reorganization of polygenic balances. In C. Barigozzi, ed., *Mechanisms of Speciation*. New York: Alan R. Liss, pp. 411–434.

Carson, H. L., and A. R. Templeton. 1984. Genetic revolutions in relation to speciation phenomena: the founding of new populations. *Annu. Rev. Ecol. Syst.* 15:97–131.

Charlesworth, B., R. Lande, and M. Slatkin. 1982. A Neo-Darwinian commentary on macroevolution. *Evolution* 36:474–498.

Cheetham, A. H. 1986. Tempo of evolution in a Neogene bryozoan: rates of morphologic evolution within and across species boundaries. *Paleobiology* 12:190–202.

Cisne, J. L., G. O. Chandler, B. D. Rabe, and J. A. Cohen. 1982. Clinal variation, episodic evolution, and possible parapatric speciation: The trilobite *Flexicalymene senaria* along an Ordovician depth gradient. *Lethaia* 15:325–341.

Cronin, T. M. 1985. Speciation and stasis in marine Ostracoda: climate modulation of evolution. *Science* 227:60–63.

Dawkins, R. 1982. *The Extended Phenotype*. San Francisco: Freeman.

——. 1986. *The Blind Watchmaker*. London: Longman.

Douglas, M. E., and C. J. Avise. 1982. Speciation rates and morphological divergence in fishes: test of gradual versus rectangular modes of evolutionary change. *Evolution* 36:224–232.

Eldredge, N. 1982. Phenomenological levels and evolutionary rates. *Syst. Zool.* 31:338–347.

——. 1984. Simpson's inverse: Bradytely and the phenomenon of living fossils. In N. Eldredge and S. M. Stanley, eds., *Living Fossils*. New York: Springer, pp. 272–277.

——. 1985. *Unfinished Synthesis*. New York: Oxford University Press.

——. 1988. *Macroevolutionary Dynamics*. New York: McGraw-Hill.
Eldredge, N., and S. J. Gould. 1972. Punctuated equilibria: an alternative to phyletic gradualism. In T. J. M. Schopf, ed., *Models in Paleobiology*. San Francisco: Freeman, pp. 82–115.
Estes, R. 1975. Fossil *Xenopus* from the Paleocene of South Africa and the zoogeography of pipid frogs. *Herpetologica* 31:263–278.
Fenster, E. J., U. Sorhannus, L. H. Burckle, and A. Hoffman. 1989. Patterns of morphological change in the Neogene diatom *Nitzschia jouseae* Burckle. *Hist. Biol.* 1:197–211.
Fryer, G., P. H. Greenwood, and J. F. Peake. 1983. Punctuated equilibria, morphological stasis and the paleontological documentation of speciation: a biological appraisal of a case history in an African lake. *Biol. J. Linn. Soc.* 20:195–205.
Ghiselin, M. T. 1974. A radical solution to the species problem. *Syst. Zool.* 23:536–544.
Gingerich, P. D. 1985. Species in the fossil record: concepts, trends, and transitions. *Paleobiology* 11:27–41.
Ginzburg, L. R. 1981. Bimodality of evolutionary rates. *Paleobiology* 7:426–429.
Gould, S. J. 1980. Is a new and general theory of evolution emerging? *Paleobiology* 6:119–130.
——. 1982. Darwinism and the expansion of evolutionary theory. *Science* 216:380–387.
——. 1984. Toward the vindication of punctuational change. In W. A. Berggren and J. A. Van Couvering, eds., *Catastrophes and Earth History*. Princeton: Princeton University Press, pp. 9–34.
——. 1985. The paradox of the first tier: an agenda for paleobiology. *Paleobiology* 11:2–12.
Gould, S. J., and N. Eldredge. 1977. Punctuated equilibria: the tempo and mode of evolution reconsidered. *Paleobiology* 3:115–151.
——. 1986. Punctuated equilibrium at the third stage. *Syst. Zool.* 35:143–148.
Hecht, M. K., and A. Hoffman. 1986. Why not Neodarwinism? A critique of paleobiological challenges. *Oxf. Surv. Evol. Biol.* 3:1–47.
Hoffman, A. 1982. Punctuated versus gradual mode of evolution: a reconsideration. *Evol. Biol.* 15:411–436.
——. 1989. *Arguments on Evolution*. New York: Oxford University Press.
Hoffman, A., and W. E. Reif. 1990. On the study of evolution in species-level lineages: controlled methodological sloppiness. *Palaeont. Z.* 64:5–14.
Hull, D. L. 1980. Individuality and selection. *Annu. Rev. Ecol. Syst.* 11:311–332.
Kelley, P. H. 1983. Evolutionary patterns of eight Chesapeake groups mollusks: evidence for the model of punctuated equilibria. *J. Paleontol.* 57:581–598.
Kirkpatrick, M. 1982. Quantum evolution and punctuated equilibria in continuous genetic characters. *Am. Nat.* 119:833–848.

Kitcher, P. 1984. Species. *Philos. Sci.* 51:308–333.

Lande, R. 1986. The dynamics of peak shifts and the pattern of morphological evolution. *Paleobiology* 12:343–354.

Larson, A. 1983. Neontological inferences of evolutionary pattern and process in the salamander family Plethodontidae. *Evol. Biol.* 17:119–218.

Lazarus, D. 1986. Tempo and mode of morphologic evolution near the origin of the radiolarian lineage *Pterocanium prismatium. Paleobiology* 12:175–189.

Levine, G. 1986. Darwin and the evolution of fiction. *New York Times Book Review*, October 5, pp. 1, 60–61.

Levinton, J. S. 1988. *Genetics, Paleontology, and Maroevolution.* Cambridge: Cambridge University Press.

Levinton, J. S., K. Bandel, B. Charlesworth, G. Muller, W. R. Nagl, B. Runnegar, R. K. Selander, S. C. Stearns, J. R. G. Turner, A. Urbanek, and J. W. Valentine. 1986. Genomic versus organismic evolution. In D. M. Raup and D. Jablonski, eds., *Patterns and Processes in the History of Life.* Berlin: Springer, pp. 167–182.

Malmgren, B. A., W. A. Berggren, and G. P. Lohmann. 1983. Evidence for punctuated gradualism in the Late Neogene *Globorotalia tumida* lineage of planktonic foraminifera. *Paleobiology* 9:377–389.

Maynard Smith, J. 1983. Current controversies in evolutionary biology. In M. Grene, ed., *Dimensions of Darwinism.* Cambridge: Cambridge University Press, pp. 273–286.

Mayr, E. 1982. Processes of speciation in animals. In C. Barigozzi, ed., *Mechanisms of Speciation.* New York: Alan R. Liss, pp. 1–20.

——. 1986. Natural selection: the philosopher and the biologist. *Paleobiology* 12:233–239.

McKinney, M. L. 1985. Distinguishing patterns of evolution from patterns of deposition. *J. Paleontol.* 59:561–567.

Moore, R. C., C. G. Lalicker, and A. G. Fischer. 1952. *Invertebrate Fossils.* New York: McGraw-Hill.

Newman, C. M., J. E. Cohen, and C. Kipnis. 1985. Neo-Darwinian evolution implies punctuated equilibria. *Nature* 315:400–401.

Penny, D. 1983. Charles Darwin, gradualism, and punctuated equilibrium. *Syst. Zool.* 32:72–74.

Petry, D. 1982. The pattern of phyletic speciation. *Paleobiology* 8:56–66.

Rhodes, F. H. T. 1983. Gradualism, punctuated equilibrium and *The Origin of Species. Nature* 305:269–272.

Rieppel, O. 1986. Species are individuals: a review and critique of the argument. *Evol. Biol.* 20:283–317.

——. 1987. Punctuational thinking at odds with Leibniz—and Darwin. *Neues. Jahrb. Geol. Palaeontol. Abh. B* 174:123–133.

Rosenberg, A. 1985. *The Structure of Biological Science.* Cambridge: Cambridge University Press.

Salthe, S. N. 1985. *Evolving Hierarchical Systems*. New York: Columbia University Press.
Simpson, G. G. 1944. *Tempo and Mode in Evolution*. New York: Columbia University Press.
——. 1953. *The Major Features of Evolution*. New York: Columbia University Press.
Sorhannus, U. 1990. Tempo and mode of morphological evolution in two Neogene diatom lineages. *Evol. Biol.* 24:329–370.
Sorhannus, U., E. J. Fenster, L. H. Burckle, and A. Hoffman. 1988. Cladogenetic and anagenetic changes in the morphology of *Rhizosolenia praebergonii* Mukhina. *Hist. Biol.* 1:185–206.
Stanley, S. M. 1979. *Macroevolution—Pattern and Process*. San Francisco: Freeman.
——. 1982a. Speciation and the fossil record. In C. Barigozzi, ed., *Mechanisms of Speciation*. New York: Alan R. Liss, pp. 41–50.
——. 1982b. Macroevolution and the fossil record. *Evolution* 36:460–473.
Stanley, S. M., and X. Yang, 1987. Approximate evolutionary stasis for bivalve morphology over millions of years: a multivariate, multilineage study. *Paleobiology* 13:113–139.
Turner, J. R. G. 1981. Adaptation and evolution in *Heliconius*: a defense of neo-Darwinism. *Annu. Rev. Ecol. Syst.* 12:99–121.
——. 1983. The hypothesis that explains mimetic resemblance explains evolution: the gradualist-saltationist schism. In M. Grene, ed., *Dimensions of Darwinism*. Cambridge: Cambridge University Press, pp. 129–169.
——. 1986. The genetics of adaptive radiation: a neo-Darwinian theory of punctuational evolution. In D. M. Raup and D. Jablonski, eds., *Patterns and Processes in the History of Life*. Berlin: Springer, pp. 183–207.
Vrba, E. S. 1980. Evolution, species, and fossils: how does life evolve? *S. Afr. J. Sci.* 76:61–84.
——. 1983. Macroevolutionary trends: new perspectives on the roles of adaptation and incidental effect. *Science* 221:387–389.
——. 1985. Environment and evolution: alternative causes of the temporal distribution of evolutionary events. *S. Afr. J. Sci.* 81:229–236.
Vrba, E. S., and N. Eldredge. 1984. Individuals, hierarchies, and processes: towards a more complete evolutionary theory. *Paleobiology* 12:217–226.
Wake, D. B., G. Roth, and M. H. Wake. 1983. On the problem of stasis in organismal evolution. *J. Theor. Biol.* 101:211–224.
Wei, K. Y., and J. P. Kennett. 1988. Phyletic gradualism and punctuated equilibrium in the late Neogene planktonic foraminiferal clade *Globoconella*. *Paleobiology* 14:345–363.
Williams, N. E. 1984. An apparent disjunction between the evolution of form and substance in the genus *Tetrahymena*. *Evolution* 38:25–33.
Williamson, P. G. 198l. Paleontological documentation of speciation in Cenozoic mollusks from Turkana Basin. *Nature* 292:437–443.

——. 1985. Punctuated equilibrium, morphological stasis and the paleontological documentation of speciation: a reply to Fryer, Greenwood and Peake's critique of the Turkana Basin mollusk sequence. *Biol. J. Linn. Soc.* 26:307–324.

Winston, J. E., and A. H. Cheetham. 1984. The bryozoan *Nellia tenella* as a living fossil. In N. Eldredge and S. M. Stanley, eds., *Living Fossils*. New York: Springer, pp. 257–265.

Wright, S. 1982. The shifting balance theory and macroevolution. *Annu. Rev. Gen.* 16:1–19.

6

Is the Theory of Punctuated Equilibria a New Paradigm?

MICHAEL RUSE

The word "paradigm" is just about the most overused in the philosophical lexicon. Professional philosophers do their best to avoid it, and today it is much more commonly used by sociologists, scientists, and others, such as the journalist Roger Lewin, who asked, in *Science* in 1980, "Will a new synthesis emerge, signaling a true paradigm shift in the Kuhnian sense?" Part of the problem is that the word "paradigm" is as slippery as the word "God." Everyone who uses it means something slightly different. Too frequently the term is used as a propaganda tool, bolstering the pretensions of some supposed major breakthrough. Paradigm founder today. Nobel Prize winner tomorrow. Burial at Westminster Abbey the day after that.

This introduction is by way of explanation and (partial) apology. Several years ago, when I first started thinking seriously about the paleontologists' theory of punctuated equilibria, I expected to find the inevitable. Some enthusiast would be hailing the theory as a new "paradigm," suggesting that all who did not jump on the bandwagon were blind to the brilliance of the new science and doomed to early extinction. The intellectual community was just waiting for the holdouts to die so it could go about its new proper business without guilt. Nor, to my morbid satisfaction, was I long disappointed. Sure enough, paradigm status was soon claimed (Stidd, 1980). The strong implication was that all those who belittled the new position—as did an eminent evolutionist of my acquaintance who contemptuously remarked: "Just hand-waving by the paleontologists, Mike!"—are irrational. All we were waiting for was the celebratory postage stamp.

As a professional philosopher with strong Darwinian sympathies, I was happy to respond vigorously and (to my satisfaction) to make mincemeat of the claim (Ruse, 1982a). In no way could one justify the conclusion that punctuated equilibrium theory is a new paradigm. Only one who is ignorant of philosophy and science could be so naive. Or so I said then. Now I think I was wrong, or at least, too quick. It is not so stupid to link together punctuated equilibrium theory and new paradigm status. It can, in fact, be positively illuminating—although whether the true link is quite what the enthusiasts have claimed is, perhaps, another matter.

In this chapter, therefore, I intend to go back and answer the question properly. In the spirit of one who wants to learn new things rather than simply settle old scores, I am not going to make judgments about truth and falsity, better and worse.[1] I ask simply: "Is the theory of punctuated equilibria a new paradigm?" And, of course, the answer depends on what is meant by "punctuated equilibrium theory" and by "paradigm." So let us start there.

What Is the Theory of Punctuated Equilibria?

In recent years, philosophers of science who are interested in the history of science have been turning, with some enthusiasm, to evolutionary theory as a model for an adequate theory of change, including scientific change (Ruse, 1986). Organisms evolve. Theories evolve. Perhaps there are connections. I am not overly excited by this particular strand of "evolutionary epistemology." I suspect that there are some severely crippling disanalogies. Yet the idea is insightful in one way. As evolutionists know only too well, there is no such thing as an average or typical organism (Mayr, 1963). In *Homo sapiens*, for example, there are big members, small members, black members, white members, bright members, stupid members. Is Stephen Jay Gould more typical than Brigitte Bardot? The question is meaningless.

Similarly, we tend to have the idea that there is such a thing as a theory, to which people subscribe. Supposedly there is a list of essential features—like the hypothetical list of essential features for species membership—and if these features are accepted, so is the theory. But not

1. The philosophers who have discussed the theory are Sober (1984) and Thompson (1983).

otherwise. The evolutionary analogy highlights what anybody who tries to write on the history of science knows too well. Theory nature—and theory allegiance in particular—is much more like the real nature of species membership. Different people believe very different things—the same people believe different things at different times—and yet rally under the same banner.

I discovered this when I wrote a book on the Darwinian revolution and wanted to characterize "Darwinian" (Ruse, 1979a). Charles Darwin accepted natural selection and sexual selection and applied his ideas to humans. Thomas Henry Huxley accepted evolution, applied it to humans, but was unenthused by selection. Alfred Russel Wallace accepted natural selection, had severe doubts about aspects of sexual selection, and pulled back from the evolution of humans. And so the story went. In the end, I had to be satisfied with some mushy sociological notion. A "Darwinian" was someone who thought of himself as a Darwinian, or some such thing.

I feel a bit the same way about characterizing punctuated equilibrium theory. Despite reassurances about uniformity and consistency—sometimes rather irritated reassurances—I see a number of different ideas and positions being staked out under the punctuationist flag. I say this in a sympathetic spirit, for one would expect a certain development of thought as people grapple to articulate a new idea. In a nondefinitive way, let me suggest that there have been three basic phases to the punctuated equilibrium theory, even in its short history.

The first phase was, fairly obviously, the form of the theory as it was announced at the beginning, most particularly in the paper by Niles Eldredge and Stephen Jay Gould, "Punctuated Equilibria: An Alternative to Phyletic Gradualism" (1972). Without going into details about whether this was truly the first articulation of the theory (Eldredge, 1971, appeared just before, although it does not seem to me to have quite the panache of the jointly authored paper), we can say that the joint paper got the attention and put the theory on the map. Also, I think we can say, with a reasonable amount of confidence, that this version of the theory was intended as a fairly straightforward extension of orthodox Darwinism (or, if you like, neo-Darwinism or the synthetic theory of evolution).

The authors noted that the fossil record does not show gradual change. Rather, it exhibits uniformity, with change occurring across gaps or breaks. They noted also that modern evolutionary theory locates much, if not most, change at times of speciation, that such speciation is allopatric (involves geographical isolation), and that Mayr's founder principle is a crucial causal element. And they concluded that the empirical material, the fossil record, is explained by the theory's modern understanding of speciation and that there is no need of subsidiary hypotheses about incomplete records and so forth. To the contrary, Eldredge and Gould professed themselves fairly happy with the fossil record. Furthermore, the authors underlined the conventionality of their position by stressing that their interpretation of natural selection is at one with that of everyone else. In particular, they wanted no truck with group or other deviant kinds of selection. Change in species comes down ultimately to change in organisms.

> The coherence of a species, therefore, is not maintained by interaction among its members (gene flow). It emerges, rather, as a historical consequence of the species' origin as a peripherally isolated population that acquired its own powerful homeostatic system. (We regard this idea as a serous challenge to the conventional view of species' reality that depends upon the organization of species as ecological uniters of *interacting* individuals in nature. If groups of nearly independent local populations are recognized as species only because they share a set of homeostatic mechanisms developed long ago in a peripheral isolate that was "real" in our conventional sense of interaction, then some persistent anomalies are resolved. The arrangement of many asexual groups into phenetic "species," quite inexplicable if interaction is the basis for coherence, receives a comfortable explanation under notions of homeostasis.) (Eldredge and Gould, 1972:223.)

This was, of course, the time when George Williams's *Adaptation and Natural Selection* was the most influential book of the day.

The second phase came as the decade drew toward its end, reaching its apotheosis in 1980 with Gould's notorious "Is a new and general theory of evolution emerging?" There had to be something new going on, for, far from portraying himself as an orthodox Darwinian, Gould assured us that the synthetic theory of evolution was effectively dead. Basically, I see a major deemphasis of the importance of organic adaptation, with a consequent downplaying of the role of natural selection. Also, Gould certainly was starting to toy with the idea of macromutations (perhaps due to chromosomal rearrangements), with species' changes occurring in one or a couple of generations. Significantly, the father figure had changed from Charles Darwin to Richard Goldschmidt (1940), one of the few English-speaking saltationists of the past half-century and intellectual rival to the synthetic theory's greatest living exponent, Ernst Mayr. (See Gould, 1979, 1980a, 1980c; Gould and Eldredge, 1977—the last-named, a bit of a hybrid paper.)

Then, fairly quickly after that we move to the theory's third, and I think to date, final form. There is a pullback from extremism, particularly with respect to the formation of new species. Thoughts of macromutation decline—indeed we are told that we were meant to think they were there in the first place—and we learn that although fifty thousand years may seem like a long time to a fruit-fly geneticist, to a paleontologist it is but a blink of the eyelid (Gould, 1982a, b, 1983a, b). Gould (1982c) presents an explicit acceptance of macromutation, although it is disassociated from punctuated equilibrium theory. "Illegitimate forms of macromutation include the sudden origin of new species with all their multi-farious adaptations intact *ab initio*, and origin by drastic and sudden reorganization of entire genomes. Legitimate forms include the saltatory origin of key features (around which subsequent adaptations may be molded) and marked phenotypic shifts caused by small genetic changes that affect rates of development in early ontogeny with cascading effects therefore" (pp. 88–89).

Although there may be pullback, there is certainly no retreat. Now we are presented with a *hierarchical* view of the evolutionary process. Down at the level of the individual organism natural selection is working away, although much of its emphasis seems to be on keeping things in line. (Below the level of the individual, at the level of the gene, there may be drift, working in ways that Japanese evolutionists suggest.) Adaptation counts, although it is only one factor, along with many constraints on development and adult form. Change generally comes at times of speciation, and although full-scale mutations are out, there is a feeling that changes in rates of development could have significant and fairly instant effects. (There are interesting parallels here with E. O. Wilson's [1975] much criticized multiplier effect.) What is important is that, as we pull back and look at species, we see that their evolution has patterns and dynamics of its own at a level higher than that of the individual. These are patterns that sit upon and are connected to the individual but are in no sense reducible to the individual. Against the well-known views of the geneticist Theodosius Dobzhansky (1951), macroevolution is *not* microevolution writ large.

In the 1980s, the prohibitions against group selection are not as strong as previously, and there are suggestions that at the species level various effects might obtain that will make differences to success or failure. But rather than group selection, more emphasis is put on the notion of "species selection," in which species act as units (this is very much in line with the philosophical claim that species are individuals and not classes) and trends are seen as working out of forces acting at this higher (i.e., above the organism) level.

I do not claim that these are three distinct theories or subtheories. Steven Stanley (1979), for instance, has been promoting species selection since the 1970s, but I do claim that looking at punctuated equilibrium theory in the way I have just sketched is not unfair to history. What is unfair is the way my discussion, from now on, will be so heavily weighted toward the ideas of Stephen Jay Gould. Given what I have said about the supporters of theories, we have no right to think that others, including Gould's regular co-author Eldredge, share his views. In fact, in Eldredge's latest books, *Time Frames* (1985a) and *Unfinished Synthesis* (1985b), I find much sympathy for what I have just characterized as the third phase of the theory. As Eldredge (1985a) explicitly states (p. 161), however, there are differences between his and Gould's beliefs, especially with respect to adaptation—and Eldredge is the less critical. In view of what I shall have to say, this is no small difference. All I can do is plead that life is short and confess to the real motive for my focus on Gould. I am a philosopher asking a philosopher's questions, and Gould more readily answers them than does anyone else.

What Is a Paradigm?

How many names are there for God? I will brew down the meanings of "paradigm" to four (Ruse, 1981). They are not accepted officially by anybody, including myself, but as with my analysis of punctuated equilibrium theory they are offered in a noncritical, constructive spirit. I do not think my meanings are unfair to the spirit of the notion as popularized by Thomas Kuhn in his *Structure of Scientific Revolutions* (1962).

First is what one might call the *sociological* sense of paradigm, the notion of a group of people who come together, feeling that they have a shared outlook (whether they do or not), and to an extent separating themselves off from other scientists. This sense of paradigm comes through in Kuhn's characterization:

> Aristotle's *Physica*, Ptolemy's *Almagest*, Newton's *Principia* and *Opticks*, Franklin's *Electricity*, Lavoisier's *Chemistry*, and Lyell's *Geology*—these and

> many other works served for a time implicitly to define the legitimate problems and methods of a research field for succeeding generations of practitioners. They were able to do so because they shared two essential characteristics. Their achievement was sufficiently unprecedented to attract an enduring group of adherents away from competing modes of scientific activity. Simultaneously, it was sufficiently open-ended to leave all sorts of problems for the redefined group of practitioners to resolve. (P. 10.)

Second is the *psychological* level to a paradigm. People in a paradigm see things in a way differently from people not in the paradigm, especially people in another paradigm. One group sees a duck. Another group sees a rabbit. One group sees dephogisticated air. The other group sees oxygen. And so forth. This is why, so often, in a change of paradigm— in a "revolution"—something akin to a conversion experience occurs. Literally a whole new way of looking at things develops.

Third is the *epistemological* level to a paradigm. Here, one's ways of doing science are bound up with the paradigm. What counts as a proper solution is defined by the paradigm, not to mention what counts as an interesting problem. For the pre-Copernican the difference between the inferior and superior planets was not a matter of great concern. For the post-Copernican, the difference was crucial, and the fact that the heliocentric theory solves the problem whereas the geocentric theory does not was taken as definitive positive evidence (Kuhn, 1957). It is this way that epistemology functions in a paradigm switch—asymmetrical from within to without the paradigm—that makes a revolution, in respects, beyond reason.

Fourth is the *ontological* aspect to a paradigm. At times, Kuhn suggests that the very world is not so much defined as created by a paradigm. What there *is* depends crucially on what paradigm one holds. For Joseph Priestly there literally was no such thing as oxygen—it is not simply that he did not believe in such a thing as oxygen. In the case of Antoine Lavoisier, he not only believed in oxygen—oxygen existed.

Others might want to put forward other notions of paradigm. These four will do for me, here.

Is the Theory of Punctuated Equilibria a New Paradigm?

As far as the sociological aspect is concerned, I see no reason to deny punctuated equilibrium theory paradigm status. I doubt that either of the founding authors would claim the same authority for "Punctuated Equilibria: An Alternative to Phyletic Gradualism" as for Aristotle's *Physica*. But Kuhn grants—especially in his later writings (for instance, the 1970 edition of *The Structure of Scientific Revolutions*)—that paradigms do not have to have earth-shattering significance. (Perhaps a slightly inappropriate metaphor in this context!) Certainly, the work by Eldredge and Gould and others has polarized evolutionists in such a way that punctuated equilibrium theory has defining paradigm properties at the social level (see, for instance, Hallam, 1977). Moreover, although I have no direct evidence, my impression (from the speed with which people took defining positions) is that the sociological factors came into place fairly early on in the theory's history. (Sociology involves human factors. It is hardly a thesis of extreme methodological reductionism to point out that this aspect of paradigmhood relates fairly directly to the brilliant literary abilities of Eldredge and Gould, combined with the journalistic interests of Roger Lewin, of *Science*, in seeing all disputes and differences in stark black and white.)

Conversely, at the ontological level, I see little reason to give the theory paradigm status. I doubt anyone seriously thinks that the theory has changed the fossil record. Or, more moderately, if one accepts some sort of realist thesis, I doubt one would be persuaded that the theory has changed the world. If one is a nonrealist, ontological change to a large extent collapses into questions of psychology and epistemology.

What, then, of the two middle categories of psychology and epistemology (which, from the paradigm point of view, are closely related)? These are more difficult questions to answer. To do so, we must seek out the reasons why the theory of punctuated equilibria has attracted support. There are three plausible directions in which this inquiry might be pursued—empirical, political, and metaphysical (or theological, in a sense).

THE EMPIRICAL ATTRACTIONS OF THE THEORY

Why would one accept the theory of punctuated equilibria? Is it because of the empirical evidence, the fossil record? Somewhat paradoxically, the founders denied that this would be a reason: between punctuated equilibrium theory and the older, "phyletic gradualism," "the data of paleontology cannot decide which picture is more adequate" (Eldredge and Gould, 1972:208). But I think we can take this with a pinch of salt. It is true that if the Duhem-Quine thesis about the relation between theory and evidence is taken seriously, some way can be found of propping up a crumbling theory,[2] but, if one is prepared to stay with science, there really does come a time when the facts start to bite. For a scientist, it is no longer reasonable to hang on to one position rather than another. With respect to punctuated equilibrium theory, Eldredge and Gould and their supporters clearly think the evidence matters. One may argue indefinitely that the gaps in the fossil record are only artifacts of incomplete fossilization, but this becomes increasingly ad hoc. And how, in any case, can stasis (lack of morphological change) be explained? One cannot seriously argue that all change occurs in the soft, nonfossilizable parts of organisms. Evolutionists today see no reason to draw distinctions between hard and soft. Why should they do so for the past?

The empirical evidence has certainly been intended as important support for the punctuated equilibrium position. By 1977 (in their follow-up paper, "Punctuated Equilibria: The Tempo and Mode of Evolution Reconsidered") Gould and Eldredge heartily denied that they had ever belittled the importance of the empirical evidence. Supposedly, all they had suggested was that the facts tend to be "theory laden," that is, we tend to look at the world from one viewpoint or another.[3]

The evidence has been intended as support, and it would be churlish to deny that it has been effective support. One thinks particularly of Peter Williamson's (1981) pertinent findings about the evolution of fish in the lakes of East Africa. Everyone seems to agree that this evidence manifests a punctuated picture of evolutionary change. On the negative side, Gould's pulling back from the extremism, the saltationism, of the second phase of the theory was in major part a function of the evidence. On the one hand, he was under fire from other evolutionists, particularly geneticists, who argued flatly that the empirical base he required

2. The Duhem-Quine thesis, or D-thesis, points out that if you have a hypothesis *H*, supposedly leading to an empirical observation *O*, but not *O* obtains, then instead of denying *H* (which is what *modus tollens* and Popper's principle of falsifiability demand) you can always deny an auxiliary hypothesis *A*, for in science you never get observations from one hypothesis alone but always in conjunction with many others (for instance, about the reliability of one's equipment).

3. This is not the only occasion in the short history of punctuated equilibrium theory when minds are changed although it is pretended that no shift has occurred. Apparently when one is fighting for one's views, any shift is taken as weakness. But I wish people would see that changing or modifying one's thinking can be a virtue as well as a fault.

does not exist. On the other hand, those same geneticists gave him an empirical escape route—or, if you like, a more fruitful path of discovery. They pointed out how much change can occur in fifty thousand years, even with slow-breeding organisms (Stebbins and Ayala, 1981).

Empirical evidence certainly earns a place in the battle over punctuated equilibrium theory. It has been clear from fairly early on, however, that—whatever the founders may have intended—the empirical evidence, particularly in the fossil record, is truly not the ultimate court of appeal. Apart from all the questions about whether our knowledge of the fossil record tells more of the fossils or of the classificatory techniques of paleontologists, there are two main problems. First, the paleontologists disagree about how often the fossil record exhibits stasis and how often it exhibits gradualism. Just about everyone agrees that there is a bit of stasis and a bit of gradualism. But how precisely one understands a "bit" is another matter. All one can say is that there may be some final answer to this debate—perhaps at some future convention, paleontologists will agree that some 85.021 percent of cases show stasis—but the final answer has not yet been given by competent paleontologists. (Gould and Eldredge, 1977, shows well the difficulties with the evidence. See also Hallam, 1977, and Gould and Eldredge, 1986.)

The second problem is that many of the supposedly crucial cases turn out to be highly ambiguous. For example, did the evolution of *Homo sapiens* exhibit punctuated equilibria? Some say yes, some say no. As Figure 6.1 (from a 1983 paper by the late Glynn Isaac) shows only too well, there is no definitive answer.

I conclude, therefore, that the empirical evidence is far from definitive, and I back this up by an (admittedly undocumented) sense that, in recent years, opponents have not been striving all-out to find crucial test cases. Such cases have tended to become a bit like prayer: comforting to believers. I am not saying that evidence could (in principle) never be effective. Moreover, empirical knowledge was crucial in the formulation of punctuated equilibrium theory. Claims about allopatric speciation are derived from our knowledge of the world. What I am saying is that we need to keep looking for the theory's hold on good biologists.

Two Political Attractions of the Theory

"Political" can mean the politics internal to science or even to biological science, or it can mean the politics of the big world, outside. Let us consider them in turn.

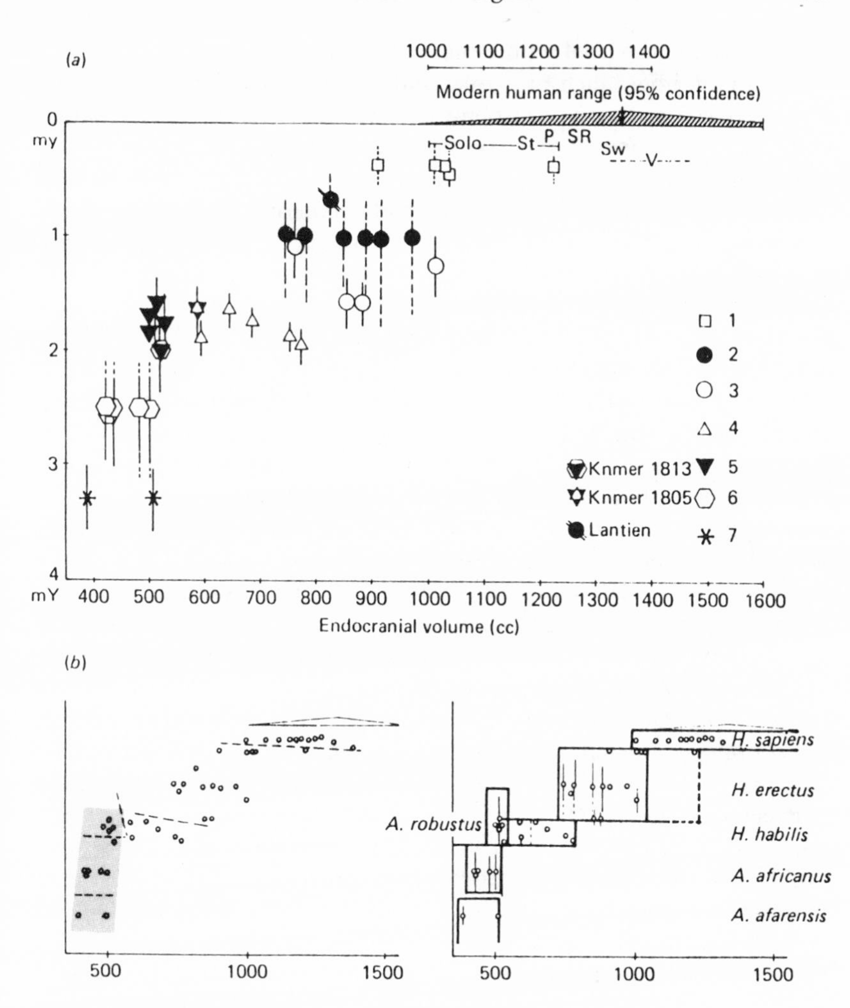

Figure 6.1. (*a*) Endocranial volumes of hominid fossil skulls plotted against a time scale. The degree of uncertainty about age is indicated by the vertical bars. 7 = *Australopithecus afarensis* from Hadar; 6 = *A. africanus*; 5 = *A. robustus* and *A. boisei*; 4 = *Homo habilis*; 3 = *H. erectus* (East Africa); 2 = *H. erectus* (Java and Lantien); 1 = *H. erectus* (Pekin). Early *H. sapiens* (the range for skulls): P = Petralona, St = Steinheim, S = Saldanha, R = Kabwe (Rhodesia man), Sw = Swanscomb, V = Vertesszöllös. (*b*) The same data (left) fitted to a phyletic gradualist model and (right) to a punctuated model. The species indicated at the right apply to both versions. Reprinted, by permission, from Glynn Isaac, "Aspects of Human Evolution," in *Evolution from Molecules to Men*, ed. D. S. Bendall (Cambridge: Cambridge University Press, 1983).

I am sure that internal politics has played (and still does play) a role. I do not know how much of a role, and I expect that the answer varies from person to person. I strongly suspect that paleontologists suffer from inferiority complexes, recognizing and resenting the fact that other evolutionists, especially geneticists, set the pace, and resenting even more that other evolutionists think this a rightful ordering of things. ("Hand-waving by the paleontologist, Mike!") Supposedly, it does not really matter what happens in the fossil record—it has got to fit in with the chromosomes, and not vice versa. When Gould toys with the notion of macromutations, he is slapped down, yet no paleontologist would dare challenge (say) William Hamilton (1964a, b) when he proposes kin selection.

I suspect that, at the beginning, there was simple pleasure in making an important point and showing its usefulness. Remember, the first phase of punctuated equilibrium theory was orthodox neo-Darwinism, and the claim to fame was that paleontology could be useful in confirming the theory. The cry seemed to be: "Don't think of us as a weak sister. We can play a full role in supporting the cause." Now, as the theory has matured to the third phase, the claim is much stronger: "There are aspects—virtually important aspects—of the evolutionary process that you just cannot understand without the aid of paleontology. You would not even recognize them without paleontology. Hence you ignore us at your peril." This is a powerful political reason for supporting punctuated equilibria. (If you think I exaggerate, look at Gould's "Irrelevance, Submission and Partnership: The Changing Role of Palaeontology in Darwin's Three Centennials, and a Modest Proposal for Macroevolution" [1983a]. The main title tells all. Significantly, this paper was given at the Darwin Centennial Conference in Cambridge, England, in 1982, when Gould—and paleontology—were on display in front of the rest of the evolutionary world.)

In the world external to science, until recently, most people would have pooh-poohed the idea that (general) politics could affect science, making some theories more attractive than others. But thanks to the work of historians (like Gould himself) we now know that this is naive thinking—perhaps such denial is itself derived from a political stance that necessitates the objectivity of science (particularly one's own science). Certainly, Gould (1981) has been one of the forerunners in arguing that aspects of contemporary evolutionary thought are impregnated with a certain political ideology. I refer to the claim that human sociobiology is capitalist libertarianism by another name (Allen et al., 1977).[4]

4. This was the work of a collective, which included Gould.

Gould has admitted, if not to being a Marxist, to having learned his Marxism at his daddy's knee (Gould and Eldredge, 1977). He has thought it pertinent to bring this fact into his scientific work (on the nature of paleontological theories), he has admitted to seeing the world as functioning according to the laws of dialectical materialism, and he has praised punctuated equilibrium theory precisely because it fits with such a dialectical worldview (Gould, 1979).

When he was questioned on these links (in print in 1981 and verbally in 1982 by the paleontological gadfly Beverley Halstead at the Cambridge Darwin Symposium), Gould tended to downplay their significance, but they were there—especially around 1980, when Gould was promoting the second phase of the theory. This, of course, was the most saltationist version and with reason might be thought the one most in tune with a Marxist worldview. Remembering also that Gould was a leader in the fight against human sociobiology—which was carried on most strenuously by left-wing groups like the Cambridge (Massachusetts) Science for People collective, of which Gould was a member (Ruse, 1979b; Segerstralle, 1986)—it is worth recalling that this fight was most intense toward the end of the last decade. And it is interesting how the sociobiological controversy and the punctuated equilibrium controversy were drawn together. If the punctuated equilibrium theory is true for humans, then essentially all of the change leading to today's forms happened in one leap right at the beginning of our species' origin. Since then, there has been little or no change, especially not change of a significant adaptive nature. Obviously, all of this blows holes in any extreme sociobiological thesis, which, through the multiplier effect, sees major adaptive changes occurring virtually overnight, within the same species. The links are there.

In 1981 Gould published *The Mismeasure of Man*, a passionate treatise on the history of hereditarian claims about intelligence, a significant theme of which is the way such views have been used to discriminate against Jews—especially in America in the first part of this century. I speak of passion in part because, of all Gould's books, this is the one that got widely criticized, particularly by historians of psychology. This perhaps suggests that strong motives frame its nature. (I should note here that I wrote a highly favorable discussion review [Ruse, 1982a].)

It does not seem to me entirely implausible that Gould's passion against human sociobiology was linked to the fear that it was yet another tool that could be used for anti-Semitic purposes. I asked Gould about this once, when we were having dinner one day in December 1981, when we were both witnesses for the American Civil Liberties

Union against the creationists in Arkansas. He did not entirely repudiate the idea but was inclined to think that the opposition stemmed more from Marxism, and as it so contingently happens, most American Marxists are from Eastern European Jewish families. Perhaps both factors were involved.

I am *not* saying that the punctuated equilibrium theory is part of a communist plot. Most of its supporters (like Eldredge) are not Marxists. Still less am I saying that critics of the theory are guilty of anti-Semitism. I have already pointed out how dangerous it is to believe automatically that one person's theory is another person's theory, even if they fall in the same camp. And I would argue this point even more strongly for motives. All I am saying is that there is a complex set of threads that bind people's motives to their support for scientific theories, and some of those motives I have just been detailing may have laid—may still lie—behind the support that some people give to punctuated equilibrium theory.

As with the empirical evidence, I do not argue that this is the entire story. I am inclined to think that, at most, we are still dealing with secondary causes. The direct connections between support for punctuated equilibrium theory and opposition to human sociobiology are frail. But they are there. Moreover, I think we are starting to hint at some of the deep reasons for the support of punctuated equilibrium theory.[5]

5. Since Gould has strongly denied significant links between his scientific position and his political beliefs, it seems fairest to give one of the more explicit discussions in which the two are linked and then a repudiation. First, from Gould and Eldredge (1977):

> When Darwin cleaved so strongly to gradualism—ignoring Huxley's advice that he did not need it to support the theory of natural selection—he translated Victorian society into biology where it need not reside. As his astute biographer W. Irvine remarks (1959, p. 98): "Darwin's matter was as English as his method. Terrestrial history turned out to be strangely like Victorian history writ large. . . . The economic conceptions [of *laissez-faire* liberalism] . . . can all be paralleled in the *Origin of Species*. But so, alas, can some of the doctrines of English political conservatism. In revealing the importance of time and the hereditary past, in emphasizing the persistence of vestigial structure, the minuteness of variations and the slowness of evolution, Darwin was adding Hooker and Burke [famous English conservatives] to Bentham and Adam Smith [equally famous liberals]. The constitution of the universe exhibited many of the virtues of the English constitution." Karl Marx, who admired Darwin greatly and once stated that the *Origin* contained "the basis in natural history for all our views," made the same point in a famous letter to Engels (1862): "It is remarkable how Darwin recognizes among beasts and plants his English society with its division of labor, competition, opening up of new markets, 'invention,' It is Hobbes' '*bellum omnium contra*

The Metaphysical Attractions of the Theory

Let me start this section with a question, the answer to which I feel confident (Ruse, 1979a). What was Darwin's big achievement, or more precisely, what did Darwin think was his big achievement? Most obviously, it was establishing the fact of evolution—and no one (except the creationists) would deny him credit for this. But in Darwin's own mind,

---

omnes,' [war of all against all] and one is reminded of Hegel's Phenomenology, where civil society is described as a 'spiritual animal kingdom,' while in Darwin the animal kingdom figures as civil society." We mention this not to discredit Darwin in any way, but merely to point out that even the greatest scientific achievements are rooted in their cultural contexts—and to argue that gradualism was part of the cultural context, not of nature.

Alternate conceptions of change have respectable pedigrees in philosophy. Hegel's dialectical laws, translated into a materialist context, have become the official "state philosophy" of many socialist nations. These laws of change are explicitly punctuational, as befits a theory of revolutionary transformation in human society. One law, particularly emphasized by Engels, holds that a new quality emerges in a leap as the slow accumulation of quantitative changes, long resisted by a stable system, finally forces it rapidly from one state to another (law of the transformation of quantity into quality). Heat water slowly and it eventually transforms to steam; oppress the proletariat more and more, and guarantee the revolution. The official Soviet handbook of Marxism-Leninism (anonymous, undated) proclaims: "The transition of a thing, through the accumulation of quantitative modifications, from one qualitative state to a different, new state, is a leap in development. . . . It is the transition to a new quality and signalizes a sharp turn, a radical change in development. . . . We often describe modern Darwinism as a theory of the evolution of the organic world, implying that this evolution covers both qualitative and quantitative changes. Leap-like qualitative changes in social life are designated by the concept of revolution. . . . The evolutionary development of society is inevitably consummated by leaplike qualitative transformation, by revolutions" (anon., pp. 88–89). It is easy to see the explicit ideology lurking behind this general statement about the nature of change. May we not also discern the implicit ideology in our Western preference for gradualism?

In the light of this official philosophy, it is not at all surprising that a punctuational view of speciation, much like our own, but devoid (so far as we can tell) of a reference to synthetic evolutionary theory and the allopatric model, has long been favored by many Russian paleontologists (Ruzhetsenv, 1964; Ovcharenko, 1969). It may also not be irrelevant to our personal preferences that one of us learned his Marxism, literally, at his daddy's knee.

A letter by Gould in *Nature* (February 26, 1981) responding to Beverley Halstead's charge, "simply silly beyond words, that punctuated equilibrium, because it advocates rapid changes in evolution is a Marxist plot," continues:

> For Halstead's second charge, I did not develop the theory of punctuated equilibrium as a part of a sinister plot to foment world revolution, but rather as an attempt to resolve the oldest empirical dilemma impeding an integration of paleontology into

at least, there was more. After all, by 1859 (the year of the publication of the *Origin*) many others had proposed the idea of evolution. For Darwin, his big achievement was his mechanism of natural selection. To know why he thought it a big achievement, we have to know what his question was. His first introduction of natural selection makes the question clear.

> Before entering on the subject of this chapter, I must make a few preliminary remarks, to show how the struggle for existence bears on Natural Selection. It has been seen in the last chapter that amongst organic beings in a state of nature there is some individual variability; indeed I am not aware that this has ever been disputed. . . . How have all those exquisite adaptations of one part of the organization to another part, and to the conditions of life, and of one distinct organic being to another being, been perfected? We see these beautiful co-adaptations most plainly in the woodpecker and missletoe; and only a little less plainly in the humblest parasite which clings to the hairs of a quadruped or feathers of a bird; in the structure of the beetle which dives through the water; in the plumed seed which is wafted by the gentlest breeze; in short, we see beautiful adaptations everywhere and in every part of the organic world.
>
> Again, it may be asked, how is it that varieties, which I have called incipient species, become ultimately converted into good and distinct species,

---

modern evolutionary thought: the phenomena of stasis within successful fossil species, and abrupt replacement by descendants. I did briefly discuss the congeniality of punctuational change and Marxist thought (*Paleobiology*, 1977, p. 145) but only to illustrate that all science, as historians know so well and scientists hate to admit, is socially embedded. I couldn't very well charge that gradualists reflected the politics of their time and then claim that I had discovered unsullied truth. But surely Halstead, who has done some statistics in his day, knows that correlation is not cause. If I may make a serious point: I grew up frightened in a leftist household during the worst days of McCarthyism in America; and I know that what seems peripheral or cranky today can become a weapon tomorrow (consider the current creationist surge in America). May we avoid red-baiting; it may not always be harmless.

There was no "plot," but there was a bit more than "correlation." Why the retreat? In part, I suspect, because punctuated equilibrium theory by the early 1980s had a high profile and did not need the "taint" of controversial ideology. In part, because the fight against human sociobiology likewise did not need such a ready weapon for the opposition. In part, because the fight against creationism was not helped by the connection. (I remember in Arkansas that, thanks to my having drawn the connection, both Gould and I were cross-examined on this by the assistant attorney general [appearing for the state in defense of the creationist law].) In part, because (as Gould says above) charges based on political ideology have such painful memories.

which in most cases obviously differ from each other far more than do the varieties of the same species? How do those groups of species, which constitute what are called distinct genera, and which differ from each other more than do the species of same genus, arise? All these results, as we shall more fully see in the next chapter, follow inevitably from the struggle for life. Owing to this struggle for life, any variation, however slight and from whatever cause proceeding, if it be in any degree profitable to an individual of any species, in its infinitely complex relations to other organic beings and to external nature, will tend to the preservation of that individual, and will generally be inherited by its offspring. The offspring, also, will thus have a better chance of surviving, for, of the many individuals of any species which are periodically born, but a small number can survive. I have called this principle, by which each slight variation, if useful, is preserved, by the term of Natural Selection. (Darwin, 1859:60.)

The point was that natural selection was to speak to *adaptation*—the hand and the eye. This was the question that Darwin wanted answered. Moreover, from all sorts of evidence, we know precisely why Darwin thought adaptation so significant a feature of the organic world. He was immersed in the natural theology of early nineteenth-century Britain. Although he was to reject the Great Designer in the sky of Archdeacon William Paley (1802), author of the authoritative texts on the subject, Darwin accepted completely the premise of the natural theologians that the organic world is *as if* designed. (Incidentally, Darwin was a Believer for most of his life, including the time of writing the *Origin*.)

This is the key point to understanding Darwin, even for one who doubts the efficacy of natural selection. Furthermore, it is the key point to understanding the thread of continuity from Darwin to the present. Consider, for example, the work of Sir Ronald Fisher, especially his *Genetical Theory of Natural Selection* (1930), or Sir Julian Huxley's *Evolution: The Modern Synthesis* (1942), or Theodosius Dobzhansky's *Genetics and the Origin of Species* (1951) (especially after what Gould has referred to as the "hardening" of the synthesis), or, in our day, G. C. Williams's *Adaptation and Natural Selection* (1966), or the sociobiological movement. Everybody thinks in terms of design—this was what Gould meant by "hardening." Williams (1966) puts the point explicitly:

Whenever I believe that an effect is produced as the function of an adaptation perfected by natural selection to serve that function, I will use terms

> appropriate to human artifice and conscious design. The designation of something as the *means* or *mechanism* for a certain *goal* or *function* or *purpose* will imply that the machinery involved was fashioned by selection for the goal attributed to it. When I do not believe that such a relationship exists I will avoid such terms and use words appropriate to fortuitous relationships such as *cause* and *effect*. (P. 9.)

Moreover, although no one (including Darwin—or Paley) would argue that all organic features are useful, the presumption is that when we are looking at features, especially complex ones, adaptation is there unless proven otherwise (which otherwise might well involve adaptation at one step removed, like past function). Richard Dawkins (1983) says it all:

> I agree with Maynard Smith (1969) that 'The main task of any theory of evolution is to explain adaptive complexity, i.e. to explain the same set of facts which Paley used as evidence of a Creator.' I suppose people like me might be labelled neo-Paleyists, or perhaps 'transformed Paleyists'. We concur with Paley that adaptive complexity demands a very special kind of explanation: either a Designer as Paley taught, or something such as natural selection that does the job of a designer. Indeed, adaptive complexity is probably the best diagnostic of the presence of life itself. (P. 404.)

Now, relate this to punctuated equilibrium theory. No one denies that natural selection plays a role. In the early versions, especially, selection plays an important role. No one denies adaptation. But if punctuated equilibrium theory does anything, it downplays the significance of natural selection and (especially) adaptation. Functions really do get squeezed down, as change is compressed into short periods. In addition—at least in some versions of species selection—when a new species is produced there is an element of randomness about adaptive value. According to Gould: "I do not claim that a new force of evolutionary change has been discovered. Selection may supply an immediate direction, but if highly constraining channels are built of nonadaptations, and if evolutionary versatility resides primarily in the nature and extent of non-adaptive pools, then 'internal' factors of organic design are an equal partner with selection" (1982a, p. 384).

This downplaying of the significance of adaptation is not a contingent by-product of the theory of punctuated equilibria. Gould, especially, has been a persistent critic of adaptationism for fifteen years. He accuses adaptationists of producing "just so" stories, like those of

Rudyard Kipling (Gould, 1981). He faults adaptationists for "Panglossian optimism" because they see design effects in nonexistent places (Gould and Lewontin, 1979). He cites examples of supposed adaptations that are not really adaptations at all. One neat example is the much-enlarged clitoris of the female hyena, which might lead one to think it functions as a quasi-penis. Gould argues that it is more plausibly a side product of testosterone levels (Gould, 1983c). And repeatedly, Gould has suggested that change might be merely contingent on change of rates of development and thus without any immediate purpose (Gould, 1977).

Is this all just negativism on Gould's part, or, at most, that he saw that his punctuated equilibrium theory downplays adaptation and so he has made a virtue out of necessity? There could well be a negative element here, especially insofar as the attack on adaptationism represents an attack on the highly adaptationist human sociobiology. This connection has been made explicitly (Gould, 1981), but it would be a bad mistake simply to think in negative terms—that punctuated equilibrium theory (particularly in its final phase) defines itself negatively by belittling a key plank in the Darwinian program. Rather, we would do better to start gathering up strands of evidence we have had thus far, particularly the European connection of which Marxism is a major intellectual outpouring, add to it Gould's expressed enthusiasm for the Germanic transcendentalist notion of *Baupläne* (blueprints for bodily structure) as well as for morphologists like D'Arcy Thompson, who were likewise keen on underlying structures (Gould, 1971), and finish off with a strong dash of Gould's reading of history, particularly in his *Ontogeny and Phylogeny* (1977), where he expresses strong admiration for various lines of European thought.

The answer that emerges is that we are faced with a thinker from a biological tradition different from the utilitarian adaptationism (to use the technical term) of the Darwinian. We are considering somebody, and in punctuated equilibrium theory somebody's theory, whose roots go back to the transcendental idealism of Goethe and other early nineteenth-century thinkers. In the words of that great historian of biology E. S. Russell (1916), the emphasis is on *form* rather than *function*. We are considering people whose emphasis is on structure, shared between animals, because of underlying laws of constraint.

I do not mean to be negative. In suggesting that punctuated equilibrium theory stands in the tradition of Goethe, I am not trying subtly to suggest that Gould is not a genuine evolutionist. The point is mechanisms; and in any case, I suggest also that Darwin stands in the tradition of Paley, far less of an evolutionist than Goethe ever was. In fact, these are the two traditions that go right across the evolutionary divide. (See Russell, 1916, and Mayr, 1982, for full details.) On the functionalist side are Paley (who goes back to the external teleology of Plato) and the French father of comparative anatomy, Georges Cuvier (who goes back to the internal teleology of Kant and Aristotle), and then the Darwinians. On the formalist side are the transcendentalists, not just the Germans, Goethe and Lorenz Oken and Karl Ernst von Baer (I would include him here), but also the French such as Etienne Geoffroy St.-Hilaire and the English, including Richard Owen, the notorious evolutionist Robert Chambers, T. H. Huxley in respects (his attacks against transcendentalism were more personal against Owen than anything else), and the Swiss-American Louis Agassiz. To this side I would now add supporters of punctuated equilibrium theory—at least, supporters of the later versions.

Virtually no one belongs to one side exclusively. Darwin was always extremely proud of his ability to explain *Baupläne*, as they manifest themselves in repeated patterns (homologies) from organism to organism (Ruse, 1979a). This "unity of type," he thought, fell out from his overall evolutionism. The point is that for Darwin these *Baupläne* are frames on which to hang adaptations. Conversely, someone like (say) Owen (1849) never denied adaptation. No more does Gould. It is all a matter of emphasis and weighting of relative importance.

Traditionally among evolutionists these two sides have been associated with different attitudes to the tempo and mode of evolution (to use G. G. Simpson's [1944] happy phrase). Darwinism can certainly accept, indeed expects, different rates of evolution, from very slow or nonexistent all the way to very rapid, which might not get recorded in the fossil record. For Darwin and his fellow thinkers, however, the essence of change has to be gradual because otherwise organisms would be liable to get out of adaptive focus. This is the real reason why Darwin repudiated saltationism. (This is evident in the early notebooks and first drafts of the theory. Darwin toyed with saltations in the early days but then realized that they pose great difficulties for adaptation. See Ruse, 1979a; Ospovat, 1981.)

Conversely, although the transcendentalist tradition has never excluded gradualism entirely, such gradualism has never been seen as essential and there has always been a place for "jumps," however these might be interpreted (Reif, 1983). Chambers (1844), for instance, was an out-and-out saltationist. And T. H. Huxley (1860) chided Darwin for tying himself too tightly to the motto "natura non facit saltum." Punctuated equilibrium theory is taking a stand well within its tradition. (This is, of course, also the tradition of the German philosophers Marx and Engels, which is why I prefer to see the various parts of Gould's thinking as connected through a general metaphysical position, rather than tacked together piecemeal as the particular controversy demands.)

In additional support of my claim that punctuated equilibrium theory's real roots lie in the transcendentalist tradition and the reason why (especially in its final form) it appeals to certain biologists is because they too turn or are attracted to that tradition, let me note that the simultaneous opposition to human sociobiology makes much sense because (apart from the direct links) such opposition is seen as part and parcel of a general attitude, given that human sociobiology is so much an extension of neo-Darwinism (Ruse, 1979b). Finally, let me quote the conclusion to one of Gould's essays (published in 1983):

> Darwin's assertion of evolution was an event of such unrivaled importance in the history of science and human society that we tend to view it as a watershed for all concepts in biology—as though everything should be discussed primarily in terms of before or after evolutionary theory. Such a perspective is inadequate, for many traditions of thought persist through evolutionary theory, emerging with a reinterpretation of causality to be sure, but intact nonetheless. In particular, certain "national styles" persisted from the eighteenth century, through Darwin's era, and into our own time. Views on adaptation provide a good example. I have said nothing about German biology because it has generally held a view of adaptation outside the scope of this essay (Rensch and other synthesists notwithstanding). Adaptation is seen as real but superficial, a kind of jiggling and minor adjustment within a *Baupläne* evolved by some other mechanism—not, in any case, a general mechanism (by extrapolation) for evolutionary change at higher levels. This viewpoint is firm in the nonevolutionist transcendental morphology of Goethe and many of the *Naturphilosophen*. It affects the entire "laws of form" tradition, underlies the pre-Darwinian evolutionism of Etienne Geoffroy de Saint-Hilaire, and persists to our time among such German evolutionists as Remane, Schindewolf, and Goldschmidt. The adaptationist tradition, on the other hand, has been an English pastime for at least two centuries. If continental thinkers glorified God in nature by inferring the character of his thought from the laws of form lining his created species, or incarnated ideas (as Agassiz maintained), then Englishmen searched for him in the intricate

> adaptation of form and function to environment—the tradition of natural theology and Paley's watchmaker. Darwin approached evolution in a quintessentially English context—by assuming the adaptation represented the main problem to be solved and by turning the traditional solution on its head. Few continental thinkers could have accepted such a perspective, since adaptation, in their view, was prevalent but superficial. The centrality of adaptation among English-speaking evolutionists in our own times, and the hardening of the synthesis itself, owes much to this continuity in national style that transcends the simple introduction of evolutionary theory itself. One might say that adaptation is nature's truth, and that we had to overthrow the ancient laws-of-form tradition to see it. But one might also say that twentieth-century panselectionism is more a modern incarnation of an old tradition than a proven way of nature. (1983b:90–91.)

One might also say that twentieth-century punctuated equilibrium theory is more a modern incarnation of an old tradition than a proven way of nature.[6]

Is the Theory of Punctuated Equilibria a New Paradigm?

I ask the question again, thinking now about the second and third senses of paradigm, the psychological and epistemological. In their original paper, Eldredge and Gould (1972) were a little bit coy. Explicitly (in a footnote) they denied that they were into the paradigm-invention business. Yet, with all their claims about the facts not deciding the issue, one senses that they would not have been brokenhearted had some-

6. Let me return again to the worry that I am dealing less with punctuated equilibrium theory per se and more with one man's (Gould's) version of it. Could it not be argued that although I may put my finger on why Gould believes in the theory, this is not necessarily the reason why others do. Indeed, if we involve the old distinction between the context of discovery and the context of justification, it would be possible to argue that just about everything I am saying is irrelevant to the reasons why the theory is accepted. I can only respond that this line of argument virtually gives the case away, by definition. Only empirical evidence is allowed to count, and that is the end of matters. I would myself challenge the discovery/justification dichotomy, and as part of my case I would note that Gould's position is stronger against adaptationism than is Eldredge's and since they both have access to the same evidence would claim that the difference has to be explained in Gould's deeper roots in Continental philosophy and science. (I am interested to note, incidentally, that although *Time Frames* contains several references to Gould's early work, there is no mention of any of the post-1977 essays. *Unfinished Synthesis* is more generous.)

one brushed aside their modesty and contradicted them. At this point in the theory's history, however, I do not think it would be appropriate to talk of a (new) "paradigm," especially not in the epistemological sense. The point about punctuated equilibrium theory when it was first introduced was that it was an extension, or rather, a correct application of an already existing paradigm, namely orthodox neo-Darwinism. It was certainly agreed that one may now see things differently. But this claim was not based on a supposed paradigm switch. Rather, the claim was that only now was one using properly the paradigm in hand.

As the theory has matured, particularly through to its third phase, I would argue that the case for paradigm status (in the second and third senses) becomes much stronger. This is especially true if one grants that the ultimate appeal of the theory lies in its transcendentalist roots. In a way, there are two metaphors that color one's picture of the organic world. On the Paley/Darwinian side is the world as an object of design, but not just any design—*artifactual* design. Organisms are seen as like artifacts. The eye is a telescope, the heart is a pump. The fins on the back of stegosaurus are heat-regulating devices akin to those found in hydroelectric stations. On the transcendentalist side, the world is still object of design, but not just any design—*crystalline* design. Organisms are seen as like crystals. There are various laws that determine possible patterns, and then these are repeated throughout various branches of the organism.[7]

All of this, it seems to me, points to something (at both psychological and epistemological levels) much like a Kuhnian paradigm difference. In looking at organisms, or their fossil remains, one is looking at different things (not necessarily in some ultimate ontological sense). And certainly, this leads to different methodological judgments and directives. The orthodox Darwinian searches for those gradual linking sequences in the fossil record and tries to understand change in terms of adaptive response to selective pressures. The punctuatated equilibrium supporter knows that these links will be rare or nonexistent, and the search for them a waste of time, and looks for other reasons behind change (or nonchange). Also, he or she thinks it worthwhile to seek answers to

7. I have spoken of "design" on the transcendentalist side. In the days when natural theology ruled, it was argued that homologies do show design, but everyone agreed that they show much less design than functions—and there were those who accepted only functions as evidence of design, who used them against design. I speak of the "crystalline" metaphor after Whewell, who is my window into much of the early nineteenth-century thinking about natural theology (1837, 1840).

questions about higher levels of the evolutionary process, which I am not sure the orthodox Darwinian recognizes as questions at all (see Maynard Smith, 1983).[8]

Grant then that there is indeed something going on that looks like a paradigm (or paradigm difference) in action. People (like my former self) who dismissed the idea were wrong—and missing something rather interesting to boot. I will conclude by making three qualifying—or clarifying—points. First, what we have are things that are rather softer than Kuhnian paradigms. For Kuhn, it is one or the other but not both. Now it is a duck. Now it is a rabbit. There are no duck-rabbits. The artifact/crystal metaphors are a little different in that there is room for overlap, with everybody at least appreciating the other side, somewhat. There is not quite the stark contrast that Kuhn supposes.

Second, the two positions have had a long parallel history. Now one position predominates, now the other. Here one position is popular, here the other. But both have been around for two hundred years. We do not get the sequential story that *The Structure of Scientific Revolutions* leads us to suspect. Indeed, I am not sure just how much support our story gives to the very notion of a "revolution" in science. I am not denying that they occur, but they are certainly not as Kuhn portrays them. The coming of punctuated equilibrium theory was not revolution, in the sense of a new paradigm. But lest I be accused of belittling the work of Eldredge and Gould and others, let me point out that the coming of the *Origin of Species* was no revolution, in this sense, either. I do not think this is a point that should be accepted in glum negative silence, but as a spur to thinking again about the nature of scientific change.

The third point follows on the other two. What of the future? May we expect a resolution? Will one paradigm or metaphor ultimately prove triumphant? I confess that I am not sure of the correct answer except to remark that no one is looking for total victory anyway. Against Kuhn, I would not rule out a priori the possibility of the empirical evidence being fairly definitive, even though it has not been so yet. Suppose some new technique of studying the fossil record yielded masses of new pertinent evidence. This might close the case, although it might not. Gel electrophoretic techniques have given us much new in-

8. Although I have much admiration for Richard Dawkins's *Blind Watchmaker* (1986) I think that he is wrong to dismiss punctuated equilibrium theory as a wrinkle on neo-Darwinism. This was true of the first version but not of the third.

formation about genetic variation, but the disputes go on (Lewontin, 1974).

Another possibility is that people will bury the hatchet and start to balance their evolutionary thinking between the two traditions. Gould, at times, seems to yearn for such a "pluralistic" account of change (see, for instance, Gould, 1983b). Given human nature, however, I wonder if this is not wishful thinking. Perhaps the two traditions will simply continue on into the future, with each side having supporters. In the realm of paleontology, there will be strict adaptationists and punctuationists (or whatever the views are then called). The consolation is that this is not necessarily bad. Indeed, the creative tension that the traditions cause can have a positively stimulating effect. The history of paleontology in the past fifteen years and the way the theory of punctuated equilibria has brought the field alive are surely proof enough.

Epilogue

Several years ago, I spent a year in the Museum of Comparative Zoology at Harvard University. On my first day there, the distinguished evolutionist who was so contemptuous of punctuated equilibrium theory showed me the ant collection. He held up a specimen of ant and started to hold forth on some intricate adaptation. "What a remarkable example of design!" he cried out. Then he stopped, looked at me sheepishly, and grinned. "For goodness sake! Don't ever tell Steve Gould that I said that."

References

Allen, E., et al. 1977. Sociobiology: a new biological determinism. In Sociobiology Study Group of Boston, eds., *Biology as a Social Weapon*. Minneapolis: Burgess.

Anonymous. n.d. *Fundamentals of Marxism-Leninism Manual*. Moscow: Foreign Languages Publishing House.

Beatty, J. 1980. Optimal-design models and the strategy of model building in evolutionary biology. *Philos. Sci.* 47:532–561.

Brandon, R. N., and R. M. Burian. 1984. *Genes, Organisms, Populations: Controversies over the Units of Selection*. Cambridge, Mass.: MIT Press.

Chambers, R. 1844. *Vestiges of the Natural History of Creation*. London: Churchill.

Darwin, C. 1859. *On the Origin of Species*. London: Murray.

Dawkins, R. 1976. *The Selfish Gene*. Oxford: Oxford University Press.

——. 1983. Universal Darwinism. In D. S. Bendell, ed., *Evolution from Molecules to Men*. Cambridge: Cambridge University Press, pp. 403–425.

——. 1986. *The Blind Watchmaker*. London: Longman.

Dobzhansky, T. 1951. *Genetics and the Origin of Species*, 3d ed. New York: Columbia University Press.

Eldredge, N. 1971. The allopatric model and phylogeny in Paleozoic invertebrates. *Evolution* 25:156–167.

——. 1985a. *Time Frames: The Rethinking of Darwinian Evolution and the Theory of Punctuated Equilibria*. New York: Simon and Schuster.

——. 1985b. *Unfinished Synthesis: Biological Hierarchies and Modern Evolutionary Thought*. Oxford: Oxford University Press.

Eldredge, N., and S. J. Gould. 1972. Punctuated equilibria: an alternative to phyletic gradualism. In T. J. M. Schopf, ed., *Models in Paleobiology*. San Francisco: Freeman, Cooper.

Engels, F. 1971. *Dialectics of Nature*. Moscow: Progress Publishers.

Fisher, R. A. 1930. *The Genetical Theory of Natural Selection*. Revised and reprinted, 1958. New York: Dover.

Ghiselin, M. T. 1974. A radical solution to the species problem. *Syst. Zool.* 23:536–544.

Gingerich, P. D. 1976. Paleontology and phylogeny: patterns of evolution at the species level in early Tertiary mammals. *Am. J. Sci.* 276:1–28.

——. 1977. Patterns of evolution in the mammalian fossil record. In A. Hallam, ed., *Patterns of Evolution as Illustrated by the Fossil Record*. Amsterdam: Elsevier, pp. 469–500.

Goldschmidt, R. B. 1940. *The Material Basis of Evolution*. New Haven: Yale University Press.

Gould, S. J. 1971. D'Arcy Thompson and the science of form. *New Lit. Hist.* 2:229–258.

——. 1973. Positive allometry of antlers in the "Irish Elk," *Megaloceros giganteus*. *Nature* 244:375–376.

——. 1977. *Ontogeny and Phylogeny*. Cambridge, Mass.: Harvard University Press.

——. 1978. Sociobiology: the art of story-telling. *New Sci.* 80:530–533.

——. 1979. Episodic change versus gradualist dogma. *Sci. Nat.* 2:5–12.

——. 1980a. Is a new and general theory of evolution emerging? *Paleobiology* 6:119–130.

——. 1980b. *The Panda's Thumb*. New York: Norton.

——. 1980c. The promise of paleobiology as a nomothetic, evolutionary discipline. *Paleobiology* 6:96–118.

——. 1981. *The Mismeasure of Man*. New York: Norton.

——. 1982a. Darwinism and the expansion of evolutionary theory. *Science* 216:380–387.

——. 1982b. Punctuated equilibrium—a different way of seeing. In J. Cherfas, ed., *Darwin up to Date*. London: IPC Magazines, pp. 26–30.
——. 1982c. The meaning of punctuated equilibrium and its role in validating a hierarchical approach to macroevolution. In R. Milkman, ed., *Perspectives on Evolution*. Sunderland, Mass.: Sinauer, pp. 83–104.
——. 1983a. Irrelevance, submission, and partnership: the changing role of palaeontology in Darwin's three centennials, and a modest proposal for macroevolution. In D. S. Bendall, ed., *Evolution from Molecules to Men*. Cambridge: Cambridge University Press, pp. 347–366.
——. 1983b. The hardening of the modern synthesis. In M. Grene, ed., *Dimensions of Darwinism*. Cambridge: Cambridge University Press, pp. 71– 93.
——. 1983c. *Hen's Teeth and Horse's Toes*. New York: Norton.
Gould, S. J., and N. Eldredge. 1977. Punctuated equilibria: the tempo and mode of evolution reconsidered. *Paleobiology* 3:115–151.
——. 1986. Punctuated equilibrium at the third stage. *Syst. Zool.* 35:143–148.
Gould, S. J., and R. Lewontin. 1979. The spandrels of San Marco and the Panglossian paradigm: a critique of the adaptationist programme. *Proc. R. Soc. Lond.* B 205:581–598.
Hallam, A. 1977. *Patterns of Evolution as Illustrated by the Fossil Record*. Amsterdam: Elsevier.
Hamilton, W. D. 1964a. The genetical evolution of social behaviour. I. *J. Theor. Biol.* 7:1–16.
——. 1964b. The genetical evolution of social behaviour. II. *J. Theor. Biol.* 7:17–32.
Hull, D. L. 1976. Are species really individuals? *Syst. Zool.* 25:174–191.
Huxley, J. 1942. *Evolution: The Modern Synthesis*. London: Allen & Unwin.
Huxley, T. H. 1860. *The Origin of Species*. (Reprinted in *Darwiniana*. London: Macmillan, 1893.)
Irvine, W. 1959. *Apes, Angels, and Victorians*. New York: Meridian Books.
Isaac, G. 1983. Aspects of human evolution. In D. S. Bendall, ed., *Evolution from Molecules to Men*. Cambridge: Cambridge University Press, pp. 509–543.
Kuhn, T. S. 1957. *The Copernican Revolution*. Cambridge, Mass.: Harvard University Press.
——. 1962. *The Structure of Scientific Revolutions*. Chicago: Chicago University Press.
Lewin, R. 1980. Evolutionary theory under fire. *Science* 210:883–887.
Lewontin, R. C. 1974. *The Genetic Basis of Evolutionary Change*. New York: Columbia University Press.
Maynard Smith, J. 1969. The status of neo-Darwinism. In C. H. Waddington, ed., *Towards a Theoretical Biology*. Edinburgh: University of Edinburgh Press.
——. 1981. Did Darwin get it right? *London Review of Books* 3 (11): 10–11.
——. 1983. Current controversies in evolutionary biology. In M. Grene, ed.,

Dimensions of Darwinism. Cambridge: Cambridge University Press, pp. 273–286.
Mayr, E. 1963. *Animal Species and Evolution.* Cambridge, Mass.: Harvard University Press.
——. 1982. *The Growth of Biological Thought.* Cambridge, Mass.: Harvard University Press.
Ospovat, D. 1981. *The Development of Darwin's Theory.* Cambridge: Cambridge University Press.
Ovcharenko, V. N. 1969. Transitional forms and species differentiation of brachiopods. *Paleontol. J.* 3:57–63.
Owen, R. 1849. *On the Nature of Limbs.* London: Voorst.
Paley, W. 1802. *Natural Theology.* In *Collected Works*, vol. 4. London: Rivington, 1819.
Reif, W.-E. 1983. Evolutionary theory in German paleontology. In M. Grene, ed., *Dimensions of Darwinism.* Cambridge: Cambridge University Press, pp. 173–203.
Ruse, M. 1979a. *The Darwinian Revolution: Science Red in Tooth and Claw.* Chicago: University of Chicago Press.
——. 1979b. *Sociobiology: Sense or Nonsense?* Dordrecht: Reidel.
——. 1981. What kind of revolution occurred in geology? In P. Asquith and I. Hacking, eds., *PSA 1978.* East Lansing, Mich.: Philosophy of Science Association 2:240–273.
——. 1982a. *Darwinism Defended: A Guide to the Evolution Controversies.* Reading, Mass.: Addison-Wesley.
——. 1982b. Review of S. J. Gould, *The Mismeasure of Man. Isis* 73:430–431.
——. 1986. *Taking Darwin Seriously.* Oxford: Blackwell.
Russell, E. S. 1916. *Form and Function.* London: Murray.
Ruzhentsev, V. Y. 1964. The problem of transition in paleontology. *Int. Geol. Rev.* 6:2204–2213.
Segerstralle, U. 1986. Colleagues in conflict. *Biol. Philo.* 1:53–88.
Simpson, G. G. 1944. *Tempo and Mode in Evolution.* New York: Columbia University Press.
Sober, E. 1984. *The Nature of Selection: Evolutionary Theory in Philosophical Focus.* Cambridge, Mass.: MIT Press.
Stanley, S. M. 1979. *Macroevolution: Pattern and Process.* San Francisco: W. H. Freeman.
Stebbins, G. L., and F. J. Ayala. 1981. Is a new evolutionary synthesis necessary? *Science* 213:967–971.
Stidd, B. M. 1980. The neotenous origin of the pollen organ of the gymnosperm *Cycadeoidea* and the implications for the origin of higher taxa. *Paleobiology* 6:161–167.
Thompson, P. 1983. Tempo and mode in evolution: punctuated equilibria and the modern synthetic theory. *Philos. Sci.* 50:432–452.
Turner, J. R. G. 1983. The hypothesis that explains mimetic resemblance ex-

plains evolution: the gradualist-saltationist schism. In M. Grene, ed., *Dimensions of Darwinism*. Cambridge: Cambridge University Press, pp. 129–169.

Whewell, W. 1833. *Astronomy and General Physics Considered with Reference to Natural Theology: Bridgewater Treatise No. III*. London: William Pickering.

——. 1837. *The History of the Inductive Sciences*. London: Parker.

——. 1840. *The Philosophy of the Inductive Sciences*. London: Parker.

Williams, G. C. 1966. *Adaptation and Natural Selection*. Princeton: Princeton University Press.

Williamson, P. 1981. Palaeontological documentation of speciation in Cenozoic molluscs from Turkana Basin. *Nature* 293:437–443.

Wilson, E. C. 1975. *Sociobiology: The New Synthesis*. Cambridge, Mass.: Harvard University Press.

Part Two

Implications for the Behavioral Sciences: The Debate Continues

7

Punctuationism in Societal Evolution

KENNETH E. BOULDING

The similarities between biological and societal evolution are very great. Both involve a taxonomy of objects into species. Both involve morphogenesis. Production in each case is how the genotype becomes a phenotype, how, for instance, the egg becomes a chicken, a blueprint becomes a house, or a production plan an automobile. Both systems are subject to ecological interaction. In the economy commodities form an ecosystem with competitive, cooperative, and predative relationships, very much like the biological ecosystem. Both systems are subject to mutation and selection, mutation through novelty in the genotype, selection through ecological interaction. It is not surprising, therefore, that the question of punctuationism applies to both systems. The evolutionary process in both is highly indeterministic. The historical record is dominated by the time when highly improbable events happen, such as the appearance of DNA or *Homo sapiens*. Evolution often goes into a new gear with such changes, sometimes very rapidly. Both systems also seem to produce "living fossils," species that have survived unchanged despite great changes in environments and in other species, such as the fish *Coelacanthus*, the Australian aborigine, the Amish, and the British Horse Guards. The evidence for punctuationism in biological evolution, which is very strong, is reinforced by the study of societal evolution.

The interactions and the parallels between biological and social systems have been observed by human beings for a very long time. The observation that humans are mammals, although peculiar ones, certainly must go back to the beginnings of the human race. From the eighteenth century on, the social and biological sciences grew up together. Adam Smith is a somewhat younger contemporary of Linnaeus, though Adam Smith is much more of a theorist and less of a taxonomist than Linnaeus. Adam Smith, indeed, has a good claim to be the first post-Newtonian evolutionary thinker, transcending equilibrium, which he also understood very well, and seeing the progress of society as an irreversible evolutionary or developmental process involving constant innovation. Economic development (the increase in wealth) he sees mainly as a result of the human capacity for learning, both learning by repetition of tasks, that is, skill, which develops from and fosters the division of labor and exchange; and, more significant, the specialized learning that comes from what he calls "philosophy" or "speculation," which, again, is subdivided so that "each individual becomes more expert in his own peculiar branch, more work is done upon the whole, and the quantity of science is considerably increased by it" (Smith, 1937:10).

Adam Smith's price theory comes close to being an ecological theory of commodities, that is, economic species, each commodity occupying what today would be called a "niche" in a total system. The relative price at which its production ("births") and consumption ("deaths") are equal is the "natural" price. If the market price is below this, production will be discouraged and consumption encouraged, the stock or population of the commodity will decline, and its price will rise. If the market price is above the natural price, production will be encouraged, consumption discouraged, the stock or population of the commodity will rise, and the price will fall. This is a concept very close to that of the equilibrium of an ecosystem. He also sees, however, that every equilibrium position is transitory and is constantly being changed by new knowledge, new discoveries, and technical change, what today we would call "mutation." He did not perhaps see quite so clearly that an equilibrium can be changed by a change in the overall environment—droughts, war, political changes, and so on.

It is not surprising, therefore, that it was when Darwin happened to read "for amusement" Thomas Malthus's *Essay on Population* that he got the idea for natural selection (Barlow, 1958:120). As an economist, Malthus was developing ideas that had already been expressed, though in less dramatic form, in Adam Smith's *Wealth of Nations*. It is perhaps no accident, also, that it was the social philosopher Herbert Spencer from whom Darwin got the term "survival of the fittest."

In the twentieth century, unfortunately, the division of labor, both within and between the social and biological sciences, became somewhat pathological in the sense that there has been a serious absence of exchange of ideas between the different subdisciplines. In Adam Smith's famous attack on the division of labor (1937:734), he says, "The man whose whole life is spent in performing a few simple operations . . . generally becomes as stupid and ignorant as is possible for a human creature to become." This perhaps is a little strong for social and biological scientists, but certainly specialization has prevented communication that could be very fruitful, as we have developed "coteries of co-citation," all the members of which quote each other and practically nobody else. Even within the biological sciences it is surprising how little communication there seems to have been between ecologists and evolutionists, or paleobiologists and evolutionists, or geneticists and evolutionists. In the social sciences, of course, the economists, sociologists, and political scientists have almost given up talking to one another, to their great loss. And communication between the social and biological sciences has shrunk almost to zero.

Nevertheless, the biosphere and the sociosphere (that is, the sphere of all human knowledge, activities, artifacts, and organizations) have become increasingly a single system simply because of the enormous capacity of the human race for its own expansion into all ecosystems and its remarkable capacity for producing artifacts. The automobile is now as much a species in the world ecosystem as is the horse. Human artifacts are species that occupy niches in the total ecosystem of the world just as biological artifacts do, and species of all kinds interact with each other. Both biological and human artifacts can have profound effects on climate, temperature, and the physical environment. Humans are profoundly changing the composition of the atmosphere and of the ozone layer. This change in itself may have an enormous effect on the biosphere. Direct human intervention, in cutting down forests, desertification, the development of agriculture, and so on, has a large and immediate effect. The human race can certainly be regarded as an ecological catastrophe, producing a rapidity of extinction of species that may well exceed any catastrophe the earth has experienced in the past, and there have been many. We cannot, therefore, regard the physical, biological, and social systems of the earth as isolated. They are all part of a single system (Boulding, 1985).

The different parts of this total system, however, do have somewhat different patterns, though they also have a great many patterns in common. All systems have component parts, at some levels the system is the sum of the parts, and the size of the system is an important limiting factor. This is the economist's concept of "scarcity." Within a system that is the sum of its parts, the more of one part, the less of all the others. In other respects, however, as the parts interact, the system is *not* the sum of its parts. A rise or fall in one part may increase the size of the system, diminish some other parts, increase other parts, and leave still other parts unaffected. Thus the great technical change in agriculture in the last one hundred years or so, by greatly increasing the amount of food produced by each agricultural worker, has greatly expanded the total output of the economy and led to a massive transfer of labor out of agriculture into other occupations. In the United States the proportion of the total labor force in agriculture fell from 21 percent in 1929 to about 3 percent today.

The component parts of the system are subject to taxonomy. Some parts are more like other parts and so form a species. Taxonomy is a very tricky business. It is often hard to tell what is alike and what is different, as we saw in alchemy, when we certainly got the elements wrong. The elements are *not* earth, air, fire, and water! Racism is another taxonomic error. It has a very weak genetic base, for the genetic differences within the races are much larger than they are between them.

Some concepts, as we shall see, are shared by all systems, especially perhaps by the biosphere and the sociosphere. All structures and systems share some of the properties of physical systems. There is no way of escaping the limitations imposed by gravity, valency, nuclear transformations, Ohm's Law, and so on. Entropy is a little tricky because the earth is an open system in regard to energy, but it still imposes limits. Within these physical limits, however, an enormous variety of biological and social structures has emerged.

Many basic concepts are found both in the sociosphere and the biosphere. One is "production," which is how the genotype turns into the phenotype, the egg into the chicken, the blueprint into the house, the idea into the institution, and so on. All production involves know-how that is embodied in some genotype. The fertilized egg of a giraffe knows how to make a giraffe; it does not know how to make a rhinoceros, though it may have some ideas about this. If we ask why there were no *Homo sapiens* a million years ago, the answer is that the "genosphere" of all genetic know-how did not know how to make us. If we ask why there were no automobiles two hundred years ago, the answer is exactly the same: we did not know how to make them. The major difference between biological and social production is that biological production never seems to have gotten above two sexes, the respective genetic know-how of which must come together to form a fertilized egg, whereas in human artifacts, the genetic material is contained in many different brains, plans, records, computers, blueprints, and so on, so that human artifacts might accurately be described as "multiparental." Just as the invention of sex speeded up biological evolution, so the development of *Homo sapiens* and human artifacts has enormously speeded up societal, and possibly biological, evolution.

Both the sociosphere and the biosphere are subject to the principles of the demography of species. Species consist of a set of individual structures similar enough to be interesting and subject to birth or conception[1] (that is, production, the beginning of new members of the species) and also to death and consumption (the destruction of old members). For a biologist, what is interesting is the capacity to reproduce, and this is the usual definition of a biological species. From an ecological point of view, however, groups of genetic species may be significant, like the lichen, a combination of two genetic species. In social systems, because of the multiparental quality of reproduction, species are much harder to define, that is, is a two-wheeled covered motorbike a bicycle or an automobile? Concepts like class, race, and nationality are vague and often misleading, and the I.Q. is a very poor measure of intelligence. In most species, aging is significant. The phenotype grows from its origin to maturity and then declines to death. In the biosphere, birth is accomplished mainly within the species itself. In the sociosphere, birth involves the cooperation of many other species. A Ford and a Chevrolet have never produced a Volkswagen. Each is produced by other species, both social and biological: factories, machines, human beings, and so on.

Another pattern common to both social and biological systems is ecological interaction, that is, ecosystems. Interaction happens because of the fundamental principle that in order for the genotype to produce the phenotype, whether biological or societal, it has to be able to capture energy in different forms, use the energy to capture materials of different kinds, and then transform the materials into the phenotype. The phenotype itself during its life also must capture energy and materials

1. The distinction between conception and birth is found in social or economic as well as in biological species. The truck is conceived in the minds of its designers, grows in the "womb" of the factory, and is "born" when it emerges from the factory.

such as solar energy, food, air, water, and so on for biological species; and air, gasoline, repairs, and the like for trucks. Usually some of these inputs are excreted. This gives rise to food chains and to the circulation of excrements, the excrement of one species being the food of another, as we see in the nitrogen cycle. The growth of the genotype also requires space and time, and these likewise are limiting factors. Space, indeed, is probably a more important limiting factor in the biosphere than is energy, which seems to be used remarkably inefficiently.

Oddly enough, the idea of a food chain, although not by that title, was familiar to the classical economists. The production of commodity species begins with corn as human food. Corn is fed into a cow, and we get meat and milk. If together with leather, corn is fed into a shoemaker we get shoes. If the farmer wears the shoes, he may be able to produce more corn so that we get feedback cycles. All this is surprisingly similar to the biological concept of the ecosystem. In both social and biological ecosystems there seem to be three basic relationships between species: (1) mutual competition, which is actually rather rare because it frequently leads to the extinction of one of the species; (2) mutual cooperation, in which increase in each species increases the niche of the other, like dung beetles and sheep, or gas stations and automobiles; and (3) predation, which also is fairly stable, though rather rare in social systems; the military and civilians are perhaps an example.

The concept of the niche is essentially the same in biological and social systems. It rests on the assumption that the birth and death rates of any species are functions of its population and that there will be some niche population at which the birth and death rates are equal and the population is at a stable equilibrium level. For populations below this level, where the niche is not filled, food is easy to find, predators find species hard to find, birth rates exceed death rates, and the population grows. If populations are too large relative to the niche, food is hard to find, predators find species easier to find, death rates exceed birth rates, and the population declines in a classic cybernetic equilibrium. This applies to automobiles just as well as it does to deer. An important concept common to both systems is that of the "empty niche" (Boulding, 1978). Biologists find this an awkward concept because they don't like to study what isn't there, but what isn't there is often more important than what is. There is no doubt that Australia had an empty niche for rabbits and that the human race had an empty niche for the AIDS virus once it appeared. An empty niche represents a population of a species that would have an equilibrium population in an ecosystem if it existed. It can come into existence, of course, either by mutation or by migration.

In both biological and social systems there are two forms of genetics. Biogenetics is DNA and all that and is certainly most important in biological systems. It is pretty strictly Mendelian, and though it is dangerous to assume that genes cannot be altered by the experience of the phenotype, such alteration is probably very small. Even in biological evolution, however, we have what I have called "noogenetic" processes. These are the learned structures which are transmitted from one generation to the next by a learning process. Biogenetically created learning systems, like brains, whatever they are, have a potential for learning which is not necessarily realized. Learning goes a long way back in the evolutionary story. It may be a little doubtful in planaria, but it seems to take place in snails. There is no question about it in birds. In monkeys it is quite significant, as the famous macaques of Japan illustrated, and in *Homo sapiens* it is completely predominant. Biogenetic change in *Homo sapiens* seems to have been very small over the last forty thousand or fifty thousand years, but the noogenetic changes have been enormous, carrying us from a handful of flint scrapers to 5 billion people, producing skyscrapers, airplanes, spacecraft, and computers.

In both social and biological systems, then, we see evolution as the process of the filling of empty niches in ecosystems, either by biogenetic or noogenetic mutation or by migration. Once an empty niche is filled, of course, it has a profound effect on all the other niches of the system. Some may shrink to zero, when their population becomes extinct; new empty niches are likely to be created. The introduction of sheep into Australia certainly opened up an empty niche for dung beetles. The increasing scarcity of whale oil and candle wax in the nineteenth century opened up a niche for kerosene and the oil industry. This produced gasoline as a by-product, which opened up an empty niche for the internal combustion engine and the automobile, which had not existed in earlier periods. The automobile opened up empty niches for supermarkets and drive-in theaters, change in sexual habits and in the family . . . and so we go on.

A feature common to both biological and social systems is that the actual history of evolution is dominated by the exact time at which improbable events happened. This is a worrying concept to those who have not made the transition into post-Newtonian science. It is a fundamental principle that any system involving information as an essential element, as both biological and social systems do, has a component of inherent and irreducible unpredictability, simply because information has to be surprising or it is not information. We cannot predict the future of human knowledge or we would know it now. DNA itself is such an improbable molecule that one of its co-discoverers, Francis Crick (1981), thinks it must have come from outer space, which certainly does not solve the problem as to where it came from. Even Edward O. Wilson, the founder of sociobiology, is puzzled by the evolution of the human brain. How did we have this long succession of mutations, apparently increasing the size of an organ that nobody was using?

All the systems of the universe are strange mixtures of stability and instability, continuity and discontinuity. The trouble is that we are able to study only a very small sample of the universe, both in time and in space, and we tend to make rash generalizations from this small sample. This is particularly true of the physical scientists, who tend to assume that what is observed to be constant here is constant everywhere. This is about like deducing the state of the world economy from a five-minute interview with a single person. It is an interesting question, to which it is hard to find an answer, as to whether the discontinuities in both biological and social history are to be seen as interruptions in a larger pattern of development which is more stable, or whether they are to be seen as basic turning points from which there is no going back. There does seem to be some evidence that there is a certain general pattern of niche structures. In the oceanic islands, separated from the mainland for a very long time, such as Mauritius or New Zealand, which the birds reached but the mammals did not, the birds seem to have evolved into some mammalian niches. In Australia the road sign that in North America has a deer jumping out has a kangaroo instead, a delightfully improbable animal. In all ecosystems there seems to be a place for larger and smaller creatures, and so on. Similarly, in social systems, though there has been a spectacular growth in the size of larger organizations, such as the corporation, the national state, and the centrally planned economy, the family and the artisan and the small business still survive, sometimes perhaps in the cracks of the system.

One of the real puzzles in both biological and societal evolution is the probability structure of both mutation and selection. Mutation is random only in the sense that change from one genotype to another has a probability, which is extremely difficult to estimate simply because of the lack of a universe of similar cases. For any one genotype certainly the probability of small changes is greater than that of large changes. It is a fundamental principle, however, that no matter how small the probability of any change, if we wait long enough it will happen. The truth is that we know very little about genetic mutation. It is very hard to observe, for the most probable thing in any case is no mutation at all in a period of observation. Science is extremely ill-equipped for dealing with the improbable. Experimental science can deal only with what is highly probable within the universe that it studies, and observational science is not in much better shape in this regard.

Another factor that has to be taken into consideration is that small changes often produce very large effects. I am tempted to call this "Browning's Law": "Oh, the little more, and how much it is! And the little less, and what worlds away!" (Browning, 1909:36, stanza 39). The fact that the overall genetic difference between the chimpanzee and ourselves is so small and yet the evolutionary effects of this small change have been enormous is a good illustration of the principle. We cannot deny the existence of what I have sometimes called "watershed systems," systems in which change goes over a watershed and we find ourselves in a completely different dynamic world. If we spit on the Continental Divide, a puff of wind may take the water to the Atlantic or the Pacific. Some watersheds, however, are more important than others. If we spit on the divide between the North and the South Platte rivers, the water, unless it evaporates, will end up in Nebraska and finally in New Orleans.

The punctuational view in evolution, for which there is a great deal of evidence, both in biological and in societal evolution, suggests that certain improbable events may produce large consequences very rapidly. The improbability of such events means that in the historical record there is also going to be a great deal of continuity, of which the "living fossils" are a good example. We find similar records in social evolution. The Australian aborigines, for instance, seem to have lived in about the same state of technical capacity with stable human artifacts for something on the order of forty thousand years before the invasion of Australia by Europeans. The record certainly suggests that all paleolithic humans, although genetically not very different from present humans, also lived in a stable society for perhaps tens of thousands of years. This is all the more surprising because the potential of the human brain of paleolithic humans was apparently much the same as the potential of humans today. There are Australian aborigines with Ph.D.'s. Why the paleolithic cultures were so stable is a mystery. It seems almost inconceivable to us that a human culture should persist for more than a thousand generations, with the children almost exactly repeating the culture of the parents. Of course, there may have been important changes in the culture which did not get in the skimpy records, changes, for instance, in language, in stories and legends, passed on from one generation to the next, in the detailed knowledge of the environment which these societies usually have, and so on. But even so, the absence of any changes in basic technology is surprising, in light of our common genetic structure, which would seem to give us curiosity and an enormous capacity for learning. One possible explanation is that the short expectation of human life under paleolithic technology meant that society had to devote all its teaching and learning resources to transmitting its culture to the young and therefore had nothing left over for innovation.

It is certainly significant that the great acceleration in the rate of change of human artifacts seems to have begun with agriculture. Agriculture, even at its most primitive levels, seems to have produced a surplus of storable food beyond what the food producer needed to sustain life and replace each generation with the next. This surplus was then available for activities beyond food production. The development of agriculture was rapidly followed by metallurgy, first bronze and then iron; weaving; house building, which permitted movement into less favorable climates; and so on. The food surplus also permitted the development of organized threat systems, first often priestly rulers, who were able to persuade the farmers that without their performances the crops would not grow. Then came the kings, as we see in the transition from Samuel to Saul, who were able to use the surplus food to feed soldiers, who could then threaten the farmers into relinquishing the surplus. This led, of course, to the age of civilization and cities. Changes in transportation are also important here, such as the development of the boat and the wheel. The domestication of livestock and horses also played a role. The size of an organized society depends on its means of transportation. It is certainly not surprising that virtually all the ancient civilizations developed on rivers or archipelagos, for water transportation has always been much easier and cheaper than that on land. A notable exception, the Incas, perhaps were able to expand because they lived above the tree line, where it was easy to make paths for runners.

What happened is clearly a speeding up of the rate of noogenetic mutation. This acceleration takes place because of a positive feedback process by which the more we know, the easier it is to know more. There are also important factors in the selective process. There are mutations in selection, just as there are in mutation itself, which biologists are slow to recognize. The filling of an empty niche, for instance, may create new empty niches and hence new opportunities for successful mutation which did not exist before. A successful mutation is precisely one that creates a phenotype that can occupy an empty niche. In societal evolution this creation of empty niches can be very rapid and open up ever new opportunities for increasing survival of mutations in the social genetic structure of know-how.

An important aspect of this developmental process is the relation between what might be called "know-what" and "know-how." This process began quite early. The success of hunting societies obviously must have depended a good deal on know-what about the behavior and habits of animals that were the prey. Likewise, the know-how that created agriculture had a background unquestionably in observation and the classification of plants. With the rise of science, of course, know-how expanded at an ever-increasing rate for five hundred years, although the impact of science on technology, that is, the translation of know-what into know-how, hardly began until the middle of the nineteenth century, with the development first of the chemical industry, which would have been impossible without chemistry; then of the electrical industry, which would have been impossible without physics; then the oil industry, which would have been impossible without geology, physics, and chemistry. As a result, the total ecosystem of the earth, including both biological and human artifacts, has almost certainly changed more in the last one hundred years than in any one hundred years in the earth's history. There has been an enormous rise of social species—automobiles, cities, airplanes, oil wells, and even domesticated animals—an unprecedented rise in the human population, and some significant declines in the number and size of biological species.

An interesting question in social systems is whether punctuational evolution is more apparent in the economy than it is in the polity, that is, the structure of political and military systems, or in what I have called the "integry," the set of human behaviors, images, and institutions which revolve around legitimacy, community, benevolence, malevolence, and so on (Boulding, 1978:189). In the economy we certainly find periods of relative stability, in which society is getting neither much richer nor much poorer, but these periods of stability do seem to be punctuated by periods of very rapid economic development. The transition from hunting-gathering societies to agriculture at any particular locality seems to set off a period of rapid economic growth. This transition was usually rather rapid and, it would seem, irreversible. I know of no example in which an agricultural society has gone back to becoming a hunting-gathering society. Perhaps this is because the niche for human beings in agriculture is much larger than is the one for hunting-gathering societies, and hence the introduction of agriculture is always followed by a population expansion to a point that would make hunting-gathering societies impossible. What might be called "imperial civilizations" shifted around a bit geographically, some centers rising and some falling, but there seems not to have been much change in their general level of riches, even in the imperial cities. Rome under the Caesars was not much more magnificent than Babylon. The collapse of an empire may produce a period of relative impoverishment, but the society soon seems to recover. After the fall of Rome, for instance, Europe started off on a process of technological improvement, much of it coming from China, with the horse collar, the stirrup, the iron-shod plow, and so on. The development of science-based technology, which began about the middle of the nineteenth century, was followed by a spectacular increase in wealth and per capita income in parts of the world which experienced it. There were, of course, Malthusian ups and downs, with impoverishment resulting from overpopulation, followed perhaps by famine, as in Ireland in the 1840s, or plague, such as the Black Death of the 1300s, which diminished population, made resources per capita more plentiful, and hence led to some enrichment. The economy perhaps has the advantage of being closer to an ecosystem than the polity or the integry and hence is more subject to evolutionary change.

There has, of course, been evolution in the polity and the integry which is very hard to measure, but it does seem to have been slower. One can certainly postulate that democratic societies represent an improvement over tyrannical ones, but tyranny persists and has certainly not become extinct. Military organizations are a curious ecosystem. They tend to be cooperative with each other and competitive with their own civilian populations. Because they have to have enemies to justify their existence, they tend to create enemies and hence do not develop into conflict-management organizations. Economic organizations on the whole provide the means of life and enrich the human race; military organizations specialize in the means of death and impoverish the human race. Nevertheless they persist. There is a good deal of evidence that war is a product of civilization because on a large scale it is conditional on an agricultural surplus. Certainly the technology of weaponry has a profound effect on political institutions, though not much, one suspects, on economic institutions. The development of the cannon certainly destroyed the feudal system and created the national state. The long-range missile and nuclear warhead have now made the national state about as obsolete as the feudal baron. A high priority for post-civilized society based on science-based technology is the abolition of war and an appropriate transformation of the national state.

In regard to integrative systems, no real substitute has ever been found for the family. Threat systems without an integrative component are fragile. Countries that nobody loves do not last very long. Economic institutions may suffer a loss of legitimacy, as the private corporations and financial markets did in the communist countries. The evolutionary dynamics of legitimacy, however, are puzzling, even though it can be argued that this is the dominant factor in social dynamics. Integrative institutions certainly possess a great deal of stability and seem hard to change. National states, churches, clubs, professional associations, and so on have a strong tendency to persist. The remarkable rise of international nongovernmental organizations (INGOs) in the twentieth century, whose number is now well over ten thousand, suggests a striking evolutionary change in the integrative system. There are indeed occasional moments of mass conversion, revolution, and reformation in these systems, which are certainly punctuational exclamation points. Once achieved, however, integrative change seems to slow down very substantially. The distribution of world religions, for instance, has not changed much in a long time, though it may be changing now with the development of the total world communication system. Christianity expanded all over the world in the last five hundred years, but this may have reflected the expansion of European population and influence. The religions of the world certainly form an ecosystem of members, but they do seem to be rather isolated ecosystems, just as biological ecosystems are isolated and depend very much on climate, altitude, land and water distribution, and so on.

The extraordinary phenomenon of acceleration in social and economic systems, especially in the last few centuries, raises the question as to whether the phenomenon of punctuation in biological evolution may not show some parallels. The critical question here is whether it is possible to have a genetic mutation that increases the rate of mutation. I know of no direct evidence for this, but our detailed knowledge of genetic mutation is extremely small. Nevertheless, it seems to me that the possibility cannot be ruled out. The stability of species may well be the result of a kind of mutational equilibrium. It would not be surprising to find that successive mutations would produce a situation in which virtually all possible further mutations were adverse. It is hard to distinguish between a situation of stable DNA, with very few possible mutations, and a situation in which virtually all DNA mutations would be eliminated by the selective process and only those reproducers that had very few mutations would survive to reproduce.

In fact, one would expect natural selection itself to favor, at least for a time, the nonmutable gene structures. What such a species would gain in stability, of course, it would lose in adaptability, and it is a fundamental principle of evolution that it is the adaptable, not the well adapted, that inherit the earth. Eventually every stable species will reach some environmental crisis simply because of the everlasting change in physical environments which goes on, such as ice ages and so on. If the change in environment is too severe, the species will become extinct. A change in environment, producing a change in the selective process, may also set off an acceleration of mutational change. It may well be, for instance, that the mammals survived whatever change in environment it was that exterminated the dinosaurs because they had great noogenetic capacity for learning new tricks to be passed on from one generation to the next without much change in their biogenetic structure. Even at very early stages, noogenetic evolution may have been considerably faster than biogenetic and hence more capable of surviving rapid changes in the environment.

One element in the overall pattern which intensifies the random character of evolution is the role of the accidental and infrequent coming together of small groups of interaction, perhaps somewhat in isolation from the larger environment. This reflects partly the importance of sex in biological evolution and correspondingly the multiparental generation of ideas and artifacts in societal evolution. The probability of survival of a large biological mutation would seem to be at first sight much smaller in sexual species. With asexual reproduction a mutant cell could divide and multiply without finding a partner. In sexual reproduction a large mutation in one sex would seem almost to demand a corresponding mutation in a mate, which seems to diminish the probability of such mutations being successful by an enormous amount. One possible solution to this problem would be the development of a mutation that did not change the phenotype much and did not interfere with sexual fertilization with mates who did not participate in this mutation, but which passed on to a succession of offspring might make further mutations easier and more widespread. I have no evidence for anything like this whatever, but the mechanism for mutation in biogenetics is very inadequately understood. I am not even sure that such "potential-creating mutation," as it might be called, is even conceivable in the DNA molecule. In noogenetic mutation, however, such as we find especially in social systems, mutations, that is, new images and new discoveries, often have this quality of increasing the potential for new mutations. The invention of printing, for instance, enormously expanded the potential for what might be called literary mutations. Mutations in the selection process itself are also highly noticeable. A striking phenomenon in intellectual history is the importance of small compatible groups that just happen to come together and reinforce each other's new ideas, which otherwise might not be accepted and hence even be given up, or at least not spread. All cultures seem to develop orthodoxies that are hostile to new ideas. This is as true in the sciences as it is in the churches. Then comes some critical mass, and the new ideas may explode. We see something like this in the Protestant Reformation. We see it in geology in the sudden acceptance of plate tectonics, an idea that was almost wholly rejected for a generation or more, and then some critical mass of experimental evidence permitted it to become an orthodoxy itself.

Another problem is that there may be competitors for an empty niche. This, I think, is probably a rare situation, but it cannot be wholly ruled out. The equilibrium of two competing species is a bit precarious, and it is possible that slight and random changes in competitive advantage can lead to the extinction of one or the other. That some countries are still predominantly Catholic and others predominantly Lutheran is such an example. The distinction between Marxist and non-Marxist societies in regard to economic thinking may be another.

The moral of all this is that evolutionary processes are highly complex and imperfectly understood in all fields. Under these circumstances interaction among the disciplines can be very helpful. Things that are fairly clear in one type of evolution may be obscure in another, although still present. One hopes that by the interactions we may be able to move toward a more general and more realistic theory of evolution than we now have.

References

Barlow, N., ed. 1958. *The Autobiography of Charles Darwin (1809–1882).* New York: Norton.

Boulding, K. E. 1978. *Ecodynamics: A New Theory of Societal Evolution*. Beverly Hills: Sage.
——. 1985. *The World as a Total System*. Beverly Hills: Sage.
Browning, R. 1909. *Poems of Robert Browning*. London: Oxford University Press.
Crick, F. 1981. *Life Itself: Its Origin and Nature*. New York: Simon and Schuster.
Smith, A. 1937. *The Wealth of Nations*. New York: Modern Library Edition.

8

The Theory of Punctuated Equilibria and Evolutionary Anthropology

SUSAN CACHEL

In the nineteenth and early twentieth centuries, evolutionary theory affected all of anthropology and was probably a major influence on the founding of the science itself. After the introduction of the idea of natural selection as a mechanism for evolution in 1859, traditional boundaries between such disciplines as anatomy, zoology, geology, paleontology, ethnology, and archaeology began to erode as the question of human evolution was raised. Recognition of the genetic relatedness or kinship between human and nonhuman animals and the subsequent novel problem of "man's place in nature" addressed by Thomas Huxley (1863) led to the founding of the Société d'Anthropologie de Paris and the Anthropological Society of London.

The comparative method was elaborated by eighteenth-century scholars to reconstruct hypothetical sequences in the development of civilization. Social anthropologists of the late nineteenth and early twentieth centuries attempted to apply the theory of natural selection to the history of culture, using the comparative method. Rather than invoking mere similarities between the behavior of modern hunter-gatherers and ancient human behavior as recorded by historians, social anthropologists could invoke evolutionary processes to reconstruct the development of society. Contemporary "primitive peoples" were presumed to represent the earliest stages of human development. Modern savagery represented civilization's past. This idea was particularly attractive because of the relative lack of human fossil evidence and an unwillingness to accept just how nonmodern the ancestors of *Homo sapiens* were. (The last consideration still affects contemporary ideas about human evolution.) Students of the comparative method also held that there might be a distinction between chronological sequence and the course of human development because of retrogression or accidental loss of traits. Furthermore, the comparative method held that one could classify societies by their structure as being primitive or advanced in development. The guarantee that higher forms of society had passed through lower forms was given by social vestiges or "survivals" that failed to evolve and continued as remnants of the past. Similarities between higher and lower forms of society could be explained by survivals. Cultural elements might be survivals or relics of past societal conditions, although now serving no discernible societal function. Kinship terminology, for example, might reflect extinct marriage practices characteristic of earlier societies. These might yield information about the origin of the family, the state, and religion (Kuper, 1988). Unequal development of human races caused variation in social organizations and institutions, although one general law of progress operated for all of humankind.

Evolutionary social theory gradually ended with the acceptance of cultural relativism and the assault on positivism generated by World War I (Burrow, 1966). Freudian ideas of humans being inherently alienated from nature and subject to irrational drives and impulses also contributed to the demise of social evolutionary ideas (Cartmill, 1983).

At the present time, therefore, physical anthropology alone retains the original emphasis on evolution which once created and organized all of anthropology. Evolutionary anthropology now is equivalent to physical anthropology. Any impact of evolutionary theory on anthropology now affects only physical anthropology; hence, the major impact of the theory of punctuated equilibria (Eldredge and Gould, 1972; Gould and Eldredge, 1977) is on physical anthropology. Nevertheless, it is possible that questions of cultural change in prehistoric archaeology may be affected by punctuated equilibrium theory.

It is ironic, however, that evolutionary theory was merely nominal in physical anthropology until the early 1950s, when Sherwood Washburn (1951a, 1951b, 1953), in a truly revolutionary restructuring of the science, argued for the application of evolutionary theory, adaptive studies, and experimentation to analyses of human and nonhuman primate evolution and variation (Hunt, 1981). Historians of anthropology refer to this revolution in the early 1950s as the emergence of the "New Physical Anthropology."

Punctuated Equilibria, Cladistics, and "The Adaptationist Program"

It is clear that discussion of rates of evolution, species discrimination in the fossil record, and the longevity and stasis of species will affect nonhuman primate evolutionary studies and paleoanthropology. Stephen Jay Gould (1984) has argued that the theory of punctuated equilibria has application to a general theory of catastrophism, reworked since the early nineteenth century, in which random factors affect geological, biological, and cultural history to a far greater extent than he presumes uniformitarianism and "phyletic gradualism" recognize or can allow. Although it is certainly true that social and historical factors are linked with scientific theory and interpretation, it is not clear that any link can be established between punctuated equilibrium theory and human social, cultural, and political history. At any rate, establishing such a link will require more than merely declaring that "the Zeitgeist is now in for an overhaul" (Gould, 1984:29). No evidence exists to show any impact of punctuated equilibrium theory on social or cultural anthropology. It also appears that the random historical events stressed by Gould in his explication of the new catastrophism would argue against the inevitable development of the political and social revolutionary ideas which he links with punctuated equilibria, because there is no guarantee that the end result of a revolutionary event will differ in the desired direction from the initial condition. Nor is there any evidence that a gradualistic, progress-oriented worldview was unwittingly fabricated by the geologist Charles Lyell and espoused by Darwin to make the world safe for the English class system and nineteenth-century liberalism (Gould, 1984), with the result that present criticism of punctuated equilibrium theory implies reactionary bias and a longing for social repose and stability on the part of the critic. Furthermore, although Gould (1984) may cast Lyell as a paragon of nineteenth-century capitalist thought and a defender of progress, he may also cast Lyell as the major protagonist in a battle against directionality in geological and biological history (Gould, 1987). Thr role played by particular major figures may shift in rhetorical embellishments, but this does not demonstrate a necessary effect of evolutionary thought on cultural or social history.

Gould (1986) has argued that the ideas of progress, gradualism, dynamic equilibrium, and adaptationism are important biases in Western thought and are influencing interpretations of geological and biological history to the detriment of the acceptance of punctuated equilibrium theory. Punctuated equilibria explicitly denigrates adaptation, but there are two other links to punctuated equilibrium theory which are less easily demonstrable—cladistic systematics and typology.

Cladistic systematics is based on the belief that classification should express the evolutionary sequence of divergence from a common ancestor. This sequence of divergence is reflected in branching (cladistic) relationships between species, shown in a cladogram constructed only from shared derived character states. In cladistic systematics, a clade is thus a monophyletic set of species descended from a certain ancestral species. A grade is a level of behavioral or structural organization that may be attained by one or a number of species during evolution. A grade may not always be a clade because characters can be attained independently by species as a result of parallel or convergent evolution. The pattern of character distribution among taxa determines their relationships. It is not necessary to examine the adaptive significance of characters or functional anatomy when drawing inferences about the direction of evolution and distinguishing primitive from derived traits. This can be done principally by examining the pattern of character distribution, or by detailed morphological, behavioral, or biochemical study, performed only to reveal the sequence of transformation from a primitive to a derived state—not to reveal function. Cladistic analysis eschews the study of grades or polyphyly, and, indirectly, the study of adaptation. For example,

> [a] flurry of theoretical work [on polyphyly] represents a sort of hypertrophication of the general view that evolution is fundamentally the transformation of intrinsic features, best explained by reference to adaptation via natural selection. Investigators cheerfully recognized the existence of taxa acknowledged not to share unity of desent. . . . Rather these taxa were based on convergence, parallelism, and, to a lesser extent, joint retention of primitive features. . . . This celebration of adaptation came directly from evolutionary theory and, inasmuch as it dwelt on the "explanation" of the evolution of phylogenetically non-extistent groups, is best regarded as a bizarre conclusion." (Eldredge and Cracraft, 1980:13–14.)

In fact, adaptive characters can evolve through parallel or convergent evolution and may distort cladistic analysis. Hence the operation of natural selection, functional analysis, and the way of life (adpative zone) or ecological niches of organisms are not studied by strict cladistic systematists. Evolutionary ecology or the grades of structural or behavioral organization which lead to adaptive radiations are not important from the point of view of cladistic systematics. Yet many systematists employ cladistic methods of searching taxa for patterns of character distribution and examining traits for sequences of development from a primitive to a derived state, without embracing or employing the strict methodology of cladistic systematists outlined below.

The methodology of cladistic systematics is based on the creation of phylogenies that consist of a series of dichotomies created by the branching of an ancestral species into two sister species with the extinction of the ancestral species at the point of dichotomy. Only the joint possession of shared derived characters can be used for character weighting in determining sister groups; character weighting by functional anatomical complex or adaptation is rejected. Any violation of these two principles leads to classifications which are not evolutionary and are therefore to be eschewed. The bifurcating origin of clades renders phyletic speciation insignificant. The cladistic scenario of strictly dichotomous branching and concomitant extinction of the ancestral species also rejects the study of grades or levels of evolutionary change because grades result if great weight is given to derived characters acquired by one sister group and not the other. Recognition of major shifts in adaptive zone might be reflected in paraphyletic groups that would express degrees of evolutionary divergence, but this precludes a classification based on cladograms.

Occasionally there is an explicit relationship between punctuated equilibria and cladistics (Eldredge and Tattersall, 1975, 1977, 1982; Eldredge and Cracraft, 1980), but at other times the relationship is implicit. Novel components of punctuated equilibrium theory lie in stimulating discussions of interspecific or species selection and macroevolution (Stanley 1979, 1981, 1982). If individual species are characterized by stasis during their existence, then any long-term trends in morphology must be the result of selection among species. The analysis of species trends is advocated to reveal nonrandom interspecific survival. Trends reveal that many lineages change directionally over long evolutionary periods. Strangely, however, use of a grade concept to examine the same evolutionary process is rejected.

Emphasis on the speciation event explains why many adherents of punctuated equilibria also practice cladistics. The strict dichotomous branching of lineages stresses speciation, not adaptation, as the central focus of evolution. Occasionally it is explicitly recognized that change in gene frequency and content, whether by natural selection or other means, contradicts the cladistic view of species as discrete entities in time and space (Eldredge and Cracraft, 1980:248, 270). The goal is to create a comparative biology based on a methodology or set of procedural rules to study the orderliness of life without reference to evolutionary ecology or functional anatomy. Transformed cladistics eschews evolution altogether. The adaptive status of species and the operation of natural selection are also avoided in discussions of punctuated equilibria dealing with modes and rates of morphological change.

The study of adaptations has been satirized as a Panglossian (all organisms are adaptively optimal) "adaptationist program" by Gould and Richard Lewontin (1979), who believe that it yields trivial research, although the study of adaptations and functional anatomy or morphology has led to many important discoveries and is a significant concern in biology (Mayr, 1983). In his recent presidential address to the American Society of Naturalists, Janis Antonovics (1987) advocates a dismantling ("dys-synthesis") of the Evolutionary Synthesis, which he links to the "adaptationist program" and simple faith in natural selection. The operation of natural selection in wild populations has recently been the focus of two major monographs (Manly, 1985; Endler, 1986), and data from one of these are retabulated by Antonovics to show that a lag in the study of natural selection occurred after the synthesis and was presumably caused by it. Doubt about natural selection as a major force of evolutionary change at the level of molecules, species, or above-species taxa is linked to doubt about the study of adaptation. Antonovics argues that two separate disciplines should emerge from evolutionary biology—one for the study of modern processes and one for the reconstruction of past events.

> The study of past events has usually come under the rubric of phylogeny reconstruction or cladistics, and the study of present-day processes has been termed population or ecological genetics. . . . A major contribution to the Dys-Synthesis has been the rise of modern cladistics, which sees each phylogeny as a hypothesis constructed by explicit methods and based on stated assumptions, rather than as a personal declaration of opinion (whose force used to depend more on the status of the investigator than on any objective criterion). Although hypotheses about the past may not be amenable to experimental test, they are nonetheless valid scientific inferences whose strength can be assessed as more lines of evidence become available. Methodologies for the study of past processes contrast with those used to study

present-day processes; these can be studied by the formulation of experimentally testable hypotheses, in a manner identical to that used in other areas of experimental science. (Antonovics, 1987:329.)

Note that the study of past events is equated with cladistics, rather than with paleontology or paleobiology, which would introduce the study of the ecology or functional anatomy (adaptations) of fossil organisms. Note also that natural selection and adaptation fall within the purview of modern processes, and they are difficult to examine. Hence it can be demonstrated that the rigorous, inflexible, orderly methodology of cladistics, statements that natural selection and adaptation are not worth studying (or at least are difficult or nearly impossible to study), and discussions of punctuated equilibrium events are intertwined.

The emphasis on morphological discontinuities and the claim that species are the units of evolution introduces typology. If species are real objects, unvarying through their existence, an essentialist or typological interpretation of species is introduced. Species are viewed as "essences" or types. Discrete species are separated by unbridgeable gaps because gradual species change through time is inadmissible, and evolution can only proceed as saltations from one essence to another. Punctuationalists may eschew typology, but stressing that species are real and nonvarying heralds the return of typology or essentialism (Cachel, 1986; Brace, 1988). Three categories of typology may be distinguished in biological thought (Levinton, 1988). Yet, whether minor variation within types or intermediate stages between types is possible, there is emphasis on discontinuity between types and a reliance on essentialism, rather than population thinking.

To demonstrate that the approach to species definition is linked to systematics and evolutionary processes, one has only to examine the phylogenetic species concept, which emphasizes that species are diagnosable taxonomic entities and evolution is the history of taxonomic differentiation. According to the phylogenetic species concept, a species is "an irreducible (basal) cluster of organisms, diagnosably distinct from other such organisms, and within which there is a parental pattern of ancestry and descent" (Cracraft, 1989:34–35). Species are defined by their ability to be differentiated; hence discontinuity is emphasized. Population thinking is eschewed because groups of organisms may be diagnosably different, though not reproductively isolated. If they are diagnosably distinct, they are distinct species. In physical anthropology, a recent proponent of the phylogenetic species concept applied to human paleontology laments that "many studies have tended to de-emphasize systematics at low taxonomic levels owing to an entrenched preoccupation with population-level evolutionary explanations of morphological/adaptive change through time" (Kimbel, 1990:250). It is clear that energy spent on morphological analysis and adaptive radiation is perceived to detract from phylogeny reconstruction.

Thus there is often (though not always) a relationship between punctuated equilibrium theory, cladistics, and a rejection of studies of adaptation. There is no necessary connection between punctuated equilibria and cladistics; nor is there any reason why functional analysis should be divorced from cladistic analysis. Yet a connection between punctuated equilbria and cladistics has been emphasized in physical anthropology. Functional analysis and adaptation have also been denigrated by practitioners of cladistics in physical anthropology. This characterization may be a matter of historical accident, resulting from the biases of individual researchers, but it nevertheless constrains recent developments in the field. Alternatively, it may be that the emergence of physical anthropology from typology since 1951 predisposes researchers to new typological approaches. Human and nonhuman primate paleontology may be especially predisposed to typology because of the relative scarcity of fossils (individual human fossils are often discussed by museum catalog numbers, for example). Another factor may be that cladistic analysis is easier to perform than functional morphology or paleoecology because it does not require knowledge of other disciplines such as biomechanics, chronometric dating techniques, or sedimentology. Cladistics enjoys the added benefit of subduing troublesome variation in extinct forms: fossils with morphological distinctions are different species. In the following sections of this chapter, the appearance of dogmatic cladistic systematics, a denigration of adaptation or natural selection, or a return to typology will be taken to reflect the influence of punctuated equilibrium theory.

Nonhuman Primate Evolution and Systematics

The influence of punctuated equilibrium theory on the study of the evolution and systematics of living or fossil nonhuman primates may be explicit (Schwartz, Tattersall, and Eldredge, 1978; Tattersall, 1982; Delson, 1983). It may also enter into discussions of nonhuman fossil primates whose evolution is finally interpreted in the traditional fashion (Rose and Bown, 1986).

Punctuated equilibrium theory has also been responsible for Philip Gingerich's (1979) creation of the stratophenetic method of paleontological analysis. Avoiding cladistic methods entirely (in fact, inveighing against these methods), Gingerich invented an approach to analysis in which ancestor-descendant relationships are inferred from similarity in phenotype and from distribution in space and time, so that morphology and stratigraphic position are of primary importance in discerning phylogeny. This emphasis on morphology and stratigraphy had previously led Gingerich (1976) to propose a new classification of primates, linking fossil adapids and all living primates except *Tarsius*, with *Tarsius* linked to fossil plesiadapiform primates.

Because of the publicity attendant on punctuated equilibrium theory, mention of it is almost obligatory in discussions of nonhuman primate evolution, as is the case for fossil mammals generally. Often cladistics is a more obvious element than punctuated equilibria, as when a major symposium on human and nonhuman primate evolution was organized to examine the "major branching points of primate evolution" (Wood, Martin, and Andrews, 1986). Work on nonhuman primates focused on functional morphology and adaptive radiations now is presented in a framework of phylogenetic relationships which includes a discussion of cladistics and sometimes other schools of phylogenetic reconstruction, such as Gingerich's stratophenetic method. Even studies directed explicitly toward primate anatomy and functional morphology are now oriented toward phylogenetic consequences of anatomical or functional morphological research (Strasser and Dagosto, 1988). A good illustration of this trend is Robert Martin's monograph (1990), which is a member of an ancient and respected lineage of endeavor in physical anthropology: the comprehensive study of the biology and evolution of nonhuman primates. Wilfrid Le Gros Clark's monograph (1971), which passed through three editions, can be compared with Martin's (1990), published nearly twenty years later. This comparison demonstrates the now pervasive influence of the search for a methodology of phylogenetic reconstruction that always includes cladistics.

Cladistic analysis using limited character sets creates problems in primate paleontology. For example, nasal bone morphology and articular patterns have been reported to be diagnostic of living great ape and human taxa. Yet over forty years of primatological literature reports extensive intraspecific cranial variation; and a recent study of the nasal region in common chimpanzees, gorillas, and orangutans demonstrates substantial deviation from patterns reported to be characteristic of individual taxa (Eckhardt, 1987). "The configurations observed were so multifarious . . . that any attempt to designate one particular pattern as representative—as a character state—obscures the extensive structural polymorphism in this region" (Eckhardt, 1987:334). Consequently, use of character states to assign particular fossils to genera and lineages can be problematic. The cladistic approach tends to underestimate within-taxon variability seen in living populations. This is another example of how cladistic systematics has exercised a profound influence on primate evolutionary studies since the 1970s, with punctuated equilibrium theory playing a more indirect role. Historians of science may someday assess that cladism (attendant upon the English translation in 1966 of Willi Hennig's *Phylogenetic Systematics*) was the fertile soil in which punctuated equilibrium theory would grow because old ideas, such as peripatric speciation, quantum evolution, and interspecific selection, were viewed as novel against the new landscape of cladistic systematics.

In a major review article on primate evolutionary studies in America during the past fifty years, Matt Cartmill (1982) predicts that the recent trend toward inventing and rigorously framing primate phylogenetic hypotheses will falter in the near future because of the demonstration that parallel evolution is a common feature in primate evolution. Parallel evolution, which creates derived characters in different lineages, is a major problem for cladists because the possession of shared derived characters is the rationale for placing species within the same lineage. Instead, Cartmill (1982) predicts that the principal concern of primate evolutionary studies in the last years of this century will focus on the origins of the four major grades of primate organization: the origins of the primate order, the origins of primates of modern aspect, the origins of higher primates, and the origins of the hominids. Future concentration on primate paleobiology and adaptive radiations would coincide with a renaissance of vertebrate functional morphological studies, integrating morphology with biomechanics, behavior, evolutionary ecology, and evolutionary processes such as convergent and parallel evolution (Hildebrand et al., 1985; Liem, 1989).

I also note that mosaic evolution is a major problem for adherents of punctuated equilibria. If different morphological or behavioral characters evolve at different rates in the same organism, that organism will be a mosaic of characters, some primitive and some derived. This concatenation of traits (perhaps evolving independently, although obviously the entire organism experiences selection pressure) represents the operation of mosaic evolution. Hence there is no coherent transformation of all or most characters in organisms. Traits may vary in rate of evolution even within the same functional system, such as the masticatory or locomotor systems, or one functional system, made up of an intricate complex of traits, may vary with respect to other systems. A typical example of mosaic evolution is seen in the human locomotor system. Human bipedal locomotion is extremely specialized, and the locomotor anatomy for bipedalism is very derived in comparison to the locomotor anatomy of the first land vertebrates. Yet there are still five digits on the human foot, which represents the primitive, ancestral condition seen in the first mammals. I stress that mosaic evolution is ubiquitous. Mosaic evolution therefore demonstrates that gaps between morphological types can actually be caused by the mere absence of appropriate anatomical areas surviving as fossil evidence. A good example of this comes from consideration of dryopithecine ape evolution.

Ever since 1926, the dryopithecine apes had been held to occupy a special place in hominoid evolution. On dental grounds, they were the ancestral group for modern hominoids (the lesser apes, the great apes, and the hominids). By the early 1970s, however, it began to be suggested that the dryopithecines were only "dental apes," because examination of surviving postcranial elements, especially wrist anatomy, revealed an Old World monkey type of morphology. This implied that all of the living hominoids diverged from each other subsequent to the existence of this "dental ape" phase because all living hominoids possess specializations of the wrist, elbow, shoulder, pelvis, and other areas of postcranial anatomy which allow anthropologists to recognize the existence of a Superfamily Hominoidea among the living primates. Punctuational events were therefore implied: extraordinarily rapid acquisition of hominoid postcranial specializations in dryopithecines and then rapid divergence of the surviving hominoid lineages. This acquisition and divergence would have had to occur within the late Miocene, between the occurrence of the last dryopithecines 12 to 8 million years ago and before the appearance of the best-documented surviving hominoid lineage, the hominid lineage, which extends back to 5.5 million years ago. Fragmentary remains of the elbow region of dryopithecines from Hungary and Pakistan dated to 12 to 8 million years ago appeared to document a rapid acquisition of typically hominoid postcranial anatomy. In the early 1980s, however, major portions of two early African Miocene species (*Proconsul africanus* and *Proconsul nyanzae*) dated to about 18 million years ago were recovered, and it was observed that forelimb features become more hominoidlike proximally. This gradient is reversed in the hindlimb, with more hominoidlike features found distally.

Adherents of punctuated equilibria often claim that too much attention is paid to the incompleteness of the fossil record, but, in the case of *Proconsul africanus*, chance recovery of only the monkeylike portions of postcranial anatomy in one specimen (KNM-RU 2036) led to speculation about punctuational events in hominoid evolution. Chance recovery of other portions of the postcranium of the same specimen nearly thirty years later upsets this interpretation.

HUMAN PALEONTOLOGY

Human paleontology has played a role in punctuated equilibrium theory since its inception (Eldredge and Tattersall, 1975; Gould and Eldredge, 1977), and hominid evolution is alternatively viewed as one of the best examples of punctuated equilibria (Eldredge and Tattersall, 1982) or one of the best examples of "phyletic gradualism" (Wolpoff, 1984) seen in mammals. Major debates often center on whether particular hominid taxa, such as *Homo erectus*, demonstrate or do not demonstrate stasis (Rightmire, 1986; Wolpoff, 1986). Yet it is becoming clear that many traits once considered essential to the definition of humanity—such as bipedal walking, use and manufacture of tools, increased relative brain size, hunting or confrontational scavenging of large animals, and control of fire—are not necessarily coincident in early hominid history. Rather, these traits, which have long been considered a major constellation in hominid origins, have evolved separately (Simons, 1989). Bipedalism is documented by fossil trackways at 3.7 million years ago, stone tools at 2.3 million years ago, and increasing relative brain size at 1.6 million years ago. Only bipedalism is left as the hallmark of hominid status. Hence hominid origin, which, until recently, may have appeared to be a punctuated event with many simultaneous major transformations, now appears to be otherwise.

Invited papers in the volume of the proceedings of the Ancestors Symposium, held at the American Museum of Natural History in 1984, deal to a large extent with the phylogenetic position of various fossils and ancestor-descendant relationships, with only scant mention of evolutionary rates and evolutionary processes (Delson, 1985). Hence I think that, as is the case in nonhuman primate evolution, cladistics per se is a more important influence than punctuated equilibrium theory. Punctuated equilibrium appears to exert influence only indirectly, through cladistic systematics. The end result is that phylogeny alone has become of principal interest to many paleoanthropologists, so that Alan Walker, discussant at the Ancestors Symposium, inquired plaintively from the dais, "What has happened to biology?"

In fact, human paleontology has become the creation of phylogenies, presented from the standpoint of cladistics (Wood, Martin, and Andrews, 1986). Only the ancestor-descendant relationships of fossil taxa are of interest. To be able to state that one fossil taxon is the putative ancestor of another, for example, is thought to encompass the most important questions in paleoanthropology. When new fossil material is discovered, attention is first focused on how phylogenies advocated by different researchers will be affected. Will one or another current phylogeny be supported, or will a major new phylogenetic review be in order? Study of functional anatomy, ecology, behavior, and what might be thought of as the natural history of fossil humans is being neglected, except insofar as these studies illuminate phylogenetic analysis. Cladistic systematists might argue that a natural taxonomy based on evolutionary events incorporates all of this information, but this information is certainly not explicit. It appears that the orderly, strict assumptions and methods of cladistics lull practitioners into phylogenetic reconstruction per se, which then becomes the principal goal of paleoanthropology.

Discussion of evolutionary ecology and evolutionary processes is also neglected. This restricts the direct influence of evolutionary theory. A typical example of this trend can be seen in an exhaustive cladistic analysis of hominid taxa (Wood and Chamberlain, 1986). Thirty-nine characters were analyzed from five cranial regions, and a cladogram was constructed for each region. Regional cladograms were then compared, but no single preeminent cladistic classification emerged. The research protocol gives great attention to taxonomic traits, outgroups, sampling, quantification, and size standardization. But functional interpretations and stratigraphy are specifically rejected in distinguishing which characters are inferred to be shared-derived from those inferred to be shared-primitive (Wood and Chamberlain, 1986:223). The discovery of KNM-WT 17000 (Walker et al., 1986) generated a flurry of notes and comments that appeared in *Nature, Science*, and the popular press (Delson, 1986; Lewin, 1986; Wilford, 1987)—another example of this trend. The principal topic of these discussions was how this new specimen of robust australopithecine affects current phylogenies. A recent symposium on the robust australopithecines was in fact initiated by phylogenetic questions created by this specimen (Grine, 1988).

Other recent developments, far more revolutionary in nature, such as the discovery that short, pongidlike ontogenetic periods may exist in many fossil hominids, do not receive equivalent notice. Yet this discovery, if confirmed, would drastically affect our reconstructions of the way of life of many fossil hominids: they would mature far quicker than modern humans. This would necessitate major analysis of prior reconstructions of early hominid social behavior. If the study of adaptations, paleoecology, or functional anatomy is trivial research, why not focus energy on the construction of phylogenies by the rigorous methodology of cladistics?

Niles Eldredge and Ian Tattersall present their case that the Evolutionary Synthesis was one of the "greatest myths of twentieth century biology" in a popular account titled *The Myths of Human Evolution* (1982:43). Here and in other major accounts which promote punctuated equilibrium theory (Gould and Eldredge, 1977), the alternative or "traditional" picture of human evolution is presented as a single lineage, progressing from *Australopithecus africanus* to *Homo erectus* to *Homo sapiens* and undergoing gradual transformation, especially with regard to brain size. Yet though it is true that one version of hominid phylogeny does interpret the evidence in a unilinear fashion, with stages of human evolution present at certain points in time widely distributed through the Old World, many other phylogenies exist and have largely displaced this version of hominid evolution. Hence to claim that punctuated equilibrium theory transforms the interpretation of human evolution is not correct. The single-species hypothesis of hominid evolution, which interprets the existence of widely divergent morphology in contemporary fossil material in terms of variation within a single species (because, at any particular time, there is only one hominid niche), has been largely abandoned since the early 1970s. In fact, evolutionary ecology has been applied to hominid evolution for well over a decade, and multilineal adaptive radiations of late Neogene hominids have been advocated in the past (Cachel, 1975). If several sympatric hominid species coexisted for long periods of time in Plio/Pleistocene Africa, then questions of niche differentiation and the limits to similarity arise. The competitive exclusion principle states that competing sympatric species cannot coexist for long periods of time if they require exactly the same resources. The problem, then, is to determine the ecological differences needed for coexistence in a competitive environment, the degree of ecological overlap that can be tolerated, how behavioral, temporal, and habitat shifts affect competition, and so on. The degree of sophistication in evolutionary theory and evolutionary ecology that exists among physical anthropologists employing a paleoecological approach can be seen in Bruce Winterhalder's discussions of the single-species hypothesis and the competitive exclusion principle in hominid evolution (1980, 1981). It may well be true, however, that this paleoecological sophistication would not be appreciated by adherents of punctuated equilibrium theory who embrace cladistics and eschew evolutionary ecology or the study of adaptation.

Often researchers face problems in defining fossil hominid taxa or in deciding what specimens form the hypodigms of these taxa. When cladistic analysis is applied to these problems, the result can be a proliferation of fossil taxa, as in the case of *Homo habilis*, where cladistic analysis of specimens attributed to this taxon yields two species (Stringer, 1986). The possibility that evolutionary change might account for the distribution of characters is rejected because a single character (brain size) appears to contradict a directional trend in a single specimen (KNM-ER 1813). Later hominids can also be subject to proliferation after cladistic analysis, with European material currently recognized as *Homo sapiens* being divisible into *Homo heidelbergensis*, *Homo neanderthalensis*, and *Homo sapiens* (Stringer, 1984).

The status of the Neanderthals has long been a problem in anthropology. Sudden, catastrophic extinction of the Neanderthals is a hoary explanation for the Neanderthal problem. Punctuated equilibrium theory fits in well with this old explanation (Stringer, Hublin, and Vandermeersch, 1984) but has also been rejected in favor of local continuity (Smith, 1984). Interestingly, a recent survey of the skeletal changes accompanying nutritional change in Sudanese Nubian populations over the last twelve thousand years documents a decrease in craniofacial robusticity, increase in cranial height with decrease in length, and rotations of the cranial vault and face, so that the face becomes positioned inferoposteriorly to the vault (Armelagos et al., 1984). In short, nutritional changes over the past twelve thousand years caused local populations to be transformed from a Neanderthal-like morphology to a more modern morphology. A strict cladistic analysis of these skeletal remains might well yield an interpretation that at least two distinct species are present (an earlier, Neanderthal-like species, and a later, more modern species), even though the remains represent evolving populations of anatomically modern humans. Loss of information about the genetic continuity of these populations and loss of archaeological evidence of food production and nutritional change would make it appear that an abrupt morphological change had occurred, instead of evolution in situ accompanying dietary shifts.

The human fossil record as characterized by Eldredge and Tattersall (1982) constantly encounters evidence that cannot be accommodated by cladistics or punctuated equilibrium theory. Parallel evolution in the human fossil record creates a problem for cladists, as it does in the nonhuman primate fossil record. Features of the cranial base and dental development which link robust australopithecines and *Homo* create such a problem (Dean, 1986). Furthermore, the notion of the abrupt appearance of robust australopthecines is challenged by the discovery of KNM-WT 17000, an early member of the *Australopithecus boisei* lineage dating to 2.5 million years ago (Walker et al., 1986). Mosaic evolution (and phyletic gradualism) is wonderfully exemplified by the discovery of KNM-WT 15000, an adolescent specimen of *Homo erectus* 1.6 million years old and startlingly complete (Brown et al., 1985). Pelvic details record a mixture of australopithecine and hominine details, with those resembling the australopithecine condition being flare of the ilium, very long femoral neck, and a small femoral neck/shaft angle. In the vertebral column, a Neanderthal-like feature occurs: the spinous processes on all recovered vertebrae are relatively longer and less inferiorly inclined than in modern hominids. The discovery of this one specimen, therefore, established links to australopithecines and later hominids.

Finally, to demonstrate that there is a strong interrelationship of cladistics and typology, Tattersall (1986) has argued that more species should be recognized in the hominid fossil record because species are real entities in time and space and should not be established by studies of intraspecific variability. Rather than focusing on intraspecific variation when determining hominid species, anthropologists should focus on interspecific variation. A static-species morphology is desired by adherents of punctuated equilibrium theory, and so intraspecies variation is downplayed. When Tattersall inveighs against a focus on intraspecific variation or study of variation in the living world, he inveighs against population thinking and the obscuring of the ideal realm of types. Hence,

> Species, reproductively isolated from the rest of the living world, acquire their discreteness as the result of an irreversible genetic and historical process. Subspecies, on the other hand, while distinguishable in some way and at some point, are in concept ephemera, unbounded by reproductive barriers;

their unique identity may at least potentially be lost at any time in merging with other subspecies populations. . . . The fact remains that, to paleontologists in general, subspecies are epiphenomena which do not merit the attention paid to species: it is species, not subspecies, that are the units of evolution, or at least of evolutionary study." (Tattersall, 1986:167–168.)

Leaving aside the problem of how to determine between-species patterns of variability for hominids if only one living species exists, one might argue that intraspecific variability is just as valid to differentiate fossil species—but then one ends up with phyletic gradualism and cannot maintain a typological approach. If species are unchanging from origin to extinction, and are more real than field biologists or systematists can see, then clearly one needs punctuated equilibria, an alternative to natural selection, which introduces types as units of evolution and which incidentally meshes well with cladistic systematics.

Furthermore, if taxonomic differentiation is the focus of human paleontology, subspecies are considered to be indistinguishable in terms of skeletal or dental anatomy (Tattersall, 1986, 1990). "It is generally true that where readily distinct bony morphs exist among closely related extant mammals speciation has already occurred" (Tattersall, 1990:306). Yet living primate subspecies do demonstrate cranial, postcranial, sexually dimorphic, and shape variation that would be detectable in the fossil record (Turner et al., 1988; Cheverud and Moore, 1990a, b).

Morphological change and speciation are currently subject to scrutiny because of species definition. H. E. H. Paterson (1981, 1982, 1986) argues that a recognition concept of species, based on a specific mate-recognition system, might be the most useful species definition for paleontologists because it does not necessitate knowledge of the genetic structure of extinct populations and gene flow between these populations. Facial pelage color and pattern (facial masks) have been examined in African monkeys of the genus *Cercopithecus* (Kingdon, 1988). These facial masks contribute to reproductive isolation and are part of specific mate-recognition systems, as discussed by Paterson. But there is no link between *Cercopithecus* species and normal, preferrred habitat types, and there is dietary competition between *Cercopithecus* species (Gautier-Hion, 1988; Oates, 1988). These are contrary to expectations of Paterson's recognition concept (1986), that there should be an association between species and a particular habitat type and that there should be a lack of interspecific competition. The relationship between speciation and morphological change within the genus *Homo* has been recently addressed using the species recognition concept and comparing hominid taxonomy with that of other mammalian families (Turner and Chamberlain, 1989). This comparison reveals that taxonomic questions sometimes plague the study of nonhuman mammal families as well. Because hominids lack specific skeletal structures (e.g., horn cores or antlers) used in sexual selection or courtship displays, identification of specific mate-recognition systems in fossil hominids will probably remain equivocal. Alan Turner and Andrew Chamberlain (1989) ultimately argue for the retention of African fossil hominid specimens within the taxon *Homo erectus*, rather than the creation of an additional species of *Homo*.

HUMAN VARIATION AND ADAPTATION

If punctuated equilibrium theory introduces typology and a denigration of the study of adaptation, then physical anthropology may return to the state it was in before Washburn reorganized the field in 1951 (Washburn 1951a, 1951b, 1953), when living and fossil humans were studied to classify individuals as to type. Human variation was studied only from a typological perspective, and nonadaptive, hence unchanging, traits were actively sought to perform this analysis (Hooton, 1926; Boyd, 1950). That this premonition is more than just speculation is suggested by a recent review article on human skin pigmentation (Quevedo, Fitzpatrick, and Jimbow, 1985), which hesitates to ascribe a functional significance to melanin pigmentation. The authors cite Gould and Lewontin (1979) to conclude that natural selection may not be as finely tuned to environmental factors as once thought and that variations in melanin pigmentation need not always be of great evolutionary significance.

Yet melanin regulation of vitamin D synthesis in the skin upon exposure to ultraviolet light has been well studied (Loomis, 1967; Clemens et al., 1982) and established to be of adaptive significance in childhood rickets caused by vitamin D deficiency and associated metabolic bone disease. Abnormal calcification creates long bone, spinal, and pelvic malformation interfering with childbirth as the pelvic inlet is reduced. Reproductive efficiency is impaired in the absence of cesarean deliveries. Adult osteomalacia causes demineralization of the spine, pelvis, and legs; body weight acting on softened bone creates bowing of the long bones, vertebral compression, pelvic distortion, and contraction of the pelvic outlet through which delivery occurs. Neonates born to adult females with osteomalacia suffer from delayed motor development and reduced bone mineralization. Rachitic tetany can occur with vitamin D deficiency in either youngsters or adults and is caused by low serum calcium levels. Low serum calcium levels resulting from low vitamin D synthesis may precipitate anxiety, psychomotor seizures, memory loss, and frenzied dissociative states. When this neurological disturbance occurs under conditions of prolonged winter darkness in high latitudes, anthropologists label it "Arctic hysteria" or pibloktok. Toxic high levels of vitamin D also create health problems, with abnormal calcification of such soft tissues as the aorta, the development of kidney stones, and impairment of kidney function. Hence a good body of evidence supports the theory that the degree of melanin pigmentation in human skin is of adaptive significance in regulating vitamin D synthesis in the skin upon exposure to ultraviolet light. The occurrence of relatively depigmented humans in temperate and high latitudes can be viewed as a response to selection pressures to promote more vitamin D synthesis in regions of reduced ultraviolet radiation; the occurrence of heavily pigmented humans in equatorial regions can be viewed as a response to selection pressures to decrease vitamin D synthesis in regions of great ultraviolet radiation. Melanin pigment is thus the natural regulator of vitamin D synthesis and may also protect against the carcinogenic effects of intense ultraviolet light on human skin in equatorial regions.

Other researchers (Branda and Eaton, 1978) have examined melanin skin pigmentation as a protection against destruction of great amounts of such vitamins as B12, E, and riboflavin upon exposure to ultraviolet light. Genetic differences in lactose absorption and tolerance affecting calcium metabolism concomitantly regulated by vitamin D have also been investigated (Johnson et al., 1981). The impact of natural selection on skin pigmentation may therefore be complex, and complex selectional interactions also appear to affect other areas of human variation (Skrikumari et al., 1987). It is, of course, notoriously difficult to measure selection intensities in wild populations, perhaps more so for humans than for most other animals; but to admit that demonstrating and measuring natural selection in the wild is difficult is not the same as claiming that it is not worth studying.

The influence of punctuated equilibrium theory on W. C. Quevedo and associates is obvious when they conclude by writing of "the new directions that evolutionary theory in general . . . take[s] during the seventies and eighties" (1985:54). Perhaps there will be a return to description in physical anthropology and an abandonment of major research projects in human adaptation to temperature and altitude (Little, 1982; Beall, 1982; Harrison et al., 1988). Study of human adaptability, integrating morphological, physiological, ecological, behavioral, and cultural approaches to human survival in various habitats, as well as consideration of the effect of natural selection on human evolution, may also be abandoned. Description of human variation, typology, and the construction of phylogenies may reign, which would herald a return to the pre-1950s state of affairs in physical anthropology.

Some recent work analyzes gene frequencies in modern Europeans and explains them in terms of demic spread of early agriculturalists from southeast to northwest through the great river valley systems of Europe as a new farming economy created population growth (Menozzi et al., 1978; Sokal and Menozzi, 1982; Ammerman and Cavalli-Sforza, 1984). This anthropological synthesis of population genetics and archaeology, emphasizing the expansive movement of farming populations in a wave of advance or demic diffusion (Ammerman and Cavalli-Sforza, 1984), is in the tradition of the New Physical Anthropology. It has been followed by a controversial attempt to synthesize the evidence of archaeology and linguistics in order to reconstruct the prehistory of Europe and the spread of Indo-European languages by demic spread of early agriculturalists (Renfrew, 1988). More recent investigation focuses on analyzing the relationship between genetic patterns (gene frequencies and cranial measurements) in modern European populations and language boundaries that impair gene flow and maintain genetic difference in spite of population admixture (Sokal et al., 1990). The interest in evolutionary processes, populations, and the dynamics of population change or maintenance is clearly part of the New Physical Anthropology. Yet such study may easily succumb to emphases of the Old Physical Anthropology.

It is possible that the beginning of a trend toward description of variation, typology, and the construction of phylogenies can be detected in the current plethora of articles on the degree of relatedness in living humans. For example, when evolutionary processes are downplayed, emphasis on the pattern and degree of relatedness of modern humans may partly negate the sophistication of the genetic, archaeological, and linguistic data generating these patterns (Cavalli-Sforza et al., 1988). In fact, natural selection must be assumed not to have occurred, if allele frequencies in modern populations are used to assess degree of relatedness. Otherwise, natural selection independently operating in various human populations can create similar allele frequencies through parallel evolution. Assuming that natural selection does not operate on these genetic loci (a patently false assumption), allele frequencies then reflect patterns of descent altering the frequencies of particular serum protein, blood group, and HLA-antigen alleles in human populations. A great amount of current research is of this sort. It focuses on modern human phylogeny and has no interest in natural selection, except to assume that natural selection has not affected genetic diversity within populations. This research analyzes global patterns of serum protein, blood group, and HLA-antigen diversity; sequence data for mitochondrial DNA; a Y-chromosome-linked DNA polymorphism; and nuclear DNA polymorphisms in the alpha- and beta-globin gene cluster (Wainscoat et al., 1986; Cann, 1988; Stoneking and Cann, 1989; Wainscoat et al., 1989). Evolutionary relationships of modern humans are being assessed, and diverse phylogenetic trees for various population groups are being generated. Pattern, rather than process, is being analyzed. The functional significance of the diversity is not the focus of study but, rather, the population size and pattern of spread of anatomically modern humans from centers of origin and the sequence of their divergence. The mitochondrial DNA analyses, which have received the bulk of attention because of implications of anatomically modern human spread from sub-Saharan Africa, suffer from a variety of problems: small sample size (less than 150 in the placental mtDNA analyses); admixture with other groups (18 of the 20 African placental samples were from African-Americans); uncertainty about mutation rates (if mutation rates are slower, the mtDNA analyses refer to the spread of *Homo erectus* out of Africa); and suspicions about the cause of relative mtDNA uniformity in humans (because comparison is made with the great apes, more variability here may be explained by relict pongid populations, bottlenecks, and small sample sizes). The spread of anatomically modern humans from sub-Saharan Africa based on genetic variation here is also flawed by the prior assumption that the vector of spread is out of Africa. These studies yield genetic distance, but there is no vector to human movement. The alternate hypothesis of colonization of sub-Saharan Africa by populations subject to extreme bottlenecks and genetic drift is also possible (Excoffier et al., 1987).

This type of analysis creates scenarios of the origin of modern humans in a single small population whose descendants migrated and spread worldwide so that local populations of *Homo erectus*, "archaic" *Homo sapiens*, or Neanderthals (*Homo sapiens neanderthalensis*) became extinct and made no genetic contribution to any living human gene pool. Christopher Stringer and Peter Andrews (1988) argue that modern *Homo sapiens* originated by speciation in a single region of the Old World, probably in sub-Saharan Africa, and replaced all other hominids. The mtDNA basis of their argument is flawed by refusal to acknowledge repeated illustrations that mtDNA types vary significantly in Old World human populations; that many mtDNA types are ancient in and unique to local populations; and that the most reliable estimation of mtDNA mutation dates a common human origin to *Homo erectus* time range (Sphuler, 1988:40). Mitochondrial DNA variability in local populations and the extended time range both favor a regional transition to anatomically modern humans, rather than a complete replacement (Wolpoff, 1989). Cladistic analysis applied to the problem (Stringer, 1989) does not allow for morphological variation and emphasizes typology (Habgood, 1989). That genetic swamping of small, local human populations is entirely possible can be demonstrated by the cline of allele frequencies in modern Europe that preserves the trace of demic spread of early farmers, some six to eight thousand years later, in spite of intense population disruption in Europe caused by warfare and disease through recorded history.

The catastrophism paradigm is not new. It has been labeled the "Shadow Man Paradigm" in paleoanthropology and has been traced back to the beginning ot the twentieth century (Hammond, 1988). The essential element of the Shadow Man Paradigm is the search for true human ancestors (which must be anatomically modern in appearance), while relegating most hominid fossil evidence to the status of evolutionary cul-de-sacs. As a result, hominid phylogenetic trees show many extinct species and genera, while preserving a true, ancient lineage of modern hominids, based on little or no fossil evidence. The perception that a major transformational event is necessary in human history is therefore not novel. What is novel is the movement of the event to very late time ranges, rather than locating it at the origin of hominids. The attempt to discover a constellation of early, typically hominid traits cannot be supported by the fossil or archaeological records (Simons, 1989) and so has been shifted to a very late period. Nevertheless, detailed analyses of the Middle Palaeolithic/Upper Palaeolithic transition in Europe and the Levant, where the archaeological record is dense, records a continuum of behavior and tool industries (Clark and Lindly, 1989a, 1989b; Mellars, 1989a, 1989b).

The catastrophism paradigm certainly meshes well with punctuated equilibrium theory, which also counsels abandonment of natural selection and adaptation studies and leads to typology and cladistics. If description, typology, dismissal of natural selection and adaptation, and construction of phylogenies become imperative in physical anthropology, punctuated equilibrium theory will have exercised a deleterious effect.

Sociobiology

The possible influence of punctuated equilibrium theory on sociobiology is very slight but might be indicated by Eldredge and Tattersall's (1982) denigration of sociobiology. They hold that only the behavior of natural groups of species descended from a common ancestor should be studied. They believe that to look at sociality beyond the limits of natural groups has only limited application and will yield only the behavioral equivalents of functional morphological study because "explanations of evolutionary history must be made for groups that evolution has produced" (1982:26). Note that aspersion is cast upon functional morphology, which is predicated upon adaptation. This coincides with other downplayings of adaptation. The principal point is that the sociobiological perspective, whether in physical anthropology or elsewhere, is denigrated, which may have some minor future influence on studies of behavior and ecology in the Order Primates.

I am not aware of any direct application of punctuated equilibrium theory to primate behavior and ecology, and I think it unlikely that any current attempts to halt the influence of sociobiological theory will succeed. In fact, a good argument can be made that, among mammalogists, primatologists have been the first and most eager to apply sociobiological theory. This can be seen in primate studies on infanticide, kin selection, nepotism, altruism, male and female strategies of reproduction, sex-ratio analysis, sexual selection, mating systems, parental investment, resource distribution and feeding competition, and theories of social evolution (e.g., Smuts et al., 1987). Nearly all of primatology (with the exception of functional anatomical or biomedical research), and some human behavioral and evolutionary ecological research as well, is pervaded by sociobiology. In short, sociobiology is now an extremely important component of physical anthropology, and it is unlikely that any present criticisms of sociobiology, whether from the realm of punctuated equilibrium theory or from other areas, will have any effect.

ARCHAEOLOGY

Eldredge and Tattersall (1982) further believe that human cultural evolution, as seen from the evidence of stone tool industries during the Paleolithic, demonstrates punctuational events. Their Paleolithic evidence is gleaned from the Olduwan and Acheulean industries. Archaeologists, however, while recognizing that very local regional differences and changes in stone tool industries did not become profound until the Upper Paleolithic, nevertheless do recognize that the Acheulean industry had variants (Kleindienst, 1961; Clark, 1982; Wynn and Tierson, 1990). Local industries can be differentiated by such things as the proportion of tool classes and the addition of flakes struck from specially prepared cores, as in the Kapthurin Acheulean industry of Kenya, where a proto-Levallois technique of core preparation is found at 230,000 years ago (Gowlett, 1984).

Eldredge and Tattersall (1982) also use Egyptian history to demonstrate long periods of cultural stasis upset by rapid major change. The four-thousand-year-long span of Egyptian history from prehistoric to Roman times has often invited descriptions of the unprecedented continuity of Egyptian culture in the full realm of human cultural variation and change. This cultural continuity is classically viewed as marred only by several major sociopolitical upsets. It would therefore appear that periods of revolutionary change disrupt ironclad continuity, which is why Eldredge and Tattersall (1982) use Egyptian history to argue for punctuational events in human cultural history. Although Egyptian culture flourished within the confines of a riverine oasis isolated by the desert and was therefore highly delimited by geography and isolated from external influences, it was far from stable or static. Karl Butzer (1984) has documented how Nile flood variation and the limited irrigation technology of pharaonic times allowed only single annual crops to be harvested within areas of the flood basin where natural conditions determined adequate water, drainage, and new silt accumulation. Fluctuations in Nile flooding caused interannual fluctuations in food supply, which could not be compensated for by traditional irrigation agriculture. The classic static picture of Egyptian culture becomes replaced by a picture of dynamic equilibrium, in which a fragile ecosystem is highly stressed by noncyclical variation in Nile flooding. An ecological perspective to Egyptian history thus illustrates that climatic variability generated insecurity and necessitated constant adjustments in a culture based on irrigation agriculture because Nile flood variation was considerable and unpredictable. Water management and control may extend back to predynastic times (Hassan, 1988). Nor were periods of major sociopolitical trauma rapid in Egyptian culture; they appear to have spanned centuries in some cases (Butzer, 1984). Complex human adaptation to profound aridity can be similarly demonstrated in the Nubian Desert across the Egyptian-Sudan border. The Nubian Desert west of the Nile has been inhabited since the Acheulean, and large, permanent Middle Neolithic communities survived under arid conditions without interruption from 7,800 to 6,200 years ago (Wendorf et al., 1985). It is, therefore, unlikely that a direct application to human cultural evolution can be made of punctuated equilibrium theory, even in areas such as Egypt, where the cultural record is long and extensive.

The nature of the archaeological record itself may contribute to the perception of rapid major change. Contact, dispersal, and exchange that appear to be instantaneous from the archaeological record may, in fact, represent events spread over a century or more. These problems are analogous to those characterizing the paleontological record.

Common archaeological practice of dividing cultural remains into stages should not be interpreted as always implying major cultural changes (punctuational events) at the boundaries of these stages. Stone tool and ceramic typologies are staples of archaeological methodology. In the early twentieth century, changes in stone tools or pottery were thought to indicate migration or conquest because boundaries between the types seemed to require external factors to explain discontinuities. An analogous problem with discontinuities confronts typologists dealing with biological material. How to explain new technology? How to explain new morphology? In both cases, careful perusal of the evidence often reveals a continuum of variability. If new stone tool types or pottery types appear in the archaeological record, they are not currently viewed as necessary indicators of the arrival of a new group of people. Major cultural complexes may emerge through local factors, perhaps a network of interaction between neighboring local populations (peer-polity interaction), rather than invasion or migration (Renfrew and Cherry, 1986). The emergence of even complex societies is now examined in the light of socioecological or subsistence factors, for example, and not by incursion or invasion. The slow buildup and transfer of population (Ammerman and Cavalli-Sforza, 1984), local creation and dispersion of ideas, and local adaptation to new ideas (Higgs, 1975) are now considered to be more significant than past archaeological emphasis on prehistoric migration and diffusion.

There is, however, one area of archaeology in which punctuated equilibrium theory may play a role, and that is in the study of Quaternary extinctions of megafauna by direct or indirect human activity. The idea was first advanced by Paul Martin in the mid-1960s, long before the advent of punctuated equilibrium theory, but a major research volume has recently been devoted to the study of Quaternary extinctions (Martin and Klein, 1984). This work mirrors recent general interest in extinction phenomena per se (Berggren and Van Couvering, 1984; Nitecki, 1984; Donovan, 1989). Martin (1984) has renovated his original hunting overkill idea so that now he recognizes blitzkrieg overkill—humans colonize land masses occupied only by naive animal species and begin hunting these vulnerable animals. The speed and intensity of human contact results in waves of extinction as humans move through the landscape. There is a relative absence of archaeological kill sites recording this process because the extinctions occur over so short a period. I see this phenomenon as perhaps reflecting punctuated equilibrium theory. Martin argues that Paleolithic humans may have caused significant Pleistocene extinctions, and he has accentuated the speed and intensity of the process since he first proposed it. Many other authors in the same volume credit climate and coextinction as being of primary importance in faunal loss. Donald Grayson (1984) has rightly pointed out that the blitzkrieg hypothesis cannot be falsified because the absence of an archaeological record indicates the operation of blitzkrieg overkill, as well as the presence of a record, by way of kill sites, butchering sites, or hearths.

CONCLUSION

The impact of the theory of punctuated equilibria on anthropology is restricted to physical anthropology because, contrary to the situation in the nineteenth and early twentieth centuries, evolutionary theory now affects only this subdiscipline of anthropology.

The major impact of punctuated equilibrium theory occurs in paleoanthropology. Here it affects discussion of rates of evolution, species discrimination in the fossil record, and longevity and stasis of species. The principal paleontological approach to species is to define them by morphological differences. This is a typological interpretation of species because unvarying types are examined. Hence species are found to be separated by unbridgeable gaps, and the speciation event itself is all-important. Emphasis on the speciation event explains why so many adherents of punctuated equilibria are also cladists. The strictly dichotomous branching of lineages and concomitant extinction of the ancestral species which occurs in cladistic analysis also emphasizes speciation. The generation and study of cladograms have become all-important in paleoanthropology, and the biology and adaptations of ancient hominids have been denigrated. Phyletic lines are dealt with as if they are the units of evolution and not populations, as can be seen in a number of recent studies of Plio/Pleistocene hominids.

Population thinking had been introduced into physical anthropology in 1951 and had caused the disappearance of both the degree of difference criterion for species and schemes of static human categories. Studies of fossil and living hominids were transformed. If studies of hominid adaptation and biology originate in the demise of typological thinking, they are endangered by the reintroduction of typological thinking by adherents of punctuated equilibria. Hominids are analyzed in terms of migration and gene flow, not adaptation to changing forces of natural selection and local population transformation in situ. Hence, for both fossil and living hominids, punctuated equilibrium theory reintroduces catastrophism as an explanation for change.

Some adherents of punctuated equilibria have denigrated sociobiology as a quest for generalizations about behavioral evolution among unnatural (noncladistic, hence nonevolutionary) groupings of organisms not descended from a common ancestor. This may affect physical anthropologists who analyze human and nonhuman primate or mammalian behavior and ecology from a sociobiological perspective.

Archaeology is still another subdiscipline of anthropology which may be influenced by punctuated equilibrium theory, although this is less evident than in physical anthropology. It has been claimed, for example, that the archaeological record shows long periods of stasis and sudden, rapid bursts of change in stone tool industries, which is the principal evidence of behavioral evolution throughout most of hominid existence. Similar claims have been made for Egyptian cultural history. Archaeologists, however, have not accepted these generalizations. Yet one area in which punctuated equilibrium theory does seem to have influenced archaeological interpretation lies in the analysis of Pleistocene and Holocene extinctions. Here the argument is that humans may have directly or indirectly contributed to the extinction of many animal species, perhaps in a catastrophic fashion.

Acknowledgments

I thank Michael Moffatt of the Department of Anthropology, Rutgers University, for discussions of possible influences of punctuated equilibrium theory on cultural anthropology. I also thank C. Loring Brace of the Museum of Anthropology, University of Michigan, for allowing me to read an unpublished manuscript.

References

Ammerman, A. J., and L. L. Cavalli-Sforza. 1984. *The Neolithic Transition and the Genetics of Populations in Europe*. Princeton: Princeton University Press.

Antonovics, J. 1987. The evolutionary dys-synthesis: which bottles for which wine? *American Naturalist* 129:321–331.

Armelagos, G. J., et al. 1984. Effects of nutritional change on the skeletal biology of Northeast African (Sudanese Nubian) populations. In J. D. Clark and S. A. Brandt, eds., *From Hunters to Farmers*. Berkeley: University of California Press, pp. 132–146.

Beall, C. M. 1982. An historical perspective on studies of human growth and development in extreme environments. In F. Spencer, ed., *A History of American Physical Anthropology, 1930–1980*. New York: Academic Press, pp. 447–465.

Berggren, W. A., and J. A. Van Couvering, eds. 1984. *Catastrophes and Earth History*. Princeton: Princeton University Press.

Boyd, W. C. 1950. *Genetics and the Races of Man*. Boston: Little, Brown.

Brace, C. L. 1988. Punctuationism, cladistics and the legacy of medieval Neoplatonism. *Human Evolution* 3:121–138.

Branda, R. F., and J. W. Eaton. 1978. Skin color and nutrient photolysis: an evolutionary hypothesis. *Science* 201:625–626.

Brown, F., et al. 1985. Early *Homo erectus* skeleton from west Lake Turkana, Kenya. *Nature* 316:788–792.

Burrow, J. W. 1966. *Evolution and Society: A Study in Victorian Social Theory*. Cambridge: Cambridge University Press.

Butzer, K. W. 1984. Long-term Nile flood variation and political discontinuities in Pharaonic Egypt. In J. D. Clark and S. A. Brandt, eds., *From Hunters to Farmers*. Berkeley: University of California Press, pp. 102–112.

Cachel, S. 1975. A new view of speciation in *Australopithecus*. In R. H. Tuttle, ed., *Paleoanthropology: Morphology and Paleoecology*. The Hague: Mouton Press, pp. 183–201.

——. 1986. *The Growth of Biological Thought* revisited. *American Anthropologist* 88:452–454.

Cann, R. L. 1988. DNA and human origins. *Annual Review of Anthropology* 17:127–143.

Cartmill, M. 1982. Basic primatology and prosimian evolution. In F. Spencer, ed., *A History of American Physical Anthropology, 1930–1980*. New York: Academic Press, pp. 147–186.

——. 1983. "Four legs good, two legs bad": Man's place (if any) in nature. *Natural History* 11:64–79.

Cavalli-Sforza, L. L., et al. 1988. Reconstruction of human evolution: bringing together genetic, archaeological, and linguistic data. *Proceedings of the National Academy of Sciences USA* 85:6002–6006.

Cheverud, J. M., and A. J. Moore. 1990a. Subspecific morphological variation in the saddle-back tamarin (*Saguinus fuscicollis*). *American Journal of Physical Anthropology* 81:204.

——. 1990b. Subspecific morphological variation in the saddle-back tamarin (*Saguinus fuscicollis*). *American Journal of Primatology* 21:1–15.

Clark, G. A., and J. M. Lindly. 1989a. Modern human origins in the Levant and western Asia: the fossil and archaeological evidence. *American Anthropologist* 91:962–985.

——. 1989b. The case of continuity: observations on the biocultural transition in Europe and western Asia. In P. Mellars and C. Stringer, eds., *The Human Revolution*. Princeton: Princeton University Press, pp. 626–676.

Clark, J. D. 1982. The transition from Lower to Middle Paleolithic in the African continent. *British Archaeological Reports International Series* 151:235–255.

Clemens, T. L., et al. 1982. Increased skin pigment reduces the capacity of skin to synthesize vitamin D3. *Lancet* 1:74–76.

Cracraft, J. 1989. Speciation and its ontology. In D. Otte and J. A. Endler, eds., *Speciation and Its Consequences*. Sunderland, Mass.: Sinauer, pp. 28–59.

Dean, M. C. 1986. *Homo* and *Paranthropus*: similarities in the cranial base and developing dentition. In B. Wood, L. Martin, and P. Andrews, eds., *Major Topics in Primate and Human Evolution*. Cambridge: Cambridge University Press, pp. 249–265.

Delson, E. 1983. Evolutionary tempos in catarrhine primates. In J. Chaline, ed., *Modalites, rythmes et mecanismes de l'évolution biologique*. Paris: Editions du C.N.R.S, pp. 101–106.

——, ed. 1985. *Ancestors: The Hard Evidence*. New York: Alan R. Liss.

——. 1986. Human phylogeny revised again. *Nature* 322:496–497.

Donovan, S. K., ed. 1989. *Mass Extinctions: Processes and Evidence*. New York: Columbia University Press.

Eckhardt, R. B. 1987. Hominoid nasal region: polymorphism and its phylogenetic significance. *Nature* 328:333–335.

Eldredge, N., and J. Cracraft. 1980. *Phylogenetic Patterns and the Evolutionary Process*. New York: Columbia University Press.

Eldredge, N., and S. J. Gould. 1972. Punctuated equilibria: an alternative to phyletic gradualism. In T. J. M. Schopf, ed., *Models in Paleobiology*. San Francisco: Freeman, Cooper, pp. 82–115.

Eldredge, N., and I. Tattersall. 1975. Evolutionary models, phylogenetic reconstruction, and another look at hominid phylogeny. *Contributions to Primatology*. 5:218–242.

——. 1977. Reconstruction of hominid phylogeny: a testable framework based on cladistic analysis. *Journal of Human Evolution*. 6:263–278.

——. 1982. *The Myths of Human Evolution*. New York: Columbia University Press.

Endler, J. A. 1986. *Natural Selection in the Wild*. Princeton: Princeton University Press.

Excoffier, L., et al. 1987. Genetics and history of sub-Saharan Africa. *Yearbook of Physical Anthropology* 30:151–194.

Gautier-Hion, A. 1988. The diet and dietary habits of forest guenons. In A. Gautier-Hion et al., eds., *A Primate Radiation: Evolutionary Biology of the African Guenons*. Cambridge: Cambridge University Press, pp. 257–283.

Gingerich, P. D. 1976. Cranial anatomy and evolution of Early Tertiary Plesiadapidae (Mammalia, Primates). *Papers on Paleontology, University of Michigan Museum of Paleontology* 15:1–140.

——. 1979. The stratophenetic approach to phylogeny reconstruction in vertebrate paleontology. In J. Cracraft and N. Eldredge, eds., *Phylogenetic Analysis and Paleontology*. New York: Columbia University Press, pp. 41–77.

Gould, S. J. 1984. Toward the vindication of punctuational change. In W. A. Berggren and J. A. Van Couvering, eds., *Catastrophes and Earth History*. Princeton: Princeton University Press, pp. 9–34.

——. 1986. Evolution and the triumph of homology, or why history matters. *American Scientist* 74:60–69.

——. 1987. *Time's Arrow, Time's Cycle*. Cambridge, Mass.: Harvard University Press.

Gould, S. J., and N. Eldredge. 1977. Punctuated equilibria: the tempo and mode of evolution reconsidered. *Paleobiology* 3:115–151.

Gould, S. J., and R. C. Lewontin. 1979. The spandrels of San Marco and the Panglossian paradigm: a critique of the adaptationist programme. *Proceedings of the Royal Society of London*. B 205:581–598.

Gowlett, J. A. J. 1984. Mental abilities of early man: a look at some hard evidence. In R. Foley, ed., *Hominid Evolution and Community Ecology*. New York: Academic Press, pp. 167–192.

Grayson, D. K. 1984. Explaining Pleistocene extinctions: thoughts on the structure of a debate. In P. S. Martin and R. G. Klein, eds., *Quaternary Extinctions: A Prehistoric Revolution*. Tucson: University of Arizona Press, pp. 807–823.

Grine, F. E., ed. 1988. *Evolutionary History of the "Robust" Australopithecines*. New York: Aldine de Gruyter.

Habgood, P. J. 1989. An investigation into the usefulness of a cladistic approach to the study of the origin of anatomically modern humans. *Human Evolution* 4:241–252.

Hammond, M. 1988. The shadow man paradigm in paleoanthropology. In G. W. Stocking, ed., *Bones, Bodies, Behavior*. Madison: University of Wisconsin Press, pp. 117–137.

Harrison, G. A., et al. 1988. *Human Biology*. 3d ed. Oxford: Oxford University Press.

Hassan, F. A. 1988. The Predynastic of Egypt. *Journal of World Prehistory* 2:136–183.

Higgs, E. S., ed. 1975. *Palaeoeconomy*. Cambridge: Cambridge University Press.

Hildebrand, M., et al., eds. 1985. *Functional Vertebrate Morphology*. Cambridge, Mass.: Belknap Press of Harvard University Press.

Hooton, E. A. 1926. Methods of racial analysis. *Science* 63:75–81.

Hunt, E. E. 1981. The old physical anthropology. *American Journal of Physical Anthropology*. 56:339–346.

Huxley, T. H. 1863. *Evidence as to Man's Place in Nature*. Reprint. Ann Arbor: University of Michigan Press, 1959.

Johnson, R. C., et al. 1981. Genetic interpretation of racial/ethnic differences in lactose absorption and tolerance: a review. *Human Biology* 53:1–14.

Kimbel, W. H. 1990. Species concepts and hominid variation in the East African Plio/Pleistocene. *American Journal of Physical Anthropology* 81:250.

Kingdon, J. 1988. What are face patterns and do they contribute to reproductive isolation in guenons? In A. Gautier-Hion et al., eds., *A Primate Radiation: Evolutionary Biology of the African Guenons*. Cambridge: Cambridge University Press, pp. 227–245.

Kleindienst, M. R. 1961. Variability within the late Acheulian assemblages in eastern Africa. *South African Archaeological Bulletin* 16:35–52.

Kuper, A. 1988. *The Invention of Primitive Society: Transformations of an Illusion*. New York: Routledge, Chapman and Hall.

Le Gros Clark, W. E. 1971. *The Antecedents of Man: An Introduction to the Evolution of the Primates*. 3d ed. Chicago: Quadrangle Books.

Levinton, J. 1988. *Genetics, Paleontology, and Macroevolution*. Cambridge: Cambridge University Press.

Lewin, R. 1986. New fossil upsets human family. *Science* 233:720–721.

Liem, K. F. 1989. Milton Hildebrand: architect of the rebirth of vertebrate morphology. *American Zoologist* 29:191–194.

Little, M. A. 1982. The development of ideas on human ecology and adaptation. In F. Spencer, ed., *A History of American Physical Anthropology, 1930–1980*. New York: Academic Press, pp. 405–433.

Loomis, W. F. 1967. Skin pigment regulation of vitamin D biosynthesis in man. *Science* 157:501–506.

Manly, B. F. 1985. *The Statistics of Natural Selection*. London: Chapman & Hall.

Martin, P. S. 1984. Pleistocene overkill: the global model. In P. S. Martin and R. G. Klein, eds., *Quaternary Extinctions: A Prehistoric Revolution*. Tucson: University of Arizona Press, pp. 354–403.

Martin, P. S., and R. G. Klein eds., 1984. *Quaternary Extinctions: A Prehistoric Revolution*. Tucson: University of Arizona Press.

Martin, R. D. 1990. *Primate Origins and Evolution: A Phylogenetic Reconstruction*. Princeton: Princeton University Press.

Mayr, E. 1983. How to carry out the adaptationist program? *Am. Naturalist* 121:324–334.

Mellars, P. 1989a. Major issues in the emergence of modern humans. *Current Anthropology* 30:349–385.

——. 1989b. Technological changes across the Middle-Upper Palaeolithic transition: economic, social and cognitive perspectives. In P. Mellars and C. Stringer, eds., *The Human Revolution*. Princeton: Princeton University Press, pp. 338–365.

Menozzi, P., et al. 1978. Synthetic maps of human gene frequencies in Europeans. *Science* 201:786–792.

Nelson, G. 1989. Species and taxa: Systematics and evolution. In D. Otte and J. A. Endler, eds., *Speciation and Its Consequences*. Sunderland, Mass.: Sinauer, pp. 60–81.

Nitecki, M. H., ed. 1984. *Extinctions*. Chicago: University of Chicago Press.

Oates, J. F. 1988. The distribution of *Cercopithecus* monkeys in West African forests. In A. Gautier-Hion et al., eds., *A Primate Radiation: Evolutionary Biology of the African Guenons*. Cambridge: Cambridge University Press, pp. 79–103.

Paterson, H. E. H. 1981. The continuing search for the unknown and unknowable: a critique of contemporary ideas on speciation. *South African Journal of Science* 77:113–119.

——. 1982. Perspective on speciation by reinforcement. *South African Journal of Science* 78:53–57.

——. 1986. Environment and species. *South African Journal of Science* 82:62–65.

Quevedo, W. C., T. B. Fitzpatrick, and K. Jimbow. 1985. Human skin color: origin, variation and significance. *Journal of Human Evolution* 14:43–56.

Renfrew, C. 1988. *Archaeology and Language: The Puzzle of Indo-European Origins*. New York: Cambridge University Press.

Renfrew, C. and J. F. Cherry, eds. 1986. *Peer-Polity Interaction and Sociopolitical Change*. Cambridge: Cambridge University Press.

Rightmire, G. P. 1986. Stasis in *Homo erectus* defended. *Paleobiology* 12:324–325.

Rose, K. D., and Bown, T. M. 1986. Gradual evolution and species discrimination in the fossil record. In K. Flanagan and J. A. Lillegraven, eds., *Vertebrates, Phylogeny, and Philosophy*. Contributions to Geology, University of Wyoming special paper no. 3, pp. 119–130.

Schwartz, J. H., I. Tattersall, and N. Eldredge. 1978. Phylogeny and classification of the primates revisited. *Yearbook of Physical Anthropology*. 21:95–133.

Simons, E. L. 1989. Human origins. *Science* 245:1343–1350.

Skrikumari, C. R., et al. 1987. Acuity of selective mechanisms operating on ABO, Rh and MN blood groups. *American Journal of Physical Anthropology*. 71:117–121.

Smith, F. H. 1984. Fossil hominids from the Upper Pleistocene of Central Europe and the origin of modern Europeans. In F. H. Smith and F. Spencer, eds., *The Origins of Modern Humans*. New York: Alan R. Liss, pp. 137–209.

Smuts, B. B., et al., eds. 1987. *Primate Societies*. Chicago: University of Chicago Press.

Sokal, R. R., and P. Menozzi. 1982. Spatial autocorrelation of HLA frequencies in Europe support demic diffusion of early farmers. *American Naturalist* 119:1–17.

Sokal, R. R., et al. 1990. Genetics and language in European populations. *American Naturalist* 135:157–175.

Spuhler, J. N. 1988. Evolution of mitochondrial DNA in monkeys, apes, and humans. *Yearbook of Physical Anthropology* 31:15–48.

Stanley, S. M. 1979. *Macroevolution: Pattern and Process*. San Francisco: Freeman.

Stanley, S. M. 1981. *The New Evolutionary Timetable: Fossils, Genes, and the Origin of Species*. New York: Basic Books.

——. 1982. Macroevolution and the fossil record. *Evolution* 36:460–473.

Stoneking, M., and R. L. Cann. 1989. African origin of human mitochondrial DNA. In P. Mellars and C. Stringer, eds., *The Human Revolution*. Princeton: Princeton University Press, pp. 17–30.

Strasser, E., and M. Dagosto, eds. 1988. *The Primate Postcranial Skeleton*. San Diego: Academic Press.

Stringer, C. B. 1984. Human evolution and biological adaptation in the Pleistocene. In R. Foley, ed., *Hominid Evolution and Community Ecology*. New York: Academic Press, pp. 55–83.

——. 1986. The credibility of *Homo habilis*. In B. Wood, L. Martin, and P. Andrews, eds., *Major Topics in Primate and Human Evolution*. Cambridge: Cambridge University Press, pp. 266–294.

——. 1989. The origin of early modern humans: a comparison of the European and non-European evidence. In P. Mellars and C. Stringer, eds., *The Human Revolution*. Princeton: Princeton University Press, pp. 232–244.

Stringer, C. B., and P. Andrews. 1988. Genetic and fossil evidence for the origin of modern humans. *Science* 239:1263–1268.

Stringer, C. B., J. J. Hublin, and B. Vandermeersch. 1984. The origin of anatomically modern humans in western Europe. In F. H. Smith and F. Spencer, eds., *The Origins of Modern Humans*. New York: Alan R. Liss, pp. 51–135.

Tattersall, I. 1982. *The Primates of Madagascar*. New York: Columbia University Press.

——. 1986. Species recognition in human paleontology. *Journal of Human Evolution* 15:165–175.

——. 1990. Recognizing hominid species in the late Pleistocene. *American Journal of Physical Anthropology* 81:306.

Turner, A., and A. Chamberlain. 1989. Speciation, morphological change and the status of African *Homo erectus*. *Journal of Human Evolution* 18:115–130.

Turner, T. R., et al. 1988. Population differentiation in *Cercopithecus* monkeys. In A. Gautier-Hion et al., eds., *A Primate Radiation: Evolutionary Biology of the African Guenons*. Cambridge: Cambridge University Press, pp. 140–149.

Wainscoat, J. S., et al. 1986. Evolutionary relationships of human populations from an analysis of nuclear DNA polymorphisms. *Nature* 319:491–493.

——. 1989. Geographic distribution of alpha- and beta-globin gene cluster polymorphisms. In P. Mellars and C. Stringer, eds., *The Human Revolution*. Princeton: Princeton University Press, pp. 31–38.

Walker, A., et al. 1986. 2.5-Myr *Australopithecus boisei* from west of Lake Turkana, Kenya. *Nature* 322:517–522.

Washburn, S. L. 1951a. The analysis of primate evolution with particular reference to the origin of man. *Cold Spring Harbor Symposia on Quantitative Biology* 15:67–77.

——. 1951b. The new physical anthropology. *Transactions of the New York Academy of Science*, ser. 2, 13:298–304.

——. 1953. The strategy of physical anthropology. In A. L. Kroeber, ed., *Anthropology Today*. Chicago: University of Chicago Press, pp. 714–727.

Wendorf, F., et al. 1985. Prehistoric settlements in the Nubian Desert. *American Scientist* 73:132–141.

Wilford, J. N. 1987. New fossil is forcing family tree revisions. *New York Times*, April 14, pp. C1–C2.

Winterhalder, B. 1980. Hominid paleoecology: the competitive exclusion principle and determinants of niche relationships. *Yearbook of Physical Anthropology*. 23:43–63.

——. 1981. Hominid paleoecology and competitive exclusion: limits to similarity, niche differentiation, and the effects of cultural behavior. *Yearbook of Physical Anthropology*. 24:101–121.

Wolpoff, M. H. 1984. Evolution in *Homo erectus*: the question of stasis. *Paleobiology* 10:389–406.

——. 1986. Stasis in the interpretation of evolution in *Homo erectus*: a reply to Rightmire. *Paleobiology* 12:325–328.

——. 1989. Multiregional evolution: the fossil alternative to Eden. In P. Mellars and C. Stringer, eds., *The Human Revolution*. Princeton: Princeton University Press, pp. 62–108.

Wood, B. A., and A. T. Chamberlain. 1986. *Australopithecus*: grade or clade? In B. Wood, L. Martin, and P. Andrews, eds., *Major Topics in Primate and Human Evolution*. Cambridge: Cambridge University Press, pp. 220–248.

Wood, B., L. Martin, and P. Andrews, eds., 1986. *Major Topics in Primate and Human Evolution*. Cambridge: Cambridge University Press.

Wynn, T., and Tierson, F. 1990. Regional comparison of the shapes of later Acheulean handaxes. *American Anthropologist* 192:73–84.

9

Periods and Question Marks in the Punctuated Evolution of Human Social Behavior

ALLAN MAZUR

Punctuated equilibrium has focused much attention on the processes of evolutionary change. This chapter examines the extent to which that theoretical framework can inform our understanding of the evolution of human social behavior. For reasons outlined below, it seems unlikely that, at this point, punctuationism can so contribute. Although it is possible to trace the physical evolution of our species, it is much more difficult to determine behavioral evolution. As long as this is the case, the issue of whether punctuated equilibrium can explain the evolution of human social behavior is moot.

The ability to date fossils has enabled paleontologists to reconstruct the broad evolutionary history of contemporary species, including humans. Relationships among living species that have been inferred from fossils and from anatomical comparisons can now be cross-checked by comparing genes across species. For example, the DNA that codes for hemoglobin in one species differs somewhat from the DNA that codes for hemoglobin in other species; presumably the greater the difference in hemoglobin DNA between two species, the more distantly they are related. On the basis of such molecular comparisons, chimpanzees appear to be the living animals most closely related to humans. It is further assumed that "neutral mutations" in DNA occur at a constant rate, which can be estimated from the fossil record, and therefore apparently humans and chimps had a common ancestor as recently as 5 million years ago (Wilson, 1985). There is still uncertainty about how the human line developed since then, but it seems to have passed through the following major types (Sherratt, 1980; Johanson and Edey, 1981; Gowlett, 1984).

Type	*Approximate Dates*
Australopithecus (various forms)	4 to 1 million years ago
Homo erectus	1.5 million to 300,000 years ago
Homo sapiens (early forms)	200,000 to 30,000 years ago
Homo sapiens (modern looking)	40,000 years ago to present

The dates shown suggest more abrupt changes than may be warranted because the sample of fossils is too small for us to be sure that the full era of each type is covered and because there are transitional specimens whose placement in one type rather than another is arbitrary.

Australopithicine hominids of various species evolved in Africa with ape-size brains but, unlike apes, walked upright and had bodies similar to those of humans. Large-brained *Homo erectus*—often regarded as the first true human—apparently evolved from Australopithicine stock, spreading through Africa, Europe, Asia, and Indonesia. Among the early *Homo sapiens* that evolved from *Homo erectus* are the famous Neanderthal people of Europe and the Middle East who were suddenly replaced by fully modern people about forty thousand years ago. It is not presently clear if modern people are descended from these Neanderthals or evolved separately from *Homo erectus* in another part of the world and then migrated into Europe.

In contrast to this empirically documented reconstruction of the physical evolution of humans, speculations about the evolution of social behavior are based on much flimsier grounds. Since behaviors do not fossilize, theorists are free to offer whatever reconstructions they like, virtually unconstrained by data that can falsify their claims (Mazur, 1983).

Anthropologists who have argued at length over the proper interpretation of a single molar will chuckle at my claim that physical evolution has firm empirical grounding. I recognize that such debates can be as empty of data as those in theology. That does not deny the firm support that modern methods of fossil dating and genetic comparisons of living species give to the big picture of species evolution—support that is lacking in reconstructions of the evolution of social behavior.

How does this speak to the implications of punctuated equilibrium? With so much controversy over the place of punctuation in physical evolution, which is relatively well documented, what can be said for social evolution, which is not? I confess at the outset that I will have little to say about punctuated equilibrium per se, for it is a refinement of evolutionary theory that requires knowledge beyond the crude understanding that we have of the rise of sociability. Indeed, there is little point to inquiries about punctuated equilibrium—whether short periods of change are preceded and followed by longer periods of relative stability—unless these periods are known with some certainty. Before asking if social evolution is punctuated, it is worth asking how we can know anything at all about social evolution.

Explaining human social behavior is at the heart of the discipline of sociology. If punctuated equilibrium theory is to be relevant for sociology, it must cast light on the evolution of social behavior. If sociologists were to gain a firm hold on why human behavior evolved as it has, then they would be in a much better position to understand why such behavior is manifest as it is. Before one can ascertain the applicability of punctuational theory, one must know how behavior has actually evolved. And this is precisely the obstacle facing evolutionary-minded sociologists who might wish to test the validity of the punctuational model: there is no clear record available to analyze.

I believe that two empirical methods show some promise of documenting the evolution of human social behavior. One compares observable behaviors across living species. This is analogous to comparing genes across species but is less satisfactory because behaviors do not always reflect genetic inheritance. The second method uses prehistoric cultural artifacts that can be dated. This is analogous to fossil dating but is again less satisfactory because artifacts are very indirect indicators of behavior, and they exist for only a few behaviors of a few species. The possibilities and problems of these two methods will be discussed in turn.

Comparative Primate Behavior

The fifty-five genera of living primates can be more or less ranked according to their physical similarity to humans, based on such diverse criteria as gross morphology, ontogeny of the nervous system, and structure of protein and DNA molecules. The apes, especially chimpanzees and gorillas, are the closest of our living relatives, then come Old World monkeys, then New World monkeys, and finally prosimians. This progression is not an evolutionary sequence; we did not evolve from chimpanzees, and chimps did not evolve from present-day monkeys. Each living species is the culmination of its own particular line of evolution. Still, this quasi-evolutionary series seems to approximate the evolutionary lineage leading to humans, and fossil evidence indicates that the earliest primate looked very much like a tree shrew, which I will place at the far end of the series. Comparative anatomists have used such series to study "evolution" of body tissue that has long since disintegrated from our true evolutionary ancestors. More important for our purposes, the behaviors of these living species can be observed to find regularities as one moves along the progression from tree shrews, through monkeys and apes, to humans.

Since most primates are not well studied, as a practical matter I represent the whole series by a small number of representative genera: tree shrew (*Tupaia*), lemur (*Lemur*, a prosimian), squirrel monkey (*Saimiri*, a New World monkey), baboon and macaque (*Papio* and *Macaca*, both Old World monkeys), chimp and gorilla (*Pan* and *Gorilla*, both apes), and human (*Homo sapiens*). Moving through this series, we see increasing physical similarity to humans, including a continuous lengthening of the various developmental periods such as gestation time and life span (Mazur, 1981).

Some features of mother-infant interaction are listed in Table 9.1 for all the genera of the primate series for which data are available. Features often thought of as human appear throughout the series or develop as we move closer to humans. Apparently, many of the nonverbal aspects of human mother-infant interaction evolved along with the physical character of the species, appearing by the ape grade or lower. These features, taken together, represent a trend toward greater dependence of the infant on its mother, a longer period of dependence, and greater caretaking by the mother.

The infant's interactions with other members of its community also have an evolved character. Among the higher primates, at least, older siblings will care for the infant. Females outside of the nuclear family—usually old juveniles or young adults without children of their own—seek out infants and play with them; the mother often lets another female take over her caretaking functions for extended periods. Though none of the primates in the series (except humans) include the father in the nuclear family, adult males of the higher primates often show a benign interest in infants and, among the apes, are permissive and playful with the young of the group. At the baboon-macaque grade and higher, infants that are orphaned are adopted and cared for by an older

	shrew	Lemur	monkey	and macaque (M)	and gorilla (G)	Human
Infant-mother interactions						
How does mother carry new infant?	Does not carry	Infant clings, or mother carries it in mouth	Infant clings to mother	Infant clings to mother; she supports it with hand	Mother supports with arms	In arms
Does mother groom infant?	No	Yes	Yes	Yes	Yes	Yes
Does mother play with infant?	No	No	No	Yes	Yes	Yes
Do isolated infants show psychic or behavioral abnormalities?	No	Yes		M: Yes	Yes	Yes
Age when infant first leaves mother on its own		2–4 weeks	3 weeks	M: 2 weeks	4–6 months	7–8 months
Age when weaning is complete	1–2 months	3–6 months	6 months	1–2 years	3–6 years	0–5 years
Cessation of special mother-child bond	2 months	<7 months		>2 years	Persists	Persists
Infant-other interactions						
Do older sibs care for infant?		No		M: Yes	C: Yes	Yes
Do females other than mother and sibs seek out and care for infant?		Yes	Yes	Yes	C: Yes G: Seldom	Yes
Do adult males interact (nonaggressively) with infants?	No	Yes	No	Yes	Yes	Yes
Are orphans adopted by older group members?				M: Yes	Yes	Yes

Source: Reprinted, by permission of Sage Publications, Inc., from Allan Mazur, "Biosociology," in *The State of Sociology*, ed. James Short, Jr. (Beverly Hills, Calif.: Sage Publications, 1981), © 1981 by Sage Publications, Inc.

group member, often a sibling. Thus community support for infants, as a backup to the mother, evolved well before the human species.

As the human infant grows into a juvenile, it follows play patterns that are typical in many ways to those of the higher primates (Symons, 1978). The formation of juvenile playgrounds (when other juveniles are available), separate from younger or older members of the community, is one such pattern. Another is that the play of young males is more rough-and-tumble than female play, with more chasing and wrestling. Perhaps this is why primates' play groups tend to be segregated by sex (though this may not be true of apes). The play of lower primates is primarily locomoter with stylized chasing and biting but little object manipulation. Play becomes less stylized and slightly more object-oriented through the monkeys and increasingly so in apes and humans, which play complex games.

By the time an animal reaches its juvenile years, it has already formed relationships that will extend into adult life. Not surprisingly, the most important of these is the relationship with its mother, which, in the higher primates, persists beyond the years of dependency. The juvenile's status in its play group, its ability to coerce or direct the activity of its peers, is correlated with the status of its mother in the community, at least among macaques and probably chimps as well; and this in turn is related to its status when it becomes an adult (Mazur, 1973). In all these ways and others, the characteristics of human behavior seem to have evolved in our primate ancestors.

There are empty cells in Table 9.1 where data are missing. These allow tests of the progressive nature of the primate series because a value for each cell is easily predicted by interpolation. If predictions from such tables are correct, the method should gain credibility. Also, as the behaviors of additional primate species are well characterized, they too can be placed in the table as a test of consistency (although there should be no expectation that all fifty-five genera of living primates will form a fully consistent linear pattern of behavior). Extinct *Australopithicus* and early *Homo* types fit into the table between the chimp-gorilla and human columns so their behaviors may also be inferred by interpolation.

It has been argued that humans have evolved a territorial nature, defending bounding plots of land against intruders (Ardrey, 1966; Van den Berghe, 1974). This claim receives no support from the quasi-evolutionary series. Prosimians, the lowest primates in the series, do defend bounded plots of land, but moving "upward," through monkeys and apes, territoriality is usually absent, although there is some ambiguity in the case of chimps (Hamburg and McCown, 1979).

I doubt that we will ever make an ironclad case for an evolutionary explanation of any human social behavior, but two criteria, taken together, appear to represent a plausible test. The first criterion is the one just discussed: the behavior, or some rudimentary form of it, must appear in the quasi-evolutionary series as we move from tree shrew toward humans. The second criterion requires that any evolved behavior be pancultural—that is, it should be common among human societies. (Since extreme environmental influences can modify any characteristic, even physical features, as in the Chinese practice of binding women's feet, we need not require that an evolved behavior be observable in every single instance.) The features of child rearing and juvenile play in Table 9.1 are pancultural. Territoriality is absent from many human societies that are or were nomadic.

I do not wish to overstate the virtues of this line of reasoning; it has serious flaws. First, we can visualize behaviors that satisfy both the pancultural and the quasi-evolutionary series criteria but are not biologically evolved characteristics in any profound sense. Consider toolmaking, for example, which was long thought to be solely a human activity until chimps were seen modifying natural objects to improve their efficiency as tools (Lawick-Goodall, 1968). Obviously chimps and humans have evolved the mental and physical abilities necessary to make tools, but is toolmaking an evolved behavior? Perhaps a common ancestor of chimps and humans learned to make tools by accident, then taught its children, who passed on the art from generation to generation. Some of this creature's present-day descendants are chimps and some are humans, all learning toolmaking. In this case, toolmaking might be better explained as an instance of cultural diffusion than of biologically evolved behavior.

On the other side, humans may have true biologically evolved behaviors that fail the quasi-evolutionary series criterion. Any characteristic that evolved after the human line split from the ape line would not show up among other primates in the series. By the test advocated here, it would be erroneously rejected as an evolved behavior.

In sum, if we are willing to accept some errors and ambiguity, we may reasonably accept or reject an evolutionary explanation of a particular human behavior if it appears or emerges in rudimentary form as we move along the quasi-evolutionary series from tree shrew toward humans, and it is common across human societies. Though this reasoning has faults, I submit that it is sounder than other forms of evolutionary method now current. I will examine two of these briefly.

First, there is reasoning by analogy between human behavior and the behavior of one or more selected species scattered throughout the animal kingdom. For example, some writers point to the brief "imprinting" period when baby birds are especially prone to follow their mother. The claim is made that human babies have a similar critical period which is crucial for the formation of later emotional attachments (Mazur and Robertson, 1972).

It is difficult to interpret the bird-human comparison. One can always form analogies between any two species, but these may not be meaningful if underlying mechanisms are very different, which is likely in species that are widely separated on the phylogenetic tree. The difficulty is partly overcome by focusing on our phylogenetically "near" relatives, the primates, but there are still pitfalls. We cannot restrict our analogies to a single nonhuman species even if it is a primate. For example, the gibbon is a territorial ape, fairly closely related to humans. Can we conclude, by analogy, that human defense of a section of homeland has the same noncultural basis as gibbon territoriality? The gibbon, like every species, has evolved its own idiosyncratic features. By looking only at gibbons, we cannot tell if territoriality is one of its idiosyncrasies or a generally evolved pattern among the higher primates. In fact, territoriality turns out to be a gibbon idiosyncrasy because most other higher primates are not territorial. Many of the disadvantages of reasoning by analogy between humans and distantly related species, or by analogy between humans and one other species, are eliminated when we use the quasi-evolutionary series of primates. The argument for imprinting in human infants would be more impressive if imprinting could be demonstrated in the several primates most closely related to humans.

There is another weak form of evolutionary reasoning which I shall call justification via natural selection. If one accepts the Darwinian view of natural selection, and if one observes that a species has some particular physical or behavioral characteristic, then one may assume that the characteristic has some relative advantage that led to its selection. One then "explains" this characteristic by providing a plausible argument telling why it is adaptive and functional. Usually a reasonable selection argument can be constructed to explain any observation, real or imagined. Dark-skinned people tend to live in climates with strong sunshine, while lighter-skinned people live in cooler or cloudy climates. This may be explained as an adaptation to ultraviolet solar radiation or optimal vitamin D production. If Africans had lighter skin than Europeans, that could be explained by the fact that light skin reflects larger amounts of heat than dark skin, making light skin adaptive in a hot, sunny climate. Such arguments, often used to explain social characteristics, are wholly speculative and essentially untestable (Mazur, 1983).

Although the quasi-evolutionary series is a superior method in some ways, it is unfortunately deficient from the standpoint of punctuated equilibrium. Since the series does not show true historic time, it gives no sense of whether an evolved characteristic emerged quickly or gradually. The next method to be discussed, which uses dated cultural artifacts, does allow inferences about the rapidity of evolution.

Inferences from Dated Cultural Artifacts

The oldest well-documented stone artifacts date from about 2 million years ago in Africa. These "Oldowan tools"—named after the type-site of Olduvai Gorge—are of two kinds: blocks of stone from which flakes have been removed and the flakes themselves; neither are easily distinguishable from naturally fractured stones (Figure 9.1). Judging from the dating and location of Oldowan tool sites, the manufacturers were advanced *Australopithicus* or primitive *Homo* types.

A new tool kit, more varied and with finer workmanship, appears about 1.5 million years ago. Called "Acheulean" after a French type-site, these tools include choppers, scrapers, awls, hammerstones, and a characteristic two-faced hand ax that is the hallmark of the kit (Figure 9.1). Acheulean tools are almost certainly the work of *Homo erectus* because artifacts and fossils have been found in close association. They form part of a culture that spread over much of the Old World, which included the use of fire, construction of dwellings with hearth places, and big game hunting.

The early tool cultures were remarkably stable over long periods of time. The constancy of the Acheulean hand-ax tradition has been especially noted, for hand axes that have been found at sites widely separated in distance and across a million years of *Homo erectus*'s existence look very similar to one another, their uniformity more striking than regional differences (Johanson and Edey, 1981; Roe, 1980). Roughly two hundred thousand years ago, when *Homo erectus* left the earth to their *sapiens* successors, they took their cultural stability with them. The newer Neanderthal culture was richer and more variable, leaving

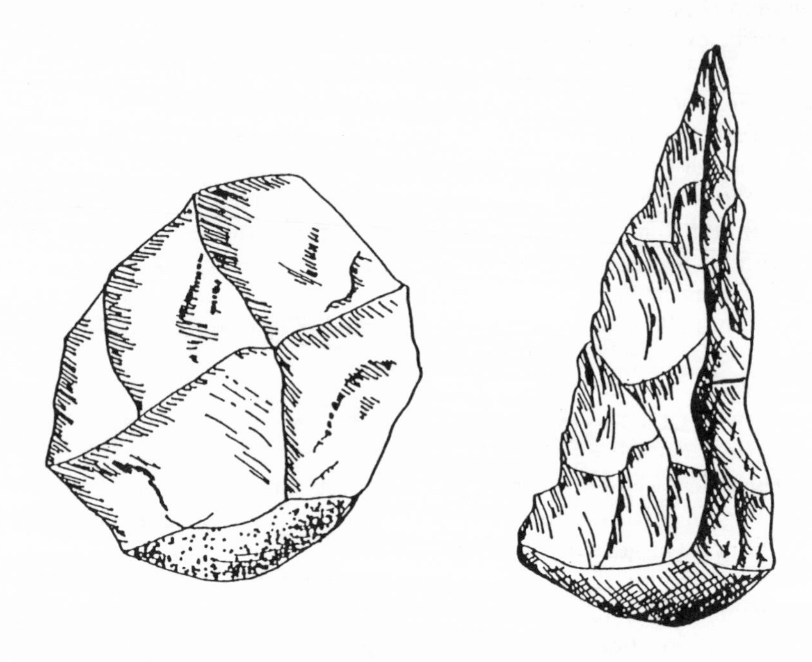

Figure 9.1. Oldowan chopper (left) and Acheulean hand ax (right). (Rendered by Rachel Mazur from specimens in Gowlett, 1984, and Roe, 1980.)

decorated artifacts, diverse collections of tools, and even signs of ritual burials where the deceased was arranged in a grave accompanied by tools, meat, and perhaps flowers. When the Neanderthals were replaced by modern-looking people about forty thousand years ago, there was a burst of new cultures, now varying from one locale to another and from generation to generation (Klein, 1980). Well-made tools appear not only in stone but also bone, antler, or ivory. Artifacts are engraved with geometric or naturalistic designs. There is sewn clothing—sometimes with bead decorations, projectile weapons, clay figurines, and cave paintings that are awesome even by modern standards.

Although the *Homo sapiens* era introduced abrupt change compared to what had gone before, it cannot be characterized as a punctuation in the sense of a short period of rapid movement sandwiched between two longer periods of stability. There has never been a follow-up period of stasis comparable to that which had gone before—at least, not yet. Rather, there was a second distinct period of dramatic change superimposed on the first one.

Beginning about ten thousand years ago, hunter-gatherer society—the normal hominid mode of life for millions of years—transformed into an agrarian mode, settled in permanent communities supported by nearby fields of grain and by animal husbandry. The animals became sources of power and transport as well as food. The populations of growing cities became differentiated into separate classes, one better off than another, with some form of king holding control, partly through hereditary right and partly through the strength of military alliances. Within several thousand years, some of these societies became "civilizations" with writing, calendars, astronomical observation, mathematics, monumental architecture, planned ceremonial and religious centers, specialization in arts and crafts, metallurgy, and intensive irrigation projects.

In one of the most remarkable coincidences of history, this transformation to agrarian civilization occurred not once but in at least six places at nearly the same time: Mesopotamia, Egypt, India, China, Mexico, and Peru (Sherratt, 1980; Gowlett, 1984). These "pristine states" seem to have arisen largely independent of one another because development in each area can be traced back to primitive indigenous roots. The coincidence is striking enough without exaggerating it. To say that all appeared at "nearly the same time" really means within five thousand years. Still, considering that nothing like this had occurred during the prior two hundred thousand years of *Homo sapiens* existence, not to mention 1.5 million years of *Homo erectus*, then their timing is very close.

These six were not the only early agrarian civilizations, but they are the best known and probably the earliest in their respective parts of the world, and all affected the cultural development of their peripheral regions. Although there are broad similarities in the development of these civilizations, they do not fit an invariant mold, and some differences are as striking as the similarities. The pristine states of Asia all occur in major river valleys, which is not true of America. Egypt was much less urbanized than Mesopotamia. The appearance of any one cultural element is variable. For example, the use of animals and the wheel for transportation was important in the Old World civilizations but not in America; metallurgy appeared in some precivilized cultures but not in all civilized ones. Thus the transformation did not take exactly the same form in each place. Nonetheless, one can hardly fail to be impressed by the degree of similarity and simultaneity that did occur and to wonder why it happened that way.

Conclusion

Modern techniques for dating fossils and for comparing genes of living species give a firm basis upon which to reconstruct the physical evolution of the human line. No comparably powerful empirical framework exists for reconstructing the evolution of human social behavior, but helpful clues come from behavioral comparisons across living primates and dated cultural artifacts.

Living primate species can be approximately placed along a gradient of more or less physical similarity to humans. Many behaviors that are common among humans occur in all other primates too, or they emerge among the species that are physically most similar to humans. Such behaviors are likely to have been inherited from our primate ancestors.

Although this quasi-evolutionary primate series helps in identifying those human behaviors with primate roots, it does not show the emergence of behaviors in real historical time and therefore has little ability to identify episodes of punctuation. Furthermore, the method is biased to favor behaviors that occur continuously across the primates or that emerge along the series in a smooth way. Suddenly appearing characteristics from the recent human past would be overlooked here as evolved features, the overall effect being to understate whatever degree of punctuation actually exists.

Dated artifacts are tied to real historic time, thus permitting the identification of abrupt or punctuated changes in the archaeological record, but there are other limitations. Artifacts become plentiful enough to provide clear trends only in the fairly recent past, and then they are associated with only a few behaviors and very indirectly at that. Nonetheless, the record of stone tools indicates that the life-style of *Homo erectus*—at least the stone toolmaking part of it—was remarkably stable for a million years. In contrast, stone tools associated with the cultures of *Homo sapiens*, within the past two hundred thousand years, show considerably more change over time and across regions. We seem justified in regarding this change from the Acheulean tool tradition to more modern designs as abrupt relative to the long period of stability that had gone before. But it was not punctuated in the sense of restabilizing, for there was no return to the prior level of constancy. Instead, cultural change continued to be rapid until about ten thousand years ago, when there was another first-order transition, this time from the hunter-gatherer life-style to agrarian civilization. This transformation apparently occurred independently in several different regions of the world at nearly the same time.

Perhaps these abrupt changes were caused by the improved mental abilities of the new types of people who enter the fossil record near the periods of transition. The articulation and transmission of culture is closely tied to language so it is tempting to guess that the Neanderthal people had more linguistic ability than *Homo erectus* and that there was a further jump in language capacity when the Neanderthals were replaced by fully modern-looking people about forty thousand years ago. The agrarian transformation followed this last change in the human stock by "only" thirty thousand years or so. Of course, one can argue the other side: since it took that long, the specific abilities of modern humans are insufficient to explain the transformation. Those who favor an environmental explanation emphasize that a period of warming climate and melting glaciers just preceded the rise of agriculture.

During the twentieth century the archaeological record of prehistory has been filled in to a degree that would have seemed implausible a century ago. What has emerged is a tentative picture of cultural development punctuated by at least two abrupt periods of change—three if one includes the recent transformation to industrialized society. Archaeological work of the next few decades should confirm this picture or show its inadequacy. If confirmed, then we have a description that begs to be explained. What caused these sudden changes? Are they the result of new kinds of people arriving with enhanced abilities, or are the explanations to be sought in environmental causes? My guess is that answers to these questions will remain at the level of untestable speculation, where they are now, unless some new empirical method that is not yet apparent can be found for probing deeper into social evolution than we can now.

As long as this is so, it is premature to ask what the relevance of punctuated equilibrium theory is for the sociologist in order to explore his or her domain. The data base is insufficient to test the theory.

References

Ardrey, R. 1966. *The Territorial Imperative*. New York: Atheneum.
Gowlett, J. 1984. *Ascent to Civilization*. New York: Knopf.

Hamburg, D., and E. McCown, eds. 1979. *The Great Apes*. Menlo Park, Calif.: Benjamin/Cummings.

Johanson, D., and M. Edey. 1981. *Lucy*. New York: Simon and Schuster.

Klein, R. 1980. Later Pleistoncene hunters. In A. Sherratt, ed., *The Cambridge Encyclopedia of Archaeology*. New York: Cambridge University Press, pp. 87–95.

Lawick-Goodall, J. van. 1968. The behavior of free-living chimpanzees of the Gombe Stream Reserve. *Animal Behavior Monographs* 1:161–311.

Mazur, A. 1973. A cross-species comparison of status in small established groups. *American Sociological Review* 38:513–530.

——. 1981. Biosociology. In J. Short, Jr., ed., *The State of Sociology*. Beverly Hills, Calif.: Sage, pp. 141–160.

——. 1983. Problems of testing inclusive fitness claims among humans, with an example on sibship. *Ethology and Sociobiology* 4:225–229.

Mazur, A., and L. Robertson. 1972. *Biology and Social Behavior*. New York: Free Press.

Roe, D. 1980. The handaxe makers. In A. Sherratt, ed., *The Cambridge Encyclopedia of Archaeology*. New York: Cambridge University Press, pp. 71–78.

Sherratt, A., ed. 1980. *The Cambridge Encyclopedia of Archaeology*. New York: Cambridge University Press.

Symons, D. 1978. *Play and Aggression*. New York: Columbia University Press.

Van den Berghe, P. 1974. Bringing beasts back in: toward a biosocial theory of aggression. *American Sociological Review* 39:777–788.

Wilson, A. 1985. The molecular basis of evolution. *Scientific American* 253 (October): 164–173.

10

Evolutionary Controversy and Psychobiology

Brian A. Gladue

One of my favorite passages from the scientific literature comes from the sociobiologist and evolutionary theoretician Richard Dawkins, whose opening lines in *The Selfish Gene* declare: "Intelligent life on a planet comes of age when it first works out the reason for its own existence. If superior creatures from space ever visit earth, the first question they will ask, in order to assess the level of our civilization, is: 'Have they discovered evolution yet?'" (Dawkins, 1976: 1).

Dawkins adds that although the theory of evolution is not open to doubt, there are numerous altercations over the interpretations and implications of what he calls Darwin's Revolution. Most scholars and scientists of all disciplines agree that evolution is here to stay, but there is less agreement about the application of evolutionary theory to fields other than anatomy, physiology, or zoology. In fact, the application of evolutionary principles of biology to behavior, as in the discipline known as sociobiology, created the greatest din of all in contemporary scholarly thought. Can social structure, political systems, selfishness, and altruism be understood or examined from a biological perspective? Dawkins says yes. Others agree, as do I.

Progress in acceptance of evolutionary theory and its incorporation into various disciplines has brought fresh debate, even controversy. This time, the argument is within the framework of evolution itself. In simple terms, this new argument draws distinctions about the scheme or mode of evolutionary change: Is it gradual (the traditional view) or "punctuated"? The latter viewpoint holds that evolutionary change, far from gradual, is characterized by local isolated species in which a sort of genetic revolution occurs, thus contributing to species change. (For a more thorough discussion of punctuated equilibrium see Eldredge and Gould, 1972; Gould and Eldredge, 1977; Eldredge, 1985; and chapters in this volume.)

Without going into detail about the merits of each viewpoint regarding the mechanisms and flow of evolution (dealt with more appropriately by experts elsewhere in this volume), it is safe to conclude that the debate is sincere and heated. The question for interdisciplinary scholars and the thrust of the present essay is, How does this debate change the way we look at evolution, and, more specifically, do we need to change the way we apply sociobiological and evolutionary principles to an understanding of human psychology? Need we rethink our use of evolutionary biology in explaining human behavior because the biologists are rethinking evolutionary mechanisms? In short, does the punctuated versus gradualism debate mean anything to behavioral and social scientists?

At the risk of irritating many colleagues who are very worried about the outcome of this evolutionary debate, I must state at the outset that I believe that the answer to each of the preceding questions is simple: don't lose any sleep over this problem.

Evolutionary Theory as It Applies to Psychology

In his 1975 text *Animal Behavior: An Evolutionary Approach*, John Alcock presented two heuristic boxes that behavioral biologists could use to sort out their clutter of explanatory questions. The "how" questions are concerned with the immediate or proximal causes of behavior. The physiology and endocrinology of competition and aggression, impact of biological rhythms on mood and decision making, influences of drugs, substances, and hormones on affect and impulse control, along with the effects of physical and biological stress on leadership are some common examples of proximal causation of human behavior.

The other inquiries scientists have regarding behavior, the "why" questions, are concerned with evolutionary or ultimate reasons why an animal does something. Within this ultimate causation there are two basic "reasons" for an animal's behavior: the ecological significance of the behavior (in terms of individual survival or fitness) and the evolutionary history of the populations from which the genes of that animal are ultimately derived. Ultimate explanations for human behavior typically follow a sociobiological model in which behavior expressed today is the culmination of forces of natural selection and adaptation upon ancestral humans, perhaps even earlier manifestations of human species.

Alcock's distinction between proximal and ultimate causations has been a powerful tool in psychobiological thought in enabling us to sort out testable and answerable questions from speculations and opinions. By employing proximal ("real time," "how") approaches to behavior, investigators have risen beyond "commonsense" schemes of explaining behavior and have used real hypotheses tested against a real framework of biology: How does this happen? Further, by taking the next step of examining why this should happen, research on behavior can enter the important predictive stage, in which an idea is evaluated according to how well the proximal explanation meshes within an overall biological-evolutionary framework.

The usefulness and applicability of evolutionary biological theories and principles to human behavior continues to be demonstrated by a variety of ethologists, psychiatrists, and psychobiologists attempting to understand the origins and mechanisms of behavior. A series of reports in a recent special issue of *Ethology and Sociobiology* addresses the value of combining psychobiological and evolutionary perspectives to the study of psychiatric populations. These interdisciplinary investigators apply ethological principles and biological techniques to studies of child psychiatric disorders (Dienske, Sanders-Woudstra, and de Jonge, 1987), the adaptiveness of anorexia nervosa related to variations in reproductive functioning (Surbey, 1987), panic disorder and fear of open places and crowds (agarophobia: Nesse, 1987), as well as to more severe human conditions such as depression (Sloman and Price, 1987) and schizophrenia (Pitman et al., 1987). Throughout, such a psychobiological approach necessitates combining the proximal understanding of the variations found in human behavior with a consideration of the ultimate origins for diversity.

A Psychological Example: Human Sexual Diversity

Of particular interest to many scholars, including myself, is the issue of what proximal and ultimate factors are associated with human reproductive behavior. Essentially, what are the mechanisms as well as the origins of human sexual diversity, especially variability in something as fundamental as preference in sexual partners, that is, sexual orientation?

The prevalence of variations in human sexual orientation, homosexuality and bisexuality, in the face of extreme cultural diversity, intolerance, and repression, is worth noting. Homosexuality is universal, occurring in all known societies, ancient as well as current (Whitam and Mathy, 1986; Whitam and Zent, 1984). Such consistency, especially when repeated efforts at explaining the diversity in sexual orientation in social-learning or psychoanalytic terms have failed, invites the notion of biological causation. And though some analysts still insist that a biological, even evolutionary, explanation for homosexuality is impossible because, as a nonprocreative reproductive strategy, same-gender sexual preferences should die out, a significant proportion of every known human society has been, is, and probably will be homosexual. Our knowledge about the development of either the core concept or subtle varieties of heterosexuality is only recently expanding (Davis and Whitten, 1987). In any case, the psychobiology of human sexuality is an area of continuing interest and importance. It is also an area best studied using both perspectives of evolutionary biology: proximal as well as ultimate causation factors.

Considerations of the origins or ultimate factors in sexual diversity have homosexuality as an expression of heterosexual reproductive strategies. As proposed by Donald Symons (1979), most if not all instances of homosexual behavior are an expression of heterosexual strategies evolved to maximize inclusive fitness. An interesting expansion of this argument by G. G. Gallup and S. D. Suarez (1983) posits that homosexuality is a by-product of selection for optimal reproductive strategies. In their scheme, homosexuality is a consequence of heterosexual frustration at acquiring opposite-gender mates or a dissatisfaction with such mates on the basis of biologically influenced gender differences in behavior. The general theme emerges that homosexual or lesbian orientation is not a personal choice or learned thing but a nonconscious behavioral outcome influenced by reproductive selection pressures acting differentially on males and females. In any such model, one has to explain variance, not only over time (as evolutionary biologists must), but among groups of individuals. Here is where proximal approaches are appropriate and often essential.

Numerous studies have attempted to identify biological and/or neural factors influential in the development of psychosexuality. These studies show that there may be, for some men and women, hormonal factors acting upon the developing body and nervous system before or shortly after birth that may predispose the individual toward same-gender sexual orientation (Gladue, Green, and Hellman, 1984; Doerner, 1985; Ehrhardt et al., 1985; Ellis and Ames, 1987; Gladue, 1987a). Other studies demonstrate chromosomal and heritable familial associations with atypical psychosexual development (Money, 1984, 1986; Money, Schwartz, and Lewis, 1984; Pillard and Weinrich, 1986). Although this evidence may seem far from overwhelming or clearly convincing, as in any newly developed field, it generates additional and varied inquiries. The idea that sexuality and sexual orientation have, at least in part, biogenetic factors instrumental in development and expression is one proximal approach to understanding human behavior.

Throughout the search for proximal factors associated with sexual development, researchers note the importance of multiple factors such as social interactions, learning, and cognitive development. Any complete understanding of erotic orientation must take into account the numerous direct (hormonal, genetic, chemical) and indirect (experience, social interactions) inputs to the nervous system that eventually result in an organized psychosexual framework. The latest addition to these studies considers sexuality to be an aspect of gender-related neurological functioning. Recent findings from my laboratory suggest differences in neuropsychological and cognitive functioning in heterosexual and homosexual men and women (Gladue et al., 1990).

Since gender differences in certain brain functions exist (algebraic reasoning, mental rotation and other visio-spatial tasks, verbal skills) and since it appears that fundamental gender differences exist for meaning in erotic or sexual images (Gladue, 1987b; Gladue and Delaney, 1990), the relationships of these differences to erotic functioning may be relevant to understanding the biological and neurological bases of sexuality itself. Simply stated, biological differences in gender may account for many behavioral and neurocognitive differences in general psychosexual development. Like language and intelligence, which require a deep neurobiological structure, psychosexual diversity may, at its core, have essential neurobiological elements. And like differences in language skills, differences in individual neurobiological functioning may account for variation in core psychosexuality and expression.

Together, proximal and ultimate perspectives result in the following psychobiological explanation for human sexual diversity. Diversity in sexual functioning and orientation in humans may result from the same evolutionary pressures and selective forces that have shaped and developed the more general neurobiopsychological variability found among members of the human species. As one scholar noted, the emotional, behavioral, intellectual, and cognitive components of being human have been modified over eons and so also psychosexual aspects of being human must have been (Mihalik, 1988). Of course, a major way of examining and testing this evolutionary argument is through the proximal approach.

All these studies have in common the use of proximal technique investigating "how" questions in the shadow of the greater Darwinian model of evolutionary "why." It is only temporarily satisfying to find out that hormones influence behavior. We all want to know why these things happen and why we have these findings and what to expect (i.e., predict). It is in this regard that the biological-evolutionary perspective has great importance for the study of human behavior. We can make predictions and objectively test them out, and each of us can do this by relying on the same unifying theme of Darwinian evolution. Though we speak with slightly different jargon and specialty interests, we all take our cues from the same school of thought: biological theory.

Why, then, do we belabor this point? All biologically oriented behavioral scientists are well aware of the value and impact of evolutionary thought in their discipline. Why retread familiar ground? Given the ensuing debate among evolutionary biologists over gradualism versus punctuationism, I wish to remind such scientists that, quite simply, their fight is not ours. We can watch and appreciate the scholarly jousting and often acrimonious rejoinders. But finally, in my opinion, the outcome of such internecine intellectual struggling among paleobiologists will not affect psychobiology.

Why the Great Debate Should Not Affect Psychobiology

Whether the evolution of traits, characteristics, and species occurred gradually (as in traditional gradualism) or in the alternative mode called punctuated equilibrium (so called because long periods of evolutionary inactivity—equilibrium—are interrupted—punctuated—by major new appearances of characteristics, thus creating new subspecies or even species) is important for the theory of evolution per se. Debate on this issue is clearly of importance for those interested in unraveling and interpreting a quirky and often incomplete historical record of life on earth. But this wrangling over mechanisms of evolution has little to do with the day-to-day practical work of psychobiological research and is perhaps even less important for theoreticians working with a grand evolutionary model of human social behavior.

There are two reasons for not being concerned about which scheme of evolution is the "better" one. The most important, practically speaking, is the time frame consideration. Second, evolutionary arguments feeding psychobiological theory and research activities are useful only when correctly applied as biological (i.e., genetic) systems. The moment evolution is used in a metaphorical sense, it ceases to serve any useful empirical function. In a sense, then, arguments over which scheme of evolution is dominant should only trouble biopolitical scientists who don't really understand biology and who are embracing biological principles merely for their metaphorical or heuristic value. A consideration of these two points follows.

Time Frame and Biological Factors in Evolution

How old is the human species, specifically *Homo sapiens sapiens*? When and how long did divergence from *Homo erectus* occur? What time scale is involved in human speciation overall? Certainly these long-standing questions continue to defy clean consensual answers. In addition to traditional physical anthropological processes for dating hominids, a new array of genetic and biochemical techniques is beginning to focus the time frame for human evolution. (For a readable overview of these techniques and their application to human evolution, see Cann, 1987.) Recent reports involving tracing genetic lineage via mitochondrial DNA of females suggests that "the first members of anatomically modern humans, *Homo sapiens sapiens*, lived somewhere between 100,000 and 200,000 years ago" (Lewin, 1987:24). Debate surrounding the precision or accuracy of these techniques results in a generally accepted time range for the appearance of common human ancestors at somewhere between "at least 50,000 but less than 500,000 years ago" (ibid.:24). Tentative interpretations of DNA data combined with the fossil record place "the transformation of archaic to anatomically modern forms of *Homo sapiens* in Africa, about 10,000 to 140,000 years ago and that all present day humans are descendants of that African population" (ibid.:26).

At the very least, then, contemporary evidence argues that modern humans were around prehistorically. One can assume that whatever human social interactions are genetically based and characteristic of the species had already been in place well before the advent of rapid cultural change as recorded in the past ten thousand years. To argue that humans, as a species, have undergone rapid natural evolution since then is to miss the point of genetic change, evolution, and speciation. If *Homo sapiens* was established by one hundred thousand years ago, we are it. All subsequent changes are minor and certainly do not alter that basic human we call our species. (This is not to say that variations among members of the species do not abound. Obviously, the varieties of human races and individuals testify to the flexibility, adaptability, and change possible in organisms under pressure from environmental and ecological forces. But the main point remains that as a species we are relatively undifferentiated from our ancestors, and as a species, it really does not matter whether we got here via gradualism or punctuated equilibrium.)

As Steven Stanley has argued (1981:206):

> The punctuational model of evolution implies a less significant impact of our culture on our biological evolution than might be envisioned by the gradualist. The fossil record shows no evidence of human evolution in Western Europe during the entire forty thousand years of our species' existence, and this is no surprise. What rate of evolution can we expect when the lineage that we call *Homo erectus* survived for upwards of a million years without altering enough for us, in retrospect, to change its name? If we restricted our analysis to major, genetically determined adaptive shifts, the punctuational model would imply that our species essentially stopped evolving the moment it was born! Fine tuning has probably taken place, and discrete races have certainly been added, but in Western Europe, no anatomical evolution significant enough to be measured has thus far been documented from the fossil record of our first forty thousand years.

Although the age of *Homo sapiens* may be debated (current estimates at one hundred thousand years instead of forty thousand years), Stanley's basic point remains. The punctuational model offers no more (or less) impact on the biology of distinctly human evolution and the associated evolution of psychobiological characteristics than does the gradualism model. The time frame is just too short.

The second, perhaps larger, implication of the punctuational model is that the "unit of selection" need not necessarily be the gene or the individual, but perhaps even the group or species. In this regard, punctuationism is likely to be hailed by group-species selectionists and denounced by gene-kin selectionists. But the point for biologically oriented social scientists, including psychologists, is not whether it is the gene-kin or group-species selectionists who are right but that, despite the outcome of the argument, in proximal terms, nothing has changed. Maybe, just maybe, in ultimate terms, it might matter if modern humans evolved through punctuationism at the group level, in which a certain isolate of archaic *Homo* became contemporary (in geologic time) man, instead of a gradual sifting and subtle modification of anatomical and physiological characteristics. This might be of intellectual interest but would not change anything in behavioral terms, either pragmatically or theoretically. Given the time periods involved for speciation, even speeded up by punctuationism, it is extremely unlikely (and in the absence of evidence, we have to go with this viewpoint) that humans have undergone substantive change (speciation) in the past one thousand to five thousand generations. The punctuationistic model remains an interesting scheme, but, factually, it has little impact on psychobiology and other social sciences.

Of course, ideas are intriguing not just for what they can offer factually or practically. There is also an elegant metaphorical charm to ideas that, in certain circumstances, can be useful heuristically in organizing a discipline's debate and intellectual expansion.

The Metaphorical Application of Evolutionary Schemes to Psychobiology

Efforts to apply biological evolution to the humanities, psychology, and the social sciences have met with mixed reviews from members of both the natural and social sciences. The biologist Carl Gans has stressed that biological examples and theories may be applied to the social sciences in two distinct ways, "as models, with the implication that they represent equivalent mechanisms, and as convenient metaphors that facilitate communication. These seemingly similar usages have profoundly different implications" (1987:225).

Gans noted that for biological theories to have modern relevance to social sciences, both systems need to employ congruent terms and principles. In particular, the relationships of genotype, phenotype, and geologic time to the evolution of human behavior (individual or social) must coexist. "The factors seen for the biological system should also govern the behavior of the social one. They can then be utilized as explanatory schemes, to explain the workings and the rules by which both systems operate" (ibid.). Ultimately, a good model must be able to predict as well as interpret.

Conversely, metaphors, according to Gans, indicate only that a set of current known aspects of the two systems is similar, but no assumption exists that the underlying rules are similar. Metaphors have limited (if any) predictive power and do not allow expectations that can be tested. Metaphors, however elegant and intuitively appealing, do not predict in the manner of theories or hypotheses.

Essentially, Gans is arguing that before one may completely embrace any new (or old) biological evolutionary approach toward explaining human behavior, evidence is needed to substantiate claims. Building a base of support for an idea or theory moves a field away from strict metaphor and toward an explanatory predictive model.

Psychology, as a discipline, can offer an example. Psychoanalytic theory has proved very ineffective in modifying or predicting the display of individual behavior. When compared with methods of behavioral modification or current techniques and perspectives in neuroscience and psychobiology, traditional Freudian psychiatry is shamefully helpless. (Of course, individuals are helped through contact with psychoanalysts from time to time, as they would be from contact with any caring, listening mental health professional. But in the aggregate, and across the breadth of mental disorders, psychoanalysis alone, without the addition of medications, behavioral modification, or adjunctive counseling psychotherapies, is comparatively ineffective.) Millions of men and women have been successfully (often permanently) treated for debilitating mental and behavioral dysfunctions through neurobiology, neurochemistry, behavioral modification regimens, or any portion of the contemporary armamentarium of modern biological psychiatry. The evidence for any effectiveness of Freudian psychoanalysis when compared with these others is appallingly slim at best. Psychoanalytic theory may be elegant and intellectually demanding, but it is essentially worthless as a predictive scheme and falls flat in its therapeutic success rate. It serves more as an academic exercise than as any real window on the reality of mind, brain, and behavior.

The revolution in psychiatry away from Freudian psychoanalysis and toward neuroscience occurred through the testing of hypotheses, a search for more information and better explanatory and predictive models, and a reluctance to worship metaphors. A slavish adherence to the beauty of Freudian theory (or any similar offshoots) is possible only when any effort to research the theory is absent. Similarly, Gans argues for evidence. If the theory of punctuated equilibria is of so much importance to behavioral scientists as a model (not as a metaphor), then it must be proved. Proof comes from evidence. Evidence comes from testing hypotheses and discerning predictions. The study of human behavior is too important for armchair speculation to be the main scholarly activity.

Conclusion

Unlike some of my colleagues, I don't believe the debate among evolutionary biologists regarding punctuated equilibrium theory can have any major impact on the various disciplines of behavioral science. The ramifications of punctuated equilibria are esoteric to this area because of the time frame involved and the lack of any empirical evidence that biological punctuational evolutionary affects human behavior. Any tested hypothesis still standing will stand until subsequent research, not theorizing, knocks it down. Psychobiologists need not worry about their findings floating about with no theoretical base upon which to rest from time to time. As one of my mentors periodically reminded me whenever I would get overly excited about some new and enticing explanation for results, "Data stand, theories fall." Repeatable observations shape and stabilize the theories upon which they stand. But a theory with no observations tips easily. Scholars and scientists in pursuit of predictable explanations of human behavior will go about their business as usual, perhaps with a bit more vigor knowing that the field has a bigger spotlight shining on it.

Punctuated equilibrium or gradualism, whichever scheme best explains an expanding collection of data from the fossil record, will have no direct altering effect on advances in the area of real-time proximal psychobiology, the "how" aspects of explaining the biological mechanisms influencing human behavior. And even in those theoretical areas in which evolutionary mechanisms or explanations are necessarily invoked, the outcome of whether a given idea is correct will ultimately depend on supportive evidence collected from here on out, not from some Devonian fossils. In the meantime, we have a lot of real-time proximal explaining to do.

REFERENCES

Alcock, J. 1975. *Animal Behavior: An Evolutionary Approach*. Sunderland, Mass.: Sinauer.

Cann, R. L. 1987. In search of Eve (a DNA trail leads to a single African woman, 200,000 years old). *The Sciences*, September–October:30–37.

Davis, D. L., and R. G. Whitten. 1987. The cross-cultural study of human sexuality. *Annual Review of Anthropology* 16:69–98.

Dawkins, R. 1976. *The Selfish Gene*. New York: Oxford University Press.

Dienske, H., J. A. R. Sanders-Woudstra, and G. deJonge. 1987. A biologically meaningful classification in child psychiatry that is based upon ethological methods. *Ethology and Sociobiology* 8:27S–46S.

Doerner, G. 1985. Sex-specific gonadotrophin secretion, sexual orientation and gender role behaviour. *Experimental and Clinical Endrocrinology* 86:1–6.

Ehrhardt, A. A., H. F. L. Meyer-Bahlburg, L. R. Rosen, J. F. Feldman, N. P. Veridiano, I. Zimmerman, and B. S. McEwen. 1985. Sexual orientation after prenatal exposure to exogenous estrogen. *Archives of Sexual Behavior* 14:57–77.

Eldredge, N. 1985. *The Unfinished Synthesis: Biological Hierarchies and Modern Evolutionary Thought*. New York: Oxford University Press.

Eldredge, N., and S. J. Gould. 1972. Punctuated equilibria: an alternative to phyletic gradualism. In T. J. M. Schopf, ed., *Models in Paleobiology*. San Francisco: Freeman, Cooper, pp. 82–115.

Ellis, L., and M. A. Ames. 1987. Neurohormonal functioning and sexual orientation: a theory of homosexuality-heterosexuality. *Psychological Bulletin* 101: 233–258.

Gallup, G. G., Jr., and S. D. Suarez. 1983. Homosexuality as a byproduct of selection for optimal heterosexual strategies. *Perspectives in Biology and Medicine* 26:315–322.

Gans, C. 1987. Punctuated equilibria and political science: a neontological view. *Politics and the Life Sciences* 5:220–244.

Gladue, B. A. 1983. Hormones, psychosexuality, and reproduction: biology, brain and behavior. *Politics and the Life Sciences* 2:86–89.

——. 1987a. Psychobiological contributions. In L. Diamant, ed., *Male and Female Homosexuality: Psychological Approaches*. Beverly Hills, Calif.: Hemisphere Publishing, pp. 129–153.

——. 1987b. Perception of sexual activity by men and women: erotic or reproductive? Paper presented at annual meeting of the Society for the Scientific Study of Sex, Atlanta, November.

Gladue, B. A., W. Beatty, J. Larson, and R. D. Staton. 1990. Sexual orientation and spatial ability in men and women. *Psychobiology* 18:101–108.

Gladue, B. A., and H. J. Delaney. 1990. Gender differences in perception of attractiveness of men and women in bars. *Personality and Social Psychology Bulletin* 16:378–391.

Gladue, B. A., R. Green, and R. E. Hellman. 1984. Neuroendocrine response to estrogen and sexual orientation. *Science* 225:1496–1499.

Gould, S. J. 1982. The meaning of punctuated equilibrium and its role in validating a hierarchical approach to macroevolution. In R. Milkman, ed., *Perspectives on Evolution*. Sunderland, Mass.: Sinauer, pp. 83–104.

Gould, S. J., and N. Eldredge. 1977. Punctuated equilibria: the tempo and mode of evolution reconsidered. *Paleobiology* 3:115–151.

Lewin, R. 1987. The unmasking of Mitochondrial Eve. *Science* 238:24–26.

Masters, R. D. 1984. Human nature and political theory: can biology contribute to the study of politics? *Politics and the Life Sciences* 2:120–126.

Mihalik, G. J. 1988. Sexuality and gender: an evolutionary perspective. *Psychiatric Annals* 18:40–42.

Money, J. 1984. Gender-transposition theory and homosexual genesis. *Journal of Sex and Marital Therapy* 10:75–82.

——. 1986. Homosexuals' genesis, outcome studies and a nature/nurture paradigm shift. *American Journal of Social Psychology* 6:95–98.

Money, J., M. Schwartz, and V. Lewis. 1984. Adult heterosexual status and fetal hormonal masculinization and demasculinization: 46,xx congenital virilizing adrenal hyperplasia and 46,xx androgeninsensitivity syndrome compared. *Psychoneuroendo* 9:405–414.

Nesse, R. M. 1987. An evolutionary perspective on panic disorder and agoraphobia. *Ethology and Sociobiology* 8:73S–84S.

Pillard, R., and J. Weinrich. 1986. Evidence of familial nature of male homosexuality. *Archives of General Psychiatry* 43:808–812.

Pitman, R., B. Kolb, S. Orr, J. deJong, S. Yadati, and M. M. Singh. 1987. On the utility of ethological data in psychiatric research: the example of facial behavior in schizophrenia. *Ethology and Sociobiology* 8:111S–116S.

Sloman, L., and J. S. Price. 1987. Losing behavior (yielding subroutine) and human depression: proximate and selective mechanisms. *Ethology and Sociobiology* 8:99S–110S.

Stanley, S. M. 1981. *The New Evolutionary Timetable: Fossils, Genes and the Origin of Species*. New York: Basic Books.

Surbey, M. K. 1987. Anorexia nervosa, amenorrhea and adaptation. *Ethology and Sociobiology* 8:47S-62S.

Symons, D. 1979. *The Evolution of Human Sexuality*. New York: Oxford University Press.

Whitam, F. L., and R. M. Mathy. *Male Homosexuality in Four Societies*. New York: Praeger Press.

Whitam, F. L., and M. Zent. 1984. A cross-cultural assessment of early cross-gender behavior and familial factors in male homosexuality. *Archives of Sexual Behavior* 13:427–439.

11

Catastrophe Theory, Evolutionary Extinction, and Revolutionary Politics

GLENDON SCHUBERT

Modern biopolitics has been touting the merits of a biological approach to politics for at least two decades. Much of the earlier professional writing had been evangelistic, exhorting conventional political scientists to recognize and accept the benefits, for the profession of political science, of a more enlightened stance that would add biological theory, methods, and data to the discipline's already well-established and culturally based stock in trade in the humanities and social sciences alike. The addition of such a life-science stance would enable—indeed, empower—political scientists to confront contemporary public policy issues with a perspective at least as broad and complex as the empirical characteristics of the issues to be analyzed and (except in the case of *real* political *scientists*) pre- or proscribed.

Some biopolitical writers have taken a more iconoclastic stance, criticizing the pretensions of animal behaviorists who described (or fantasized) animal politics by shooting from the hip as barefoot theorists instead of attempting to acquire even the rudiments of political science (or political anthropology, political economics, political sociology, or political psychology) theory of human political behavior (see Schubert, 1982a, 1983a, 1986b).

Biopolitical writers have also criticized sociobiological evolutionary theorists for their naive endeavors (in regard to human nature) to model cultural evolution on the basis of biologically deterministic (and unselfconsciously Libertarian political economics) selfish gene/demes (White, 1981; Losco and Baird, 1982), an enterprise in which a few philosophers of science, economic anthropologists, and other social scientists have joined—but with the support of precious few zoologists, ethologists, or evolutionary theorists per se (but see West Eberhard, 1976; Eldredge and Tattersall, 1982:23–26, 161–162, 176, 178–179; Ginsberg, 1982, 1988). Most biopolitical writers have ranged from neutrality to enthusiasm in their response to the suggestion that the human urge to maximize one's direct or collateral progeny is matched only by humans' selfishness as individual consumers—except for trade-offs (such as, for example, wool for wine). No doubt real sociobiologists can be excused for their apparent ignorance of twentieth-century French demographics; but their biopolitical apologists certainly know better.

Until recently there has been even less conspicuous reason for biopolitical scientists to question the orthodox teachings of the neo-Darwinian synthesis, to say nothing of the implications of those teachings for political theory. After all, those political scientists (with very few exceptions, i.e., Peter Corning, Steven Peterson, and Joseph Losco) had not been socialized as graduate students into that Modern Synthesis; to the contrary, most have bootstrapped their belated and self-taught knowledge of evolutionary theory as early-to-mid-career post-doc professionals (see Schubert 1986a:663). Except for a symposium essay by a behavioral biologist, "Punctuated Equilibria and Political Science" (Gans, 1987), and on which several political scientists commented (ibid.:229–240), there appear to have been only four instances (Masters, 1981; Corning, 1983; Peterson and Somit, 1983; Schubert, 1985:237–239, 252) in which political scientists have directed attention to "punctuational" theory per se, and *none* to catastrophe theory; although other dissents from Modern Synthesis evolutionary theory have been voiced (Schubert, 1981a, 1981b, 1985). In particular, *species* epigenesis theory (i.e., at the phylogenetic rather than ontogenetic level) fits hand-in-glove with punctuational/catastrophe theory, which relies on r-selection theory and epigenetic processes as the modus operandi for the sudden and rapid strong selection (Geist, 1978:256–264) that characterizes the punctuation of the virtual stasis, of the paradigm of evolutionary change, of the Modern Synthesis.

This chapter first presents a brief summary of neo-Darwinian orthodoxy in evolutionary theory. Then it discusses major heresies from the Modern Synthesis: Sewall Wright's theories of genetic drift and of strong selection in small populations; C. H. Waddington's theory of "canalized" growth and epigenetic selection; punctuationism as applied to speciation and hominid evolution; and the rapidly accelerating acceptance (among paleobiologists, meteorologists, geologists, and astronomers) of the emerging hybrid of punctationism/catastophism. The chapter then analyzes extinction theory, which interfaces with both the normal (continuing) phenomenon of *background* punctuational species extinction, and also the exceptional (stepwise) phenomenon of sudden and drastic ("catastrophic") mass extinction of many species at the same (geological) time—as clades, plus their higher taxa such as genera, families, orders, and even classes. Finally, it discusses the facts of the biosocial threats for the mass extinction of the human species; the highly conservative implications of gradualist (orthodox neo-Darwinian) evolutionary theory for both the structures and the behaviors of human politics; the sharply different implications of punctuationism, for both international and domestic politics; and the revolutionary political implications of catastrophe theory.

EVOLUTIONARY CHANGE

Gradualism

The synthetic theory of evolution is preeminently one of gradualism: genetic change is normal, but it takes place at a glacial pace. As Niles Eldredge and Ian Tattersall (1982:37, 39) have summarized the theory (and cf. Eldredge and Gould, 1972:94–96; Eldredge, 1985a:chap. 4, and 1985b:196–201):

> The prevailing view of the evolutionary process is called the 'synthesis' because it integrated, in the 1930's and 1940's, the seemingly disparate data of genetics, systematics . . . and paleontology into a single, coherent theory. . . . Then population genetics . . . developed a mathematical theory showing how frequencies of alternative forms of genes (alleles) could change within populations through time, given certain mutation rates and intensities of natural selection. . . . [Thus] natural selection, working on a groundmass of genetic variation, changes gene frequencies each generation. Mutations are the ultimate source of the variations, but it is selection, working to perfect [*sic*] adaptations or to keep a population in step with changing times, that is the real agent of generation-by-generation change. Over long periods of time—millions of years—such minute step-by-step change will have large effects. So large, in effect, that all the evolutionary patterns seen by systematists working on living plants and animals, as well as the fossils seen by paleontologists, are nothing more than the results of these small-scale processes summed up over geologic time.

(See also Fisher, 1930; Haldane, 1932; Dobzhansky, 1937; Huxley, 1942; Simpson, 1944, 1953.)

Nonconformism

There are three principal components of a new evolutionary paradigm. All three had their beginnings fifty years or so ago, when neo-Darwinism was new and just gaining acceptance as the dominant evolutionary paradigm. The first of these is Sewall Wright's (1940) small population and genetic drift model (Stanley, 1979:24, 165–168). The second is the Waddington/Geist model of epigenetic/dispersal change. And the third is punctuationism/catastrophism.

I have discussed the evolutionary ideas of Wright (1968, 1977, 1978) in Schubert (1981a:202, 208–210); Geist (1978) in Schubert (1981b); and Waddington (1957, 1960, 1975) in Schubert (1985); but not previously those of punctuationism/catastrophism. Therefore I deal here only briefly with the first three but at greater length with the latter theory.

Wright suggested that genetic change takes place very rapidly and with dramatic effects in colonizing and other small populations detached from the habitats of their main species. He proposed also a theory of adaptive landscapes and genetic peaks that supports an interpretation of group rather than individual selection in human populations. Waddington emphasized the importance of behavioral response to environmental stress as a device for reorganizing the expressive effects of genes, which in turn bring positive feedback to bear upon possible behavioral response to further and frequently different environmental stress; and see also Ginsberg (1978:13–14), quoted in Schubert (1989: 327–328) and Eldredge (1985b:74) on developmental gene amplification of effects. Similarly, phenotypic structures vary in development following the transactional interplay between behavior and environment. Waddington emphasized that it is populations of organisms that evolve, and Valerius Geist (agreeing with Waddington) added that the population selection Waddington heralded occurs only under conditions of resource abundance and therefore for small colonizing populations of what Geist calls the "dispersal phenotype" that such animals exhibit. "Therefore," he says, "new adaptations, diagnostic features, which differentiate the colonizing from the parental population, can arise only during dispersal" (Geist, 1978:122).

Punctuationism

Punctuationism begins with the work of Ernst Mayr (1954) and was more imminently anticipated by Eldredge (1971), but was expanded and didactically stated by Eldredge and Stephen Jay Gould (1972); on the punctuational origins of the revival of punctuational theory, in the graduate student relationship at Columbia University during the 1960s, between Eldredge and Gould (inter alia), see Eldredge (1985a:49): "I was a member of the group at the American Museum of Natural History that provided the context and stimulation for the studies which produced the notion of 'punctuated equilibria.'"

Bobby Joe Williams (1986:8) points out: "In it's [*sic*] inception the theory or doctrine of punctuated equilibria represented either an analogy or a direct extension of the Marxist idea that human society changes in major ways only by revolution, not by slow and continuous evolution. This may be a slight bowdlerization of Marx's idea but that this was the origin of punctuated equilibria is not in doubt (Gould and Eldredge, 1977). Regardless of its origin, and regardless of whether it is analogy or not, the punctuational interpretation of fossil sequences stands or falls on it's [sic] own internal logic, and on evidence."

Steven M. Stanley (1981:77–78) discusses the development of the theory and asserts:

1. New species arise by the splitting of lineages.
2. New species develop rapidly.
3. A small subpopulation of the ancestral form gives rise to the new species.
4. The new species originates in a very small part of the ancestral species' geographic extent—in an isolated area at the periphery of the range.

Alternatively, Williams (1986:9) stipulates as the defining characteristics of punctuationism:

1. Large species cannot evolve rapidly enough to produce changes seen in the fossil record. They evolve little or none ("stasis").
2. Rapid change ("punctuation") occurs at the time of speciation.
3. This change occurs as a result of gene drift in a small population, usually allopatric.
4 Speciation and associated changes are, therefore, geologically instantaneous.
5. The parent population does not change and usually becomes extinct.
6. Speciation is random with respect to fitness.
7. Long-term trends in the fossil record are a result of species selection.

By two decades later there were available many additional contributions to punctuationism (Stanley, 1979, 1981; Eldredge, 1985a, 1989; Gould, 1984; and Green, 1986:163–183; and more briefly Goldsmith, 1985:148–149; Gould, 1985:242; and Elliott, 1986:38; and with particular application to hominid evolution Eldredge and Tattersall, 1982; Stanley, 1986). But there were also dissenters (Levinton and Simon, 1980; Templeton, 1980; Charlesworth, Lande, and Slatkin, 1982; Schopf, 1982; and Kitchell and Mac Leod, 1988).

Stanley (1979:183; and cf. Eldredge and Tattersall, 1982:64–65) contrasts *micro*evolution (intraspecies) with *macro*evolution (interspecific), as follows (and for his extended discussion of how and why macroevolution is different, see his pp. 63, 99–100, 141–142, 211–212; Eldredge, 1989; and Green, 1986:168):

Evolutionary mechanisms

Macro	Micro
Phylogenetic drift	Genetic draft
Directed speciation	Mutation pressure
Species selection	Natural selection

Stanley (1979:209) remarks: "Evidence for the punctuational model becomes a double-edged sword, undercutting two of [gradualism's] premises. In the first place, it presents a strong empirical case that natural selection within established species is somehow stifled, forcing us to look beyond phyletic evolution to account for large-scale transition. In the second place, it shows that quantum speciation [and see ibid.:179; Eldredge, 1985b:79–80; and Goldsmith, 1985:148–149] is a real phenomenon and a source of great variability, thus opening the way, on a theoretical plane, for rapid macroevolution via species selection." Furthermore, "genetic recombination . . . is crucial to the process of quantum speciation: new body plans and behavior patterns must develop rapidly, with genetic rearrangements, and re-pairings playing an important role . . . divergent speciation would be difficult without genetic recombination" (Stanley, 1981:198).

Stanley emphasizes (1981:181) that "in the punctuational scheme speciation is seen as the focus of evolutionary change . . . [although] at any time, the direction of the next event of speciation will be heavily dependent upon unpredictable [environmental] and genetic accidents"; and see Eldredge (1989:106–153). However, "no single species ever becomes very highly diversified . . . diversification proceeds by [adaptive radiation] from already established species . . . [in] the punctuational scheme of evolution" (Stanley, 1981:94). But "sex can have a profound effect only within small interbreeding populations of the sort that are involved in speciation" (Stanley, 1979:221). And not least, "The punctuational view implies . . . that evolution is often ineffective at perfecting the adaptation of animals and plants: there is no real ecological balance of nature; that most large-scale evolutionary trends are not produced by the gradual reshaping of established species but are the net result of many rapid steps of evolution, not all of which have moved in the same direction" (Stanley, 1981:5). The negative consequence of this for paleobiology is that most such change "has taken place so rapidly and in such confined geographic areas that it is simply not documented by [the] fossil record" (ibid.:5).

All three of the leading protagonists of punctuationism have indicated (Stanley, 1979, 1981, 1986; Eldredge, in Eldredge and Tattersall, 1982, passim; and Gould, 1985:198) their conviction that punctuational evolutionary theory best explains hominid evolution. Stanley, for example, states (1979:63) that "phyletic evolution in the Hominidae (human family) has apparently been even slower than that for other groups of the Mammalia, implying a punctuational pattern in the ancestry of *Homo sapiens*"; and (1981:206) "the fossil record shows no evidence of human evolution in Western Europe during the entire forty thousand years of our species' existence [and see Gould, 1985: chap. 12] . . . [earlier] *Homo erectus* survived for upwards of a million years without altering." Indeed, Eldredge and Tattersall demonstrate (1982: the front and rear inside cover two-page chart of the last 4 million years of hominid speciation; and cf. Walker, 1984:121) and discuss at considerable length (chaps. 5 and 6, pp. 67–159; and see Walker, 1984:120–123, 130–133) that four hominid species, three Australopithecines and one *Homo*, each survived for a million years or more, and each without significant evolutionary change (within-species selection) taking place. These were *Australopithecus afarensis* (4.0–3.1 million years ago), *A. africanus* (3.0–2.0 million years ago), *A. boisei* (2.2–1.2 million years ago), and *Homo erectus* (1.6–0.4 million years ago). *A. africanus* was perhaps the successor to *afarensis* (and on Australopithecine behavior, from 4 to 2 million years ago, see Zihlman, 1983, and Isaac, 1978); but beginning about 2 million years ago when *boisei* appeared, there were two other species, an Australopithecine (*robustus*) and *Homo habilis*. All three coexisted (although probably not, and certainly not necessarily, in the same places at the same times) in Africa for more than half a million years (from about 1.9 to about 1.4 million years ago).

Both *A. afarensis* and *A. africanus* were gracile (i.e., not heavy-set), as were two of the successor species of the *Homo* genus, *habilis* and *sapiens*. But both of the other Australopithecines (*boisei* and *robustus*) were robust, and so were all of the other *Homo* species (*erectus, archaic* sapiens, and *neanderthalenis*). The lack of directional progress in evolution is illustrated by the fact that *H. erectus*, a quite robust species, succeeded gracile *H. habilis* and also preceded (with the intervention of *archaic sapiens* for half a million years) the reappearance of gracility in modern *Homo sapiens sapiens*, at least by 40 thousand years ago and possibly as many as 200 thousand years ago (Kurtén and Anderson, 1980:355; Leakey and Walker, 1985; Rensberger, 1988; Stringer, 1988).

Eldredge and Tattersall comment (1982:140–142) that "the period of 2 million to 1 million years ago does provide us with the first absolutely unequivocal evidence that we have the co-existence of at least three separate hominid lineages, and also with a nice example of stasis in the hominid fossil record. For sites all over Africa have yielded evidence not only of gracile hominids but also of robust ones." As an example of quantum speciation, they note that "*boisei* springs forth fully fledged and hyper-robust at its first appearance in the geological record and continues unchanged as far as one can tell for the million years until its last occurrences. Its South African contemporary, *robustus*, is more conservative than *boisei*, and it is even possible that the latter species was derived from an early [but thus far undiscovered] population of *robustus*." The placement of *Homo erectus* at 1.6 million years ago was at the beginning of the Pleistocene when the current ice age accelerated with "a worldwide climatic impact, and the environmental fragmentation which this must have involved presumably provided plenty of opportunity for the isolation of populations on which speciation so largely depends."

Other writers further discuss the late Pleistocene *Homo* speciation (since 500 thousand years ago), from the robust *erectus* to the robust *archaic sapiens*, and then allopatrically to the two modern *sapiens* subspecies (or species, a question that remains in dispute), to the robust *neanderthalis*, but also to the gracile *H. sapiens sapiens* (i.e., us) (Reader, 1981; Reichs, 1983; Leakey and Lewin, 1977; Leakey and Walker, 1985; Young, Jope, and Oakley, 1981; Kurtén, 1986:chap. 13, on our Neanderthal cousins; and Kurtén and Anderson, 1980:355–356, on *H. s. s.* settlement of North America).

Catastrophism

According to Gould (1985:242), there are "two distinct levels of explanation—punctuated equilibrium for normal times, and the *different* [emphasis added] effects produced by separate processes of mass extinction. Whatever accumulates by punctuated equilibrium (or by other process) in normal times can be broken up, dismantled, reset, and dispersed by mass extinction. If punctuated equilibrium upset traditional expectations . . . mass extinction is far worse. Organisms cannot track or anticipate the environmentalist triggers of mass extinction. No matter how well they adapt to environmental ranges of normal times, they must take their chances in catastrophic moments . . . [mass] extinctions can demolish more that 90% of all species, [so] groups [are being lost] forever by pure dumb luck among a few clinging survivors designed for a [prior] world." For a sophisticated discussion focusing on the Pleistocene in North America, see Björn Kurtén and Elaine Anderson (1980:chap. 19); and for a major example of a local catastrophic extinction event, see Kurtén's (1986:chap. 5) description and discussion of the flooding of the Mediterranean (Tethys Sea) basin about 5 million years ago, after it had been completely dried up for a million years. More generally, for a beautifully illustrated and detailed review of the theories and evidence on mass extinctions throughout the last 650 million years, see Stanley, 1987, which includes an excellent annotated bibliography of more than a hundred references on mass extinctions and catastrophes.

To exemplify his point that catastrophic mass extinction has a key influence on the history of life on earth, Gould (1985:408–409) invokes that best-studied catastrophe, the Cretaceous era, to generalize on the significance of catastrophism:

> The great world-wide Cretaceous extinction . . . snuffed out about half the species of shallow water marine invertebrates . . . [as well as the dinosaurs that] had ruled terrestrial environments for 100 million years and would probably reign today if they had survived the debacle. Mammals . . . spent their first 100 million years as small creatures inhabiting the nooks and crannies of a dinosaur's world. If the death of dinosaurs had not provided their great opportunity, mammals would still be small and insignificant creatures. We would not be here, and no consciously intelligent life would grace our earth. Evidence gathered since 1980 . . . indicates that the impact of an extraterrestrial body triggered this extinction. What could be more unpredictable or unexpected than comets or asteroids striking the earth literally out of the blue. Yet without such impact, our earth would lack consciously intelligent life. Many great extinctions . . . have set basic patterns in the history of life, imparting an essential randomness to our evolutionary pageant.

Extinction and Catastrophe

Extinction

According to Donald Goldsmith (1985:48), there are tens of millions of extant species today. (Most of them are insects.) But of the estimated "ten to twenty *billion* different species [that] have existed on Earth during the nearly four billion years since life first arose," only a few hundred thousand—perhaps one-five hundredth of 1 percent of that total—have been found as fossils. The presently living species constitute only two-tenths of 1 percent of that total; 99.8 percent have become extinct.

Of course, individual animals become extinct, by the millions or perhaps billions (counting the insects and other microfauna) every day. Paleobiologists, who eventually deal with the consequences of the deaths of individual macroanimals, distinguish two types of extinction. The ordinary "background" extinction of species goes on all the time, usually after a prolonged period of "success" during which neither the species nor its niche changes significantly. The niche changes first because of (for whatever reason) substantial modification of the environment in some respect that becomes critical to a narrow-range species whose individual members cannot change themselves sufficiently—or sufficiently rapidly—to remain adapted. Then that species dies out. Goldsmith (1985:147–148) states: "Mass extinctions occur every few tens of millions of years, removing a large fraction of . . . families of . . . organisms and an even larger fraction of . . . species, [playing] a key role in the evolution of life on Earth." Unlike the random disappearance of species "gradually" through background extinction, whose causes and effects are presumed to entail incremental reciprocal changes in biota and their environments, the effect of mass extinction is to bring about (Goldsmith, 1985:47) "cataclysmic changes in the distribution and numbers of species." Gould states (1985:231–232) that "mass extinctions must, by four criteria, be reinterpreted as ruptures, not the high points of continua. They are more *frequent*, more *rapid*, more *profound* (in numbers eliminated) and [have effects] more *different*" than those of normal times. And he proposes (1985:448) that "mass extinctions may be the primary and indispensable seed of major changes and shifts in life's history. Destruction and creation are locked in a dialectic of interaction . . . [while] mass extinction . . . strikes *at random*" (emphasis added).

Paleobiologists define the major geologic eras and their principal periods (which in turn are divided into stages) in terms of biotic extinction peaks. Most discussion focuses on the most recent half-billion years or so, beginning 570 million years ago with the Cambrian Period of the Paleozoic Era; the Mesozoic Era extends from 232 until 65 million years ago; our present era is the Cenozoic. Goldsmith (1985:60) reports a table based on the best-accepted empirical chronology, which identifies a dozen extinction peaks over the course of the most recent quarter of a billion years (for which the data are best), beginning with the Dzhulfian Stage of the Permian Period 248 million years ago and ending with the Miocene Stage of the Tertiary Period 11.3 million years ago. Evidently the average time between these twelve extinction peaks was 21.5 million years. David Elliott (1986:185) indicates five major mass extinction events since the beginning of the Paleozoic, as follows: (1) end of the Ordovician Stage (435 million years ago); (2) late Devonian Stage (345 million years ago; (3) Permain/Triassic boundary (232 million years ago); (4) Triassic/Jurassic boundary (195 million years ago); and (5) Cretaceous/Tertiary boundary (65 million years ago). The intervals between these mass extinction peaks are 90, 87, 37, and 130 million years, respectively; and the average time between these major mass extinctions is 86 million years.

The end of the Devonian clearly was one of the greatest mass extinctions: but the next one, the Permian/Triassic, was the most drastic of all, eliminating more than 95 percent of all living species 232 million years ago, mostly marine invertebrates. Goldsmith (1985:114) remarks that "mass extinctions, especially those at the Permian-Triassic and Cretaceous-Tertiary boundaries . . . occurred [at] times of significant changes in global environments." Elliott (1986:200) states that the end-Permian mass extinction may have been due to "decreased habitable shelf area leading to decreased population sizes, which increase species' vulnerability to stochastic extinction processes." Many other commentators, however, propose more rapid and dramatic catastrophic causes.

Protodinosaurs (Thecodonts) had evolved from other reptiles at the Permian/Triassic boundary of 232 million years ago, to overlap during the Upper Triassic with the several dinosaur orders that replaced them (Elliott, 1986:185; Goldsmith, 1985:14). Then at the end of the Triassic only 10 million years later, both mammals and birds evolved independently from different reptilian orders, to coexist with the dominant dinosaurs for the next 150 million years, but with the difference that the class of birds continued to expand while mammals stabilized at the level they had reached after an initial rapid but relatively brief radiation until the Cretaceous/Tertiary boundary, when dinosaurs became extinct relatively abruptly. Then birds stabilized while mammals began the massive radiation that continued up to their curtailment by humans during the Recent 15 thousand years or so.

Stanley (1971:220–221) gives examples of *H. s. s.* depredations of megafauna and their habitats, beginning at least by the Late Pleistocene:

> At present it seems that the earliest episode of [Late Pleistocene] large-scale extinction occurred in Africa at some time roughly 60,000 years ago. (The same may be true of tropical Asia but this has not yet been studied in similar detail.) The men of this time were, culturally, late Acheulians—that is to say, they had a highly developed hunting culture of the hand-axe tradition.
>
> They also used fire, and it has been thought that repeated firing was the main agency by which the animal world was depleted. There is evidence of forest firing in Europe as early as in the Holsteinian interglacial, and the deforestation of much of America and Australia may be due to the "peripatetic pyromania" of aboriginal groups. By burning off the vegetation of wide areas, probably just in order to drive game, early man destroyed the environment and so wrought an immensely greater havoc than that caused by hunting alone . . . [and] almost forty per cent of the genera of larger mammals in Africa became extinct. . . . The remaining, impoverished fauna of large mammals . . . is essentially the one that is still in existence in Africa.
>
> In Europe a somewhat similar story was enacted but at a later date—roughly during the interval 30–10,000 years ago. This . . . is the time of the *Homo sapiens* hunting cultures in Europe. . . . The great majority of the animals that became extinct in Europe belonged to the mega-fauna; only about fifty per cent of the species of the larger mammals remained alive.

Stanley's hypothesis has been much more notoriously related to humans during the transition from a gatherer-hunting adaptation to a pastoral-agricultural one at the Pleistocene/Holocene boundary, at about the beginning of the Recent Period 14 thousand years ago (Schubert, 1986a), when *H. s. s.* constituted a vastly accelerated threat to other large mammals and systematically hunted them down to feed for a short time longer an expanding global human population that no longer could be sustained by gathering and traditional symbiotic hunting practices, before the transition to domestication of both plants and animals was forced on humans anyhow. (The appearance of the Clovis hunters in North America about eleven thousand years ago coincides with the transition in the Fertile Crescent of the Middle East to pastoralism and agriculture.) There is no doubt that humans on a global basis are now in the process of completing the extinction of all other large species of mammals, birds, and amphibia, a selected few by direct predation but the vast majority "in the old-fashioned way" by habitat destruction (i.e., of our own along with theirs). Most paleobiologists think that the (relatively small, compared to others discussed above) mass extinctions of the end of Pleistocene reflected a similar human role to that which is occurring now. Indeed, fourteen thousand years is such an extremely brief time, in relation to the time span of the earlier extinction events discussed above, that what happened 14 thousand years ago and now should be viewed as the *same* extinction event when looked at from the longer-range perspective otherwise used.

Goldsmith (1985:41–42) describes the end-of-Pleistocene megafaunal losses as including saber-toothed cats, 14 thousand years ago; woolly mammoth, 10; giant ground sloth, 8.5; giant armadillo, 7.8; mastodon, 6; and a dozen others during the period 14 to 6 thousand years ago; and Paul S. Martin (1984:362–363) lists thirty-three large mammal species as having become extinct by the end of the Rancholabrean 10 thousand years ago, including cheetah, horses, camels, and peccary (in North America). The principal protagonist of the "Pleistocene overkill" thesis has been and remains Martin (1967, 1984), who for the past three decades has been (appropriately enough) ensconced in the heart of Goldwater country. But S. David Webb (1984:206) asserts that the North American extinction episodes of the last 10 *million* years "were correlated with terminations of glacial cycles, when climatic extremes and instability . . . reached their maxima" and especially for "the late Rancholabrean [megafaunal extinction] and the Wisconsin [ice age] termination" about 10 thousand years ago. Elliott (1986:110) adds: "Because most phyla did not suffer more than background extinctions, the late Pleistocene will not qualify as a mass extinction event. It was, nevertheless, catastrophic for the large herbivore and large carnivore [species] of American and Australian land vertebrates." Data and discussion by C. Vance Haynes (1984:346, 350–351) indicate that the North American Clovis hunters of 11 thousand years ago applied the coup de grace to animals that were already rapidly becoming extinct because of the terminal deglaciation and consequent aridity of the environment; the human hunters moved into the scene just as environmental degradation made the large mammals easiest to pursue.

According to Webb (1984:189), "In North American during the middle Myocene (12 to 15 million years ago), major radiations were in progress among native ungulates like the horses, camels, and pronghorn . . . [but by] the late Myocene [beginning about 9 million years ago] a series of declines began that continued through the late Pleistocene extinctions. . . . [The late Myocene] extinctions appear to form a string of related events culminating in the best-documented one about ten thousand years ago." Thus, "in the broader scale of the late Cenozoic . . . man appears too late to be the major cause . . . [arriving] in the New World only for the last of a series of extinctions . . . *climatic deterioration* becomes the primary causal hypothesis for the late Cenozoic large mammal extinctions" (ibid.:192). Richard Klein (Martin's co-editor) concludes (1984:567) that the role of Stone Age people in the extinction process "was subtle, as part of the wider evolving system to which the extinct species failed to adapt." Peter Ucko and G. W. Dimberley (1968) and Kurtén and Anderson (1980:361) agree with Klein that "a major cause of Blancan extinctions was increasing aridity. The forest savanna on the Great Plains, for example, was replaced by vast grassland. Forest dwellers and browsers died out, surviving grazers developed strongly hypsodont teeth, rodents underwent an evolutionary explosion, and small carnivores multiplied. Along the Gulf Coast, climatic deterioration was not so severe, and this area was a refuge for many South American immigrants. Many typical Pliocene mammals disappeared at the end of the Blancan, some leaving descendant species, others, representing the end of their lineage, vanishing forever. Competition was probably intense between the native fauna and invading species better adapted to the deteriorating climatic conditions."

Catastrophes

John and Katherine Imbrie (1979:189–191) conclude their discussion of the ice ages with a summary of the last billion years of earth climate. There were three glacial ages: the earliest and largest was the late Precambrian (from 800 to 600 million years ago); the next, the late Paleozoic, was much shorter (from 320 to 250 million years ago); and what they describe as the "Cenozoic climate decline" commenced 54 million years ago, although the present glacial age dates from only about 10 million years ago (and judging from the other two, even though this one may be the shortest of the three we are nevertheless still entering it; for a sophisticated discussion of the effects on mammals of the Pleistocene Ice Age [1.8 million years ago to now], see Kurtén, 1971:chap. 9).

The Imbries suggest a tectonics theory to explain the glaciations:

> An unstable mass of ice accumulates in high latitudes whenever a substantial portion of the earth's land area is located near the poles. In general, the facts

> about Permo-Carboniferous [the late Paleozoic] glaciation fit this theory, for in those remote times the earth's land masses were assembled in a single supercontinent know as Pangaea. . . . During the 200 million years that followed the Permo-Carboniferous Glacial Age, the earth returned to a non-glacial regime, and was often considerably warmer than it is today. This condition apparently resulted from a northward movement of Pangaea so that its southern tip no longer included the South Pole. . . . [The present] trend is associated with a gradual breaking-up of Pangaea into the separate continents we know today. . . . As more and more land became concentrated in the high latitudes of both polar hemispheres, surface reflectivity increased and climate cooled. . . . Three million years ago . . . once the [continental] ice sheets in this hemisphere formed, they were sensitive to astronomical variations—and they began a long and complex series of fluctuations. . . . It is these cycles that are explained by the astronomical theory of the ice age.

Alfred Fisher (1984:132, 146–147) likewise proposes that tectonics best explains the glaciations of the past billion years, but his theory is one of mantle convection, which depends for its mechanism on volcanism in relation to solar radiation. The glaciations melt during periodic but lengthy cycles of volcanism caused by the plate movements; these result in excess carbon dioxide, with consequent greenhouse effects, gradual melting of the ice sheets, and flooding of the continental shelves, "until the increased intensity of weathering once again balanced the inflow from volcanism." His summary of the past half-billion years is that

> the icehouse state of the latest Precambrian gave way to greenhouse conditions that were coincident with a major sea level rise. Except for a brief excursion at the Ordovician-Silurian boundary, this Mid-Paleozoic greenhouse state persisted up to the Late Devonian, when climates inverted to the Late Paleozoic-Triassic icehouse. Greenhouse conditions were reestablished at the end of Triassic times, and persisted through most—or all—of Eocene times, after which the present icehouse commences. [Indeed, Kurtén (1971:192) asserts that "despite slight temperature oscillations—like the short-range warming trend of the twentieth century—we are on our way towards a new glaciation. The Pleistocene is still going on."] Greenhouse and icehouse states thus alternated at intervals of about 150 [million years], in response to a cycle in mantle convection of about 300 [million years]. The next major [natural] greenhouse episode lies more than 100 [million years] in the future, though man's return of fossil carbon into the atmosphere might bring about an artificially induced—and, hopefully, temporary—condition in the next century.

Decades sooner than was predicted only a few years ago, the artificial culturally induced greenhouse already had arrived (Dateline Canada, 1988).

Although climatic responses to tectonics provide what seem to be a major cause of both continuing background extinctions and the periodic catastrophic biotic effects in mass extinctions, there are more drastic intermittent causes related to an even grander stage than that provided by our own solar system (and see the Velikovsky Affair: Velikovsky, 1950, 1952, 1955; de Grazia, 1966; Greenberg and Sizemore, 1977; Goldsmith, 1977; Bauer, 1984). The Shiva theory is that mass extinctions are caused by periodic cometary bombardments. The current revival of this theory in its modern form dates from less than a decade ago, with the discovery of persuasive evidence (Gould, 1985:chap. 28) that a comet with a head 10 to 11 kilometers in diameter and an estimated mass of 400 trillion kilograms (Goldsmith, 1985:102) hit the earth at the time of the Cretaceous boundary extinction 65 million years ago (for details, see Goldsmith, 1985; Allaby and Lovelock, 1983; Lampton, 1986; United Press International, 1987b; Smith, 1988; Chandler, 1988; and Randall, 1989). As stated by Goldsmith (1985:8, 122–128; and see Gould, 1985:chap. 30, and 1983: chap. 25), the theory includes several interrelated hypotheses, in addition to the one already stated, about what happened 65 million years ago:

1. Mass extinctions occur cyclically with a frequency of about 30 million years ± 10 percent.
2. The cycle involves perturbation in the orbits of one or more subsets of the trillions of potentially available comets that orbit our solar system.
3. Perturbations of cometary orbits are a consequence of gravitational effects caused by giant molecular clouds, traversed by our solar system in its own circular orbit of 240 million years around the galactic center of the Milky Way. (An alternative hypothesis, involving a postulated but as yet undetected brown-dwarf "companion" star to our sun, also is discussed in Goldsmith, 1985:129–146; but cf. Phinney, 1988.)

Either perturbation effect is hypothesized to bring large numbers of comets into the gravitational field of Neptune, and according to Goldsmith (1985:100), if a perturbation

> suddenly (in astronomical terms) sends a billion comets into the planetary region, the result would be a "pulse" of possible colliders just as soon after

> the perturbation as it takes the comets to arrive—about a million years, for comets perturbed at a distance of 20,000 A.U. from the sun. [One A.U. (Astronomical Unit, or the average distance of the earth from the sun) equals about 92,600,000 miles.] This pulse would decline sharply as most of the comets were again expelled, but a few would acquire short-period orbits, remaining in Earth's vicinity for hundreds of thousands, or even a few million, years. Throughout this period the number of those comets, some of them capable of acquiring Earth-crossing orbits, also would "tail" off. This "tail" provides a way to reconcile the record of mass extinctions, which does not show any perfect periodicity, with a theory that cometary orbits are perturbed at intervals of 26 to 33 million years. The perturbations may occur with complete regularity, but their effects—impact on Earth—would have a time delay that spreads, and by a varying amount, over as much as one to two million years.

Goldsmith then calculates (1985:101) that the expected frequency of a comet with a diameter equal to 10 kilometers impacting earth is once every "few" hundred million years. A 10-kilometer-diameter comet (or asteroid) would produce from 10 to 100 billion tons of dust and smoke, thereby provoking the Nuclear Winter studied by TTAPS (with "S" for Sagan: Gould, 1985:chap. 29) exponentialized in effect to at least the tenth and possibly the 10^{10} power.

Political Change

Catastrophic Problems

Ecological

The prospects for present and future human life are foreboding no matter what we do. Doing nothing, or virtually nothing (as we as a species have been responding to the technologically based destruction of our niche throughout the past two centuries; see Green, 1986), is of course the gradualist response to catastrophic problems. But the ecological problems that are absolutely certain to bring about the extinction of our species in very proximate geological time (Ehrlich and Ehrlich, 1981)—unless catastrophes do it first by punctuating our false sense of equilibrium—are ones that are themselves continuing to worsen slowly but surely, in complete accord with the gradualist model. These include the consequences of environmental degradation such as the chronic dumping of poisonous chemical wastes, in the most cost-efficient way to the polluters (Kapp, 1950), into rivers and once-great lakes and oceans and even beneath the playgrounds of our public schools (Levine, 1982) and through meteorological processes into the atmosphere (to circle the globe together with the escalating proliferation of "space junk"—Broad, 1987) of poisonous rain, fog, and air to be breathed. Also, the new global epidemic of skin cancers is directly caused by the industrial by-production of chloroflurocarbons plus their direct production for aerosol packaging of cosmetics for ourselves and our machines, in the ultimate destruction of our planet's shield from ultraviolet radiational keratosization of the human species' highly vulnerable derma (Peterson, 1987).

Our increasing practice (as during the past decade) of dumping not merely sewage and garbage and medical offal but also liquid fossil fuels into the oceans kills (usually, considerably more than) two birds with one spill: the fuels are no longer available as usable energy, and the dissipation contributes to the destruction of life in the seas as well as at their littoral. But except as it feeds back in the form of permanently unpotable *American* municipal water supplies—no longer just European and elsewhere in the world—the processes described above are less immediate in their impact than the consequences for Americans of the technological artificialization of the "conditioned" (with Legionnaire's bacteria) air they breathe, the ersatz (in content as well as packaging) food they eat, and the (exploding or highly flammable) clothes that even their infants wear. Most persuasive and damaging of all, in what has become the very short run of less than another century, is the elimination of the sensory stimulation that human bodies in their presently evolved form *must* respond to if our physiological senses are to develop normally. New public housing projects or industrial complexes or concrete fortresslike university buildings quickly degrade into asphalt jungles, and the proximate extinction of our species' competitors for food, space, and freedom (i.e., *all* other animals large enough for humans to see) will eliminate all other mammals (at least) before the end of the twenty-first century—with extraordinarily malevolant consequences for human mental health and behavior (Shepard, 1978). These are a few of the most obvious examples of how gradualistic biotic-cum-cultural evolution is eliminating the human species niche from the biosphere. And the probable vast acceleration in the synergistic impacts of the "slow death" background extinction suggested above will convert our hypothetical denouement into an empirical mass extinction—necessarily for our genus and family as well as for our species. There will be no surviving zoologists to record the event, but even if there were, another hundred years or so (to extinguish even remnant populations) is a mere microsecond in geological time. We have no clade; so quantum speciation to replace us will occur (and cf. Day, 1936), but not from within our own family.

Political

Most Americans could take in their stride the execution of a few thousand Palestinians in their Lebanese or Israeli concentration camps, or a few thousand Hindus killed or mutilated by the explosion of a cyanide chemical plant in Bhopal. Even American Three-Mile Island was discounted as analogous to an ordinary and usual Amtrak accident (see Morone and Woodhouse, 1986). But not Chernobyl. Kiev has functioned as our peacetime Hiroshima; the next "peaceful" nuclear-energy-powered plant accident may well befall New York City, and the casualties a year down the line from *that* catastrophe can be tens of millions of Americans dwelling in the Northeast corridor—evidently, it was not a mere few thousand migrant Lapps and their reindeer that suffered from Chernobyl: *perestroika* has confirmed (Los Angeles Times Service, 1990) that there remain between 3 and 4 million victims of the fallout still living in the vicinity of the burnout, still eating locally grown contaminated food, and still dying from hyperexposure to nuclear radiation.

Even so, wartime uses of nuclear explosive devices—whether initiated by France, Libya, Israel, Pakistan, India, Iran, or some other American or Soviet or Chinese-supplied munitions consumer not yet in the public eye—are likely to be more quickly catastrophic to the species than even the Chernobyls of the future. The past twenty years have seen increasing reliance on improved technical delivery capacity for nuclear weapons, with conventional ground military systems best symbolized by extremely expensive but nonoperational World War I–type weapons—not named the Sergeant York for nothing. But why does it matter whether our tanks were duds? Our Western European allies had no hope of defeating Soviet ground forces, and NATO relied on our cruise missiles and *opposed* nuclear detente until the Soviet economy finally confronted its Götterdämmerung, raising the Iron Curtain and generating the politics of *glasnost* and *perestroika*. But there are still at least a dozen (and an increasing number) of minor-league Dr. Strangelove countries with aspirations to own nuclear weapons. From MAD to SDI, the first nuclear war is apt beautifully to exemplify the punctuational model: a considerable period of disuse terminated by a sudden and extremely rapid deployment of nuclear warhead missiles—both offensive and "defensive"—all detonated in the upper atmosphere with consequences roughly equivalent to what would happen if the earth were to experience a direct hit by an ordinary, average-sized comet.

Goldsmith has explained (1985:106) that, of course, the use of nuclear warheads will entail "the effects of radioactive debris, of ultraviolet radiation [and] of gamma radiation [plus chemical] rain [but these] pale in significance beside the chief effect produced both by thermonuclear bombs and by a large extraterrestrial impact—the blotting out of the sun" that will trigger the Nuclear Winter (Gould, 1985:chap. 29) and mass extinction, if not directly through blast and burning, then indirectly through radiation and starvation. (Large mammals like humans are maximally vulnerable to starvation because of their high food requirements.) Stanley (1981:207) closes his book on a wistful but hopeful note, speculating that nuclear war will only "depopulate the Earth, leaving small, insular, and perhaps [*SIC!*] genetically altered populations"—the very ones fit, ceteris paribus, for quantum speciation. But his is the most favorable and least probable scenario, even after ignoring his monumentally unrealistic presumption of genetic stasis (see Frank, 1947). Gould (1985:426) is much more realistic: "Full nuclear exchange would probably generate the same kind of dust cloud and darkening of the skies that may have wiped out the dinosaurs." But this time it will be us humans.

Gradual Progress

Ever since its own "speciation" as a distinctive subject of higher academic instruction more than a century ago (Somit and Tanenhaus, 1967), political science has identified with a static conception of political structure isomorphic with the Constitution of the United States preordained by classical Newtonian physics (Landau, 1961). That Constitution is a design for a political machine (Schubert, 1967) with a predictable organization based on laws that define, limit, and constrain political processes and operations. Such a polity is intended to endure ("for ages") with changes confined to the replacement of worn-out parts and occasional preventive maintenance (to make the component parts last as long as possible; cf. Artigiani, 1987). The best environment for such a machine would be a sterile factory—itself a kind of machine to assure constant conditions of temperature, energy supply, lubrication, waste removal, and product encapsulation and removal—in short, a Skinner box analogous to the one that Burrhus constructed for his own infant daughter.

Evolutionary gradualism is a model for dictatorships, oligarchies, and constitutional democracies alike, thereby cutting across and applying to all of the traditional categories in terms of which political scientists conventionally distinguish among basic types of polities. From a gradualist perspective all are alike because, once established, the focus of political analysis is on the maintenance of their stability, with slow, moderate, and gradual change postulated as the expected (and desired) norm.

Gradualist theory insists (Fisher, 1930; Hamilton, 1964; Williams, 1966) that natural selection operates on individual animals, not on social groups of them (or at least, not on larger groups than closely related "kin"), and most certainly not on entire species, genera, families, or orders. The mechanism postulated by orthodox gradualism is genetic competition between and among animals of the same species; and the aggregation of changes in the gene pool for a species as a result of the differential breeding success of individuals constitutes natural selection. Mutation functions at a minute pace to favor fortuitously occasional individuals while disadvantaging most others: but see R. Goldschmidt's "punctuational" theory of "macromutations" (Eldredge, 1985a:74).

Gradualism glorifies stasis and justifies it on the basis of genetic competition among individual humans, and the only use for such an ethic as applied to twentieth-century industrial and postindustrial politics (Green, 1986) is ideological, as a purported justification for conservative exploitation of the masses and for failure to take concerted political action—which requires considerably more synergy than the residue from a host of individual human breeding choices—to avoid or avert Apocalypse Tomorrow. Surely the chronic problems of niche degradation cannot be resolved by gradualistic solutions. Certainly, too, the questions of nuclear power and weaponry cannot wait ten years, or twenty years, or (like our national debt) for our children and grandchildren to solve, or for the future to heal: they demand revolutionary political change right now, in the immediate present. Biological gradualist evolutionary theory as the model for human politics is a blueprint for the Doomsday of the human species by reinforcing the politics of certain failure to change either enough or in time (see the political pablum—euphemistic social optimism—proffered by two social-engineering technologists: Morone and Woodhouse, 1986). It is not surprising that nineteenth-century Darwinian theory (in the guise of Social Darwinism) was rejected as a model for social, economic, and political change and that the twentieth-century neo-Darwinian modern synthetic theories of evolutionary gradualism and sociobiology have had virtually no attraction for political scientists or use in serious modeling of politics (and for the *de minimus* exceptions, see Kaufman, 1985; Kort, 1983; and cf. Kort, 1986).

Eldredge and Tattersall remark (1982:161–162) that "the prevailing view among historians . . . remains an expectation of slow, steady, gradual change—in a word, progress. Historians . . . have suffered from the same sort of Victorian myths that have afflicted [gradualist] evolutionary biologists." This view contrasts sharply with Eldredge and Tattersall's impression "that human history over the past 6,000 years . . . reveals a continuation of the same basic pattern [as that of the prehistoric record of both genetic and cultural punctuationism] of rapid innovation followed by far longer periods of little or no change." Eldredge and Tattersall quote sociologist Kenneth Bock (1980) in explicit support of their own stated position, and they note that Bock "goes on to argue even more strenuously against the idea that there is [any] inevitability underlying social change—[either] biologically based or [otherwise]." Bock reports that "far from being unilinear, graded progressive, and inevitable, social change is a rare phenomenon, the result of specific events. When change happens, it happens quickly. For the most part, history is nonchange unless and until something happens."

The relation between biological and cultural evolution is then discussed by Eldredge and Tattersall (1982:176–179, 181):

> The traditional response of cultural evolutionists to biologists who, like sociologists, seek to reduce cultural evolution to the terms of biological evolution, justifiably emphasizes inheritance. . . . But [the punctationist] idea that once a way of being is invented it will persist until change is forced on the system is hardly a metatheory of evolution. . . . [Punctuationism] sees species as basic units in evolution: species are the ancestors and descendants of the evolutionary process. The pattern of change in the fossil record strongly suggests that most change in the evolutionary process is related to the origins of new species. . . . That patterns of genetic and cultural differentiation are not coextensive should be enough . . . to deter biological determinists (like sociobiologists) who seek to reduce cultural evolution to biological terms. . . . Sociobiology is an astonishing anachronism.

And they add that "geographic patterns—especially isolation—can foster change and consequent stability in both evolutionary systems. Why the change need be rapid . . . is [perhaps] 'sink or swim' [at least] for biological systems."

Eldredge and Tattersall (1982:172) attribute the illusion of progress to "misperception of pattern" rather than to "social ideology"; but this is a distinction without a difference. Similarly, in his introduction to the pilot Eldredge and Gould paper on punctuationism (1972:83), their editor Thomas Schopf said that "the larger and more important lesson" of their work is not punctuationism, but rather their demonstration that "*a priori* theorems often determine the results of 'empirical' studies before the first shred of evidence is collected . . . [the] idea that theory dictates what one sees." About half a century earlier, according to quantum physicist Werner Heisenberg (1977:5), Einstein had remarked to him that "it is the theory which decides what can be observed." This means that it is always culture that corrupts biology, rather than vice versa (as in the notorious mystique of "biological determinism" associated with the Social Darwinism of yesteryear, or of today's counterpart in feminist fears of biological corruption and contamination of their goal of social, economic, and political equality). The psychobiology of the brain makes it possible for a culture—any culture—to be expressed; but the culture supplies the content of what is thought, and biology supplies the process of thinking. Progress and punctuationism are equally cultural; neither is biological.

Punctuating Political Stasis

Clearly, gradualist evolutionary theory has little relationship to international relations among polities because competition there is at the equivalent biological level of species—not of individual animals—and gradualism denies that species are selected other than through competition among their individual members. From the perspective of philosophy of science, the ascribed posture of political international relations, as "explained" by evolutionary gradualism, is methodological individualism with a vengeance.

Punctuationism rejects the monotheistic individualism that characterizes exclusive reliance on gradualism as the mechanism for change. Punctuationism thus takes a very different stance toward progress than gradualism. As Gould (1985:44) puts it, "No internal dynamic drives life forward. If environments did not change, evolution might well grind to a virtual halt. . . . [The] primary 'struggles' of animals are with changing climates, geologies, and geographics, not with each other. Competition . . . [is] a sporadic and local interaction . . . not . . . [life's] driving force." And Stanley (1979:203) also has questioned gradualism's monomaniacal preoccupation with competition as the universal deus ex machina of natural selection: he makes the point that vulnerability to predation must also be considered: "Disease may have caused the extinction of many species . . . [but] this special form of predation is seldom detectable in the fossil record."

Punctuationism emphasizes species selection as the principal mechanism for speciation and evolutionary genetic change, which are postulated to occur very rapidly in small populations detached from the main species population and in habitats that demand rapid phenotypic adjustment to dynamically accelerating ecological disturbances (Stanley, 1979, 1981; Eldredge, 1989). New species result from stress, by invoking from among fortuitously available epigenetic combinations of structures and behaviors some that fit the new demands better than the older species adjustment did or could. Some of the new species and their clades are better adapted than are others; those better adapted may flourish while maladapted ones may rapidly become extinct under the stress of drastic ecological change.

If we are to analyze international political change from the comparative perspectives of punctuationism instead of gradualism, then we must clarify the presumptions of indulging in the metaphor that *polities* can be conceptualized as species. An animal species is a population of relatively highly genetically interrelated individuals capable of the sexual pairwise reproduction of fecund offspring of the same average degree of relatedness as the other animals constituting the population. A polity is a population of relatively highly political-culturally interrelated individuals capable of speaking the same maternal language(s) (see Masters, 1970) as the other humans constituting the population. An animal species must exploit for sustenance a niche to which it becomes and remains successfully adapted, often for millions of years and with repeated generations so that, with a million successive population cohorts, the millionth will be very much the same as the hundredth. A polity controls the exploitation of a geographically defined space that provides sustenance for the population during a culturally defined series of regimes (political generations). A polity typically persists until its subject population is displaced, absorbed, or extinguished by a different and successor population of humans who define a new polity. For large mammals the genetically defined biological life span of a cohort may be many decades; for humans it is approximately eighty years (Fries and Crapo, 1981). The longevity of regimes usually varies from a few years to several decades, and polities sometimes persist for centuries.

Many consequences follow from the punctuational assumption that polities are equivalent to different species in competition for the same or overlapping resource bases in the same time/spatial environment. Many species predate upon others, which function as vital resources for the predators; and even sympatric (i.e., of the same genus) species may relate to each other in environmental settings in such a way that one competitively excludes (i.e., extinguishes) the other. There are, of course, many political analogues of biological competitive exclusion, as exemplified by the extermination of Polish Jews during World War II by both Nazi Germany and Stalinist Russia or the displacement of most of the Moslem population from Palestine when the post–World War II Israeli polity was established.

Punctuationism can also be used to model domestic politics, through cohort analysis of generational change and differences in political culture, attitudes, and behavior. The American generation that came of age in the early 1950s, as contrasted to the one maturing in the late 1960s, can serve as an example: products of the McCarthy era versus those of the Vietnam era. Evolutionary theorist George Williams has suggested that each new generation be viewed as colonists trying to establish themselves in terra incognita; this is Sewall Wright's concept redefined so that habitats vary according to *time* as well as *place*. By also conceptualizing each cohort as the cultural analogue of a biological species, the metaphor to punctuationism is evident.

Revolutionary Change

Stanley (1981:205) suggested a punctuational analogy "between the evolutionary stagnation of large, complex species and the resistance of large institutions to change." A book by political scientist Herbert Kaufman (1985) analyzes bureaucratic stasis from an evolutionary perspective, but the author's innocence of punctuational theory left him to rely on gradualism, which compounded rather than helped to resolve his problems. At a time when I likewise was unaware of punctuationism, I commented ironically on the prospects for achieving "global peace" through a single "one-world" superstate, asserting that in the case of such a "hypothesized one-world totalitarian state, with its international civil service, it would be a challenge to political scientists to devise a method of nonviolent orderly systemic political change that might avert the periodic necessity of having the evolutionary alternative of revolution asserted" (Schubert, 1982b:77). Clearly, from a punctuational point of view, the method would be neither nonviolent nor orderly nor systemic; the necessity would be aperiodic; and revolution, which is not an alternative sanctioned by neo-Darwinism, is the natural process for making needed change possible. Political punctuationism has no need to invoke Marxism or extraterrestrial intervention; it is the political ideology of the American Revolution of more than two centuries ago, in the spirit of Thomas Jefferson's well-known (but little followed) advice about not letting twenty years pass *without* a rebellion—so that, in his colorful biological analogy, the tree of liberty can be nourished with its natural manure: the blood of patriots.

Basing his work on a more modern source (Ilya Prigogine's [1980] quantum theory of dissipative structures), Robert Artigiani (1987:252) proposes an independent but highly compatible theory of revolutionary political change: "Seen from the perspective of [Prigogine's] model, revolutions occur when a society is driven, either by internal perturbations or altered environmental inputs, through a 'catastrophe' to a bifurcation point, where it makes a non-linear *leap to a new level of stability*" (empasis added).

Instead of continuing to practice political statics, political scientists should recognize (Schubert, 1982b:77) that

> evolution signifies change in organisms (including their behavior) due to uncontrollable, and what heretofore have been virtually unpredictable, environmental changes; and revolution (like "catastrophe") is an index to one of the many possible rates of change. Hence political revolutions demonstrate—they most assuredly do not contradict—that changes are taking place in political evolution. Our traditional static approach to political theory presumes that change is abnormal, undesirable, and a sign that something is wrong; but a punctuational evolutionary approach will assume that change is a function of "environmental stress" (Waddington, 1960:95), and that the task of political science is to understand what is happening in the sociopolitical as well as the natural environment that stresses humans in particular political demes, and how and why political behavior is changing in a variety of responses to such perceptions of environmental stress.

By directly invoking evolutionary catastrophe theory, the point can be stated more simply and directly. Let us begin with Gould's assertion (1980:84) that "evolution is the differential survival and deployment of punctuations," which I redefine for application to politics as that *political* evolution is the result of changes brought about through *revolutions* and other catastrophic political events.

Classic examples (such as the American Revolution of 1776–81, the French Revolution of 1789, and the Russian Revolution of 1914–20) and many more recent ones (such as the partitioning of India in 1947, or the Vietnam War of 1946–75, or the Iranian Revolution of 1978–79) all brought about catastrophic political solutions to gradualist political problems of long standing. Other catastrophic political events that brought about massive cultural changes in this country include the three assassinations that clustered in the 1960s: those of John F. Kennedy, Martin Luther King, and Robert F. Kennedy. Like the fall of the Romanov dynasty half a century earlier, these political murders occurred on days that shook the world. And given the collapse of the Berlin Wall (and the rest of the Iron Curtain) and the differential emancipation of Eastern European countries beginning in the autumn of 1989, it is already past time for us to consider (i.e., in the spring of 1991) the catastrophic consequences of the hyperescalated reunification of Germany, the Soviet rejection of independence for the Stalinist satrapies of the Baltic, and the political as well as economic disintegration of the Russian Empire.

The great advantage of catastrophism in political theory is that it proffers an ideology, premised in the biology of our species, that can be used to legitimize the revolutionary solutions demanded by the catastrophic changes already well advanced in our biospheric niche; in our phenotypic selves; and in our chances of having either the time or the capacity to preclude a macroevolutionary megacatastrophe such as has happened to our planet many times in the past (although even the most recent was long before hominids evolved). Such "natural" catastrophes are highly likely to happen in a future as relatively remote—in relation to the probable duration of our species—as the relevant past; but as for the present, a nearby supernova may already have exploded without yet quite having manifested itself to earthlings (and see Washington Post Service, 1987a, 1987b; Detjen, 1987; United Press International, 1987b; and see Lindley, 1988, on the yearling Supernova 1987a, plus Phinney, 1988, on its red "speck" companion). Goldsmith (1985:113) suggests that "if a star within 50 light years of the sun [i.e., about 290 trillion miles]—one of the thousand or so nearest stars—were to become a supernova its cosmic-ray flux would be sufficient to produce catastrophic effects." There is no point worrying about such matters because they are utterly beyond our control or influence; what we should worry and do something revolutionary about is, first, the preclusion of nuclear holocaust (Joyce, 1987), and second, the reclamation of a niche that we can adapt to while still remaining human for yet a while longer (see Yanarella, 1984).

The acceptance of the implications for ourselves of catastrophic evolutionary theory would certainly constitute a major paradigm shift for all of the social sciences (Kuhn, 1962; Rodman, 1980; Schubert, 1983b). Shifting from gradualism to punctuationism and catastrophism might be presumed to have no automatic effect on the manifold public policy problems that confront us and (both collectively and as individuals) imperil our continued existence as an animal species—however some individuals may continue to save their souls. But the assumption that how we think about problems does not change their nature is fundamentally wrong. How we first perceive and then conceptualize them makes all the difference in what we can imagine doing about problems and, indeed, even to be able to define them as "problems," not to mention what might seem possible and appropriate for us to do about them. So the paradigm shift is itself the first step to be taken in trying to do something to postpone the extinction of our species.

References

Allaby, M., and J. Lovelock. 1983. *The Great Extinction*. London: Secker and Warburg.

Artigiani, R. 1987. Revolution and evolution: applying Prigogine's dissipative structures model. *Journal of Social and Biological Structures* 10:249–264.

Bauer, H. 1984. *Beyond Velikovsky: The History of a Public Controversy*. Urbana: University of Illinois Press.

Berggren, W. A., and J. A. Van Couvering, eds. 1984. *Catastrophes and Earth History*. Princeton: Princeton University Press.

Bock, K. 1980. *Human Nature and History: A Response to Sociobiology*. New York: Columbia University Press.

Broad, W. J. 1987. Growth of orbiting junk threatens space missions. *International Herald Tribune*, August 6 (European ed.), p. 7.

Chandler, D. L. 1988. Extinction theories. *Honolulu Star-Bulletin and Advertiser*, October 30, p. D-1.

Charlesworth, B., R. Lande, and M. Slatkin. 1982. A neo-Darwinian commentary on macroevolution. *Evolution* 36:474–498.

Corning, P. 1983. *The Synergism Hypothesis: A Theory of Progressive Evolution*. New York: McGraw-Hill.

Dateline Canada. 1988. Harmful prospects noted. *Honolulu Advertiser*, June 18.

Day, C. 1936. *This Simian World*. New York: Knopf.

de Grazia, A., ed. 1966. *The Velikovsky Affair*. New York: University Books.

Detjen, J. 1987. Supernova gives a new peek at the universe. *Honolulu Star Bulletin and Advertiser*, March 15, p. A-17.

Dobzhansky, T. 1937. *Genetics and the Origin of Species*. New York: Columbia University Press.

Ehrlich, P., and Ehrlich, A. 1981. *Extinction: The Causes and Consequences of the Disappearance of Species*. New York: Random House.

Eldredge, N. 1971. The allopatric model and phylogeny in Paleozoic invertebrates. *Evolution* 25:156–167.

——. 1985a. *Time Frames: The Rethinking of Darwinian Evolution and the Theory of Punctuated Equilibrium*. New York: Simon and Schuster.

——. 1985b. *Unfinished Synthesis: Biological Hierarchies and Modern Evolutionary Thought*. New York: Oxford University Press.

——. 1989. *Macro-Evolutionary Dynamics: Species, Niches, and Adaptive Peaks*. New York: McGraw-Hill.

Eldredge, N., and S. J. Gould, 1972. Punctuated equilibria: an alternative to phyletic gradualism. In T. J. M. Schopf, ed., *Models in Paleobiology*, San Francisco: Freeman and Cooper, pp. 82–115.

Eldredge, N., and I. Tattersall. 1982. *The Myths of Human Evolution*. New York: Columbia University Press.

Elliott, D. K. 1986. *Dynamics and Extinction*. New York: Wiley.

Fisher, A. G. 1984. The two phonerozoic supercycles. In W. A. Berggren and J. A. Van Couvering, eds., *Catastrophes and Earth History*. Princeton: Princeton University Press, pp. 119–150.

Fisher, R. A. 1930. *The Genetical Theory of Natural Selection*. Oxford: Clarendon Press.

Frank, P. 1947. *Mr. Adam*. London: Victor Gollantz.

Fries, J. F., and L. M. Crapo. 1981. *Vitality and Aging*. San Francisco: W. H. Freeman.

Gans, C. 1987. Punctuated equilibria and political science. *Politics and the Life Sciences* 5:220–244.

Geist, V. 1978. *Life Strategy, Human Evolution, Environmental Design: Toward a Biological Theory of Health*. New York: Springer-Verlag.

Ginsberg, B. E. 1978. What will students in political science have to know about biology to understand the new dimensions of their discipline and to advance the frontiers of knowledge? Paper presented at the American Political Science Association meeting, New York.

——. 1982. Review of *Sociobiology and Human Politics*. *Politics and the Life Sciences* 1:75–79.

——. 1988. The evolution of social and political behavior: sociobiology and alternative explanations. Paper presented at the American Political Science Association meeting, Washington, D.C.

Goldsmith, D., ed. 1977. *Scientists Confront Velikovsky*. Ithaca: Cornell University Press.

——. 1985. *Nemesis: The Death-Star and Other Theories of Mass Extinction.* New York: Walker.

Gould, S. J. 1980. *The Panda's Thumb.* New York: Norton.

——. 1983. *Hen's Teeth and Horse's Toes.* New York: Norton.

——. 1984. Toward a vindication of punctuational change. In W. A. Berggren and J. A. Van Couvering, eds., *Catastrophes and Earth History.* Princeton: Princeton University Press, pp. 9–34.

——. 1985. *The Flamingo's Smile.* New York: Norton.

Gould, S. J., and N. Eldredge. 1977. Punctuated equilibria: the tempo and mode of evolution reconsidered. *Paleobiology* 3:115–151.

Green, H. P. 1986. *Power and Evolution: The Disequilibrium Hypothesis.* Columbia: Institute of International Studies, University of South Carolina.

Greenberg, L. M., ed. 1982. *Evolution, Extinction, and Catastrophism.* Glassboro, N.J.: Kronos Press.

Greenberg, L. M., and W. B. Sizemore. 1977. *Velikovsky and Establishment Science.* Glassboro, N.J.: Kronos Press.

Haldane, J. B. S. 1932. *The Causes of Evolution.* London: Longman.

Hamilton, W. J. 1964. The genetic theory of social behavior: I and II. *Journal of Theoretical Biology* 7:1–32.

Haynes, C. V. 1984. Stratigraphy and late Pleistocene extinction in the United States. In P. S. Martin and R. G. Klein, eds., *Quaternary Extinctions: A Prehistoric Revolution.* Tucson: University of Arizona Press, pp. 345–353.

Heisenberg, W. 1977. Remarks on the origin of the relations of uncertainty. In W. Price and S. Chrissick, eds., *The Uncertainty Principle and Foundations of Quantum Mechanics.* New York: Wiley.

Huxley, J. S. 1942. *Evolution: The Modern Synthesis* London: Allen & Unwin.

Imbrie, J., and K. P. Imbrie, 1979. *Ice Ages: Solving the Mystery.* Short Hills, N.J.: Enslow.

Isaac, G. 1978. The food-sharing behavior of protohuman hominids. *Scientific American* 284 (4): 90–108.

Joyce, A. A. 1987. The nuclear arms race: an evolutionary perspective. Paper presented at the American Political Science Association meeting, Chicago.

Kapp, W. 1950. *The Public Costs of Private Enterprise.* Cambridge, Mass.: Harvard University Press.

Kaufman, H. 1985. *Time, Chance, and Organizations: Natural Selection in a Perilous Environment.* Chatham, N.J.: Chatham House.

Kitchell, J., and N. MacLeod. 1988. Macroevolutionary interpretation of symmetry and synchroneity in the fossil record. *Science* 240:1190–1193.

Klein, R. G. 1984. Mammalian extinctions and Stone Age people. In P. S. Martin and R. G. Klein, eds., *Quaternary Extinctions: A Prehistoric Revolution.* Tucson: University of Arizona Press, pp. 553–573.

Kort, F. 1983. An evolutionary-neurobiological explanation of political behavior and the Lumsden-Wilson "thousand year rule." *Journal of Social and Biological Structures* 6:219–230.

——. 1986. Considerations for a biological basis of civil rights and liberties. *Journal of Social and Biological Structures* 9:37–52.

Kuhn, T. S. 1962. *The Structure of Scientific Revolutions*. Chicago: University of Chicago Press.

Kurtén, B. 1971. *The Age of Mammals*. London: Weidenfeld and Nicolson.

——. 1986. *How to Deep Freeze a Mammoth*. New York: Columbia University Press.

Kurtén, B., and E. Anderson. 1980. *Pleistocene Mammals of North America*. New York: Columbia University Press.

Lampton, C. 1986. *Mass Extinctions: One Theory of Why the Dinosaurs Vanished*. New York: Franklin Watts.

Landau, M. 1961. On the use of metaphor in political analysis. *Social Research* 28:331–353.

Leakey, R., and R. Lewin. 1977. *Origins*. New York: Dutton.

Leakey, R., and A. Walker. 1985. *Homo erectus* unearthed. *National Geographic* 168:624–629.

Levine, A. G. 1982. *Love Canal: Science, Politics and People*. Lexington, Mass.: Lexington Books.

Levinton, J. S., and C. M. Simon. 1980. A critique of the punctuated equilibria model and implications for the detection of speciation in the fossil record. *Systematic Zoology* 29:130–142.

Lindley, D. 1988. Supernova 1987A: one year old. *Nature* 331:567.

Los Angeles Times Service. 1990. Thousands [millions] living amid fallout from Chernobyl. *Honolulu Advertiser*, April 23, p. D-1.

Losco, J., and D. D. Baird. 1982. The impact of sociobiology on political science. *American Behavioral Scientist* 25:335–360.

Martin, P. S. 1967. Prehistoric overkill. In P. S. Martin and H. E. Wright, Jr., eds., *Pleistocene Extinctions: The Search for a Cause*. New Haven: Yale University Press, pp. 75–120.

——. 1984. Prehistoric overkill: the global model. In P. S. Martin and R. G. Klein, eds., *Quaternary Extinctions: A Prehistoric Revolution*. Tucson: University of Arizona Press, pp. 354–403.

Masters, R. D. 1970. Genes, language, and evolution. *Semiotica* 2:295–320.

——. 1981. The values—and limits—of sociobiology: toward a revival of natural right. In E. White, ed., *Sociobiology and Human Politics*. Lexington, Mass.: Lexington Books, pp. 135–165.

Mayr, E. 1954. Change of genetic environment and evolution. In J. S. Huxley, A. C. Hardy, and E. B. Ford, eds., *Evolution as a Process*. London: Allen & Unwin, pp. 157–180.

Morone, J. G., and E. J. Woodhouse. 1986. *Averting Catastrophe: Strategies for Regulating Risky Technologies*. Berkeley: University of California Press.

Nitecki, M. W., ed. 1984. *Extinctions*. Chicago: University of Chicago Press.

Peterson, C. 1987. Skin cancers become an epidemic in U.S. Washington Post Service News Report, *Honolulu Advertiser*, March 10.

Peterson, S. A., and A. Somit. 1983. Punctuated equilibria, sociobiology, and politics. Paper presented at the Southern Political Science Association meeting, Birmingham, Ala.

Phinney, E. S. 1988. The case of the speckled partner. *Nature* 331:566–568.

Prigogine, I. 1980. *From Being to Becoming*. San Francisco: W. H. Freeman.

Randall, T. 1989. The dinosaurs' demise: Iowa crater linked to extinction. *Honolulu Advertiser*, August 6, p. E-1.

Reader, J. 1981. *Missing Links: The Hunt for Earliest Man*. Boston: Little, Brown.

Reichs, K. 1983. *Hominid Origins*. Washington, D.C.: University Press of America.

Rensberger, B. 1988. Modern human older than Neanderthal? *Honolulu Advertiser*, February 18, p. 1.

Rodman, J. 1980. Paradigm change in political science: an ethological perspective. *American Behavioral Scientist* 24:49–78.

Schopf, T. J. M., ed. 1972. *Models in Paleobiology*. San Francisco: Freeman, Cooper.

——. 1982. A critical assessment of punctuated equilibria. I. Duration of taxa. *Evolution* 36:1144–1157.

Schubert, G. 1967. The rhetoric of constitutional change. *Journal of Public Law* 16:16–50.

——. 1981a. The sociobiology of political behavior. In E. White, ed., *Sociobiology and Human Politics*. Lexington, Mass.: Lexington Books.

——. 1981b. Glaciers, neoteny, and epigenesis: a review essay. *Journal of Social and Biological Structures* 6:65–80.

——. 1982a. Infanticide by usurper Hanuman langur males: a sociobiological myth. *Social Science Information* 20:199–244.

——. 1982b. Political ethology. *Micropolitics* 2:51–86.

——. 1983a. The structure of attention: a critical review. *Journal of Social and Biological Structures* 6:65–80.

——. 1983b. The evolution of political science: paradigms of physics, biology, and politics. *Politics and the Life Sciences* 1:97–124.

——. 1985. Epigenetic evolutionary theory: Waddington in retrospect. *Journal of Social and Biological Structures* 8:233–253.

——. 1986a. Scientific creation and the evolution of religious behavior. *Journal of Social and Biological Structures* 9:241–260.

——. 1986b. Primate politics. *Social Science Information* 25:647–680.

——. 1989. *Evolutionary Politics*. Carbondale: Southern Illinois University Press.

Shepard, P. 1978. *Thinking Animals: Animals and the Development of Human Intelligence*. New York: Viking Press.

Simpson, G. G. 1944. *Tempo and Mode in Evolution*. New York: Columbia University Press.

——. 1953. *The Major Features of Evolution*. New York: Columbia University Press.

Smith, J. Y. 1988. Luis Alvarez, Nobel-winning atomic physicist, dies. *Washington Post*. September 3, p. B-6.

Somit, A., and J. Tanenhaus. 1967. *The Development of Political Science*. Boston: Allyn and Bacon.

Stanley, S. 1979. *Macroevolution: Patterns and Process*. San Francisco: W. H. Freeman.

——. 1981. *The New Evolutionary Timetable: Fossils, Genes, and the Origin of Species*. New York: Basic Books.

——. 1986. Is Human evolution punctuational? In B. J. Williams, ed., *On Evolutionary Anthropology*. Malibu, Calif.: Undena Publications, pp. 77–89.

——. 1987. *Extinction*. New York: Scientific American.

Stringer, C. 1988. The dates of Eden. *Nature* 331:565–586.

Templeton, A. R. 1980. Macroevolution. *Evolution* 34:1124–1227.

Ucko, P. J., and G. W. Dimberley, eds. 1968. *The Domestication and Exploitation of Plants and Animals*. Chicago: Aldine.

United Press International. 1987a. New data support theory on dinosaurs. *Honolulu Advertiser*, May 8, p. 5.

——. 1987b. "Enormous event": supernova's neighbor stuns astronomers. *Honolulu Advertiser*, May 22, p. A-20.

Velikovsky, I. 1950. *Worlds in Collision*. New York: Doubleday.

——. 1952. *Ages in Chaos*. New York: Doubleday.

——. 1955. *Earth in Upheaval*. New York: Doubleday.

Waddington, C. H. 1957. *The Strategy of the Genes*. London: George Allen & Unwin.

——. 1960. *The Ethical Animal*. London: George Allen & Unwin.

——. 1975. *The Evolution of an Evolutionist*. Ithaca: Cornell University Press.

Walker, A. 1984. Extinction in hominid evolution. In M. Nitecki, ed., *Extinctions*. Chicago: University of Chicago Press, pp. 119–152.

Washington Post Service. 1987a. Mysterious supernova isn't behaving as astronomers expected. *Honolulu Advertiser*, March 7 p. A-12.

——. 1987b. Supernova casts doubt on exploding-star theory. *Honolulu Advertiser*, April 15, p. A-25.

Webb, S. D. 1984. Ten million years of mammal extinction in North America. In P. S. Martin and R. G. Klein, eds., *Quaternary Extinctions: A Prehistoric Revolution*. Tucson: University of Arizona Press, pp. 189–210.

West Eberhard, M. J. 1976. Born: sociobiology. *Quarterly Review of Biology* 51:89–92.

White, E. 1981. Political socialization from the perspective of generational and evolutionary change. In E. White, ed., *Sociobiology and Human Politics*. Lexington, Mass.: Lexington Books.

Williams, B. J., ed. 1986. *On Evolutionary Anthropology*. Malibu, Calif.: Undena Publications.

Williams, G. 1966. *Adaptation and Natural Selection*. Princeton: Princeton University Press.

Wright, S. 1940. The statistical consequences of Mendelian heredity in relation to speciation. In J. S. Huxley, ed., *The New Systematics*. Oxford: Clarendon Press, pp. 161–183.

——. 1968. *Genetic and Biometric Foundations*. Chicago: University of Chicago Press.

——. 1977. *Experimental Results and Evolutionary Deductions*. Chicago: University of Chicago Press.

——. 1978. *Variability within and among Natural Populations*. Chicago: University of Chicago Press.

Yanarella, E. J. 1984. Slouching toward the apocalypse: visions of nuclear holocaust and eco-catastrophe in contemporary science fiction. Paper presented at the American Political Science Association meeting, Washington, D.C.

Young, Z. J., E. M. Jope, and K. P. Oakley, eds. 1981. *The Emergence of Man*. London: Royal Society and British Academy.

Zihlman, A. 1983. A behavioral reconstruction of australopithecus. In K. J. Reichs, ed., *Hominid Origins*. Washington, D.C.: University Press of America, pp. 207–238.

12

Gradualism and Discontinuous Change in Evolutionary Biology and Political Philosophy

ROGER D. MASTERS

POLITICAL BIAS AND EVOLUTIONARY BIOLOGY

Both gradualist and punctuated equilibrium models of evolution have been attacked on the grounds that they reflect political bias. Such charges are particularly important because theories of evolution can have direct implications for our understanding of human life. It is therefore both legitimate and important to question whether models of gradual or discontinuous rates of change in evolution necessarily have political implications.

According to the generally accepted view of scientific method, theories or hypotheses that can be tested by competent specialists should not be equated with the motivations or passions of an individual researcher. Objective demonstration of theoretical bias requires that partisanship or political motivation arise from the scientific concepts as distinct from the scientists proposing them. The validity of a scientific theory or finding is not contradicted merely because of its political implications in a specific historical context. Although Galileo's astronomical observations and theories had theological as well as political consequences, they were in themselves scientific. Any biologist tempted to dismiss the difference between politics and science would do well to reflect on the careers of Lysenko and his critics.

To apply this distinction to recent debates on evolutionary theory, we need to distinguish the structure of a scientific explanation from its use by particular scientists. If a concept or theory is inherently biased, this political or ideological element should be evident in the pattern and implications of its use over time. When an approach is used by diverse thinkers with varied political or religious consequences, we can presume that the political implications stemmed from the social context or the individual scientist's motivations rather than from the scientific concept or theory per se.

A model of evolutionary change would be vitiated by political bias only if it logically entails concepts of social change that favor some groups or political interests over others. If the concept of gradualism in biology necessarily implies the likelihood or desirability of gradual or imperceptible rates of change in human history, or if the model of punctuated equilibria in the evolution of species always justifies revolutionary change in politics, perhaps the biological theory could be criticized as a covert statement of political preference. In contrast, should the concepts of both gradual and discontinuous rates of change in biology be combined with diverse political or social theories, ideological commitment would presumably not be due to these scientific concepts as distinct from the way they are used.

In this chapter I survey the antecedents of gradualist and punctuated equilibrium models of evolution in Western philosophy and science. The goal is not completeness. Even a limited sample should show whether there have been divergent political uses of these views of biological evolution. As we will see, the evidence shows that neither gradual nor discontinuous concepts of evolution has intrinsic political implications independent of the historical context.

Ancient Biology and History: Aristotle and Empedocles

It is widely believed that contemporary scientific theories have few if any historical antecedents in antiquity worth considering (e.g., Bowler, 1984). More often than not, as discussions of modern physics suggest, the opposite is the case (Heisenberg, 1958; Capra, 1975). Although Darwinian biology was contrary to the religious doctrines that had prevailed in the seventeenth and eighteenth centuries, evolutionary theories clearly existed in other times and places. We know, for example, that the ancient Greeks explored the origin, persistence, and modification of species in ways that are directly relevant to the issues of modern science.

To be sure, ancient thinkers should not be treated as if they were precursors of modernity in a linear development; such a reading of the past introduces, within the history of science, the very assumptions that are at issue. Stephen Jay Gould was therefore correct, on methodological grounds, to "reject an approach to the history of science that rapes the past for seeds and harbingers of later views; such a perspective only makes sense within the abandoned faith that science progresses by accumulation towards absolute truth" (Gould, 1977:16). But one cannot go to the other extreme and claim that all prior thought is irrelevant without implying that each epoch is unique. Such extreme historical relativism is self-defeating because it, too, can be traced to classical antiquity. It is as inappropriate to assume that the history of science conforms to a punctuated equilibrium model of change as it is to adopt implicitly a gradualist and progressive image of intellectual history.

To explore the structural relationships between concepts of biological change and political attitudes in the past, one needs to avoid factual errors and prejudices. Ernst Mayr has asserted, for example, that Plato and Aristotle adopted a "typological" view of animal species which is antithetical to evolutionary principles: "Typological thinking is the other major misconception that had to be eliminated before a sound theory of evolution could be proposed. . . . The concepts of unchanging essences and of complete discontinuities between every *eidos* (type) and all others make genuine evolutionary thinking well-nigh impossible. I agree with those . . . who claim that the typological philosophies of Plato and Aristotle are incompatible with evolutionary thinking" (1963:5). Without contesting Mayr's immense contribution to biology, this account of traditional philosophy will not stand scrutiny.

Although Plato's biology is more relevant to contemporary thought than has been recognized (Masters, 1989a), it will be useful to focus on Aristotle because he dealt directly with an issue akin to punctuated equilibria versus gradualism. But to understand his position, one must consider the writings of Aristotle rather than the medieval theologians or nineteenth-century antievolutionists who may have cited him.

The Character of Aristotle's Biology

Aristotle wrote extensively and descriptively about animals but rarely if ever with a concern for what we call evolution. His primary focus was the analysis of form and function—or what today would be called the study of adaptation, ontogeny, and systematics. As one careful student of Aristotle's concept of *eidos* (form or type) concludes: "I believe that Aristotle is not committed to Noah's Ark Essentialism, to 'typology', or to put it most paradoxically, he is not committed to the taxonomic theory which is sometimes called Aristotelian Essentialism" (Preus, 1979:81).

Why has there been such a massive misunderstanding of Aristotle? First, Aristotle's goal in such works as *Parts of Animals, History of Animals,* and *Generation of Animals* is to establish biology as an empirical science. In particular, he seeks to dismiss the approach to classifying species, found for example in Plato's *Statesman*, that is based on simple "dichotomies" (*Parts of Animals*, 1.2–4.642b–644b [1984:I, 1000–1003]). In so doing, Aristotle's usual method can be described, in contemporary terms, as "phenetic" classification—that is, he based his concept of species on visible or phenotypical traits rather than on genetic origin (Preus, 1979; Arnhart, 1990).

Second, and equally important, Aristotle was not particularly interested in the "origin of species." Because his primary concern was the relationship between form and function, Aristotle sought to explain what we call "adaptation" as a natural process, making living beings amenable to scientific analysis. Though often speaking of species as if they had existed from all time, Aristotle was quite open to the emergence of new forms (e.g., *Generation of Animals*, 2.2.746a–b [1984:I, 1158]) as well as to the existence of intermediate forms inconsistent with what Anthony Preus calls "Noah's Ark Essentialism" (*Parts of Animals*, 4.5.681a [1984:I, 1062–1063], cited by Preus, 1979:81).

In contemporary terms, Aristotle sought to understand ontogeny rather than phylogeny. As is evident from his interest in reproduction and embryos, he was concerned with the functional adaptation of traits rather than their evolutionary origin and change. When Aristotle speaks of the *telos* or end of a structure, he is not presuming that animals demonstrate a divine plan of the sort accepted in the Judeo-Christian tradition; rather, Aristotle seeks empirical reasons for the observed shapes and characteristics of living beings on the assumption that only such "final causes" would explain the difference between animate and inanimate beings (e.g., *Physics*, 198b–199a [1984:I, 339]). Aristotle's concept of functionalism is thus close to what is today called "teleonomy" (Pittendrigh, 1958; Mayr, 1974) and should not be confused with the theological uses of teleology.

Because Aristotle was more concerned with the maintenance and functional importance of species-specific traits than with phylogenetic reconstruction, he stressed the question of the resemblance between parents and offspring in a single generation rather than the process of speciation. As Max Delbruck (1971, 1977) points out, Aristotle could be described as the discoverer of the principle of heredity because he was the first biologist of note to develop a coherent theory for the transmission of bodily traits from one generation to the next. And though Aristotle suggests analogies between a stage of development and a kind of organism, it is misleading to see his thought as a precursor of Ernst Haeckel's theory of ontogeny as the recapitulation of phylogeny: "Aristotle's comparisons are analogies, drawn to reinforce his belief in the epigenetic nature of development" (Gould, 1977:16).

There were sound intellectual reasons for Aristotle's focus, because he confronted objections to a scientific biology that were the opposite of those facing Darwin. In the mid-nineteenth century, the main obstacle to a scientific study of living beings came from the theological doctrine that all animals were the result of God's plan at the Creation. For Aristotle, the issue arose because earlier thinkers had attributed everything to chance and thus denied that what we call a species is "natural." More specifically, Aristotle was concerned to show that the theories of Empedocles could not explain the transmission of traits from parents to offspring and hence were unsatisfactory as a foundation for biology.

It can be objected, as Gould notes in summarizing Empedocles' theory, that "any comparison between this scheme and later evolutionary thinking is purely gratuitous; Empedocles' system . . . was for another time and another purpose" (Gould, 1977:413). But here we are comparing concepts of change in evolutionary biology and in politics—as Gould himself explores the relationship between ontogeny and phylogeny—without assuming that Empedocles was in some way a forerunner of modern evolutionary theory. What is the difference between Empedocles, as an early exponent of a biological theory structured in terms of discontinuities followed by periods of stability, and Aristotle, whose method leads to a stress on continuities? Are their different approaches to biological change associated with political bias?

Why Did Aristotle Reject Empedocles' Biology?

Although we lack extensive texts by Empedocles, Aristotle gives enough citations and references to permit us to reconstruct the issue. Like a number of Greek philosophers (notably some of the pre-Socratics, Heraclitus, and Epicurus), Empedocles saw the visible world as a process of continual change. In this flux, Empedocles held that accident or chance rather than "nature" determines not only what a human being perceives and thinks (*Metaphysics*, 4.5.1009b [1984:II, 1594]) but how things themselves are constituted. Hence, according to Aristotle (ibid., 5.4.1015a [1984:II, 1602]), Empedocles says:

> Nothing that is has a nature,
> But only mixing and parting of the mixed,
> And nature is but a name applied to them by men.

For Empedocles, species are arbitrary categories imposed by humans rather than naturally existent groups of animals.

Although Empedocles' theory resembles Darwinian biology in postulating a process of random variation and a kind of natural selection, the components that vary are entire morphological structures. According to Aristotle, "Empedocles accounts for the creation of animals in the time of his Reign of Love, saying that 'many heads sprang up without necks', and later on these isolated parts combined into animals" (*Generation of Animals*, 1.18.722b [1984:I, 1122]). Mere accident gave rise to different living forms, most of which—like a "man-faced ox-progeny"—perished; those animals that were adapted to the world survived, as a result of what Aristotle calls "necessity" rather than "nature" (*Physics*, 2.5–7.196b–198b [1984:I, 335–339]).[1] Empedocles could therefore be said to have a concept of time not unlike the Gould-Eldredge model of punctuated equilibria: periods of rapid change, in which random events produce a wide variety of forms, lead to the selection of types that tend to be conserved over long periods. Though crude by comparison with contemporary biological theory, in part because of the simplicity of his image of "hopeful monsters," Empedocles thus provides us with an early instance of nongradual change as the basis of speciation.

Aristotle's biology responds to Empedocles, treating the existence of functional adaptation as a legitimate and serious question:

> Why should not nature work, not for the sake of something, nor because it is better so, but just as the sky rains, not in order to make the corn grow, but

1. Aristotle uses the concepts "necessity" and "chance" in a way that differs fundamentally from modern usage: what he calls chance is an unintended or accidental result within the domain of purposive behavior. What we today call "chance" is, for Aristotle, an instance of "necessity" (see Masters, 1989a, 1989b). Many modern commentators have misunderstood Aristotle profoundly because they have failed to attend to this crucial difference between his concepts and those of our own day.

> of necessity? (What is drawn up must cool, and what has been cooled must become water and descend, the result of this being that the corn grows.) Similarly if a man's crop is spoiled on the threshing-floor, the rain did not fall for the sake of this—in order that the crop might be spoiled—but that result just followed. Why should it not be the same with the parts in nature, e.g. that our teeth should come up of necessity—the front teeth sharp, fitted for tearing, the molars broad and useful for grinding down the food—since they did not arise for this end, but it was merely a coincident result; and so with all other parts in which we suppose that there is purpose? Wherever then all the parts came about just what they would have been if they had come to be for an end, such things survived, being organized spontaneously in a fitting way; whereas those which grew otherwise perished and continue to perish, as Empedocles says his "man-faced oxprogeny" did. (Ibid., 2.8.198b [1984:I, 339].)[2]

Aristotle's reason for rejecting such a biology lies not in phylogeny but in ontogeny—and not in theory but in observation.

In trying to account for the generation of animals, Aristotle notes "that it is thought that all animals are generated out of semen, and that the semen comes from the parents. That is why it is part of the same inquiry to ask whether both male and female produce it or only one of them, and to ask whether it comes from the whole of the body or not from the whole" (*Generation of Animals*, 1.17.721b [1984:I, 1120]). Confronted by such uncertainty, Aristotle was forced to assess questions like the inheritance of acquired characteristics (e.g., "the argument that mutilated parents produce mutilated offspring"), which would be evidence that the entire soma or body of both parents is the source of reproductive material: "First, then, the resemblance of children to parents is no proof that the semen comes from the whole body, because the resemblance is found also in voice, nails, hair and way of moving, from which nothing comes. And men generate before they yet have certain characters, such as a beard or grey hair. Further, children are like their more remote ancestors from whom nothing has come, for the resemblances recur at an interval of many generations, as in the case of the woman in Elis who had intercourse with a negro; her daughter was not negroid but the son of that daughter was" (ibid., 1.18.722a [1984:I, 1121]).

2. Darwin cites this passage explicitly in the "historical sketch" that he added at the outset of *The Origin of Species* (n.d.:xiii, note). As is evident from Darwin's comments on the passage, he did not understand the context and misinterpreted Aristotle's point.

These arguments are easily restated in contemporary terms: Aristotle noted that not all parent-offspring phenotypic similarities need be inherited, that some resemblances are due to ontogenetic patterns rather than to the somatic state of parents at the time of procreation, and that the transmission of "recessive traits" would be impossible without a material basis of inheriting physical characteristics (cf. *Politics*, 2.3.1262a [1984:II, 2002–2003]). As was remarked above, Aristotle's emphasis on these aspects of reproduction has led to the suggestion that he be credited with the discovery of the principle of DNA (Delbruck, 1977; Masters, 1989b).

The functional integrity of organisms and the regularity of ontogenetic development led Aristotle to reject both direct somatic inheritance and Empedocles' account of the origin of species in a chaotic epoch in which "isolated parts" were "combined into animals": "Now that *this* is impossible is plain, for neither would the separate parts be able to survive without having any soul or life in them, nor if they were living things, so to say, could several of them combine so as to become one animal again. Yet those who say that semen comes from the whole of the body really have to talk in that way, and as it happened during [Empedocles'] Reign of Love, so it happens according to them in the body" (*Generation of Animals*, 1.18.722b [1984:I, 1122]; italics in original). Observable phenomena in ontogeny, rather than speculative phylogeny, provide Aristotle with basic insights into the nature of animals.

This contrast between Aristotle and Empedocles is equally evident in their accounts of gender. Empedocles had claimed that sex is determined by the environment during fetal development: "It is shed in clean vessels; some wax female, if they fall in with cold." Aristotle retorts with empirical evidence suggesting that sex must be determined by hereditary material: "It is plain that both men and women change not only from infertile to fertile, but also from bearing female to bearing male offspring, which looks as if the cause does not lie in the semen coming from all the parent or not, but in the mutual proportion or disproportion of that which comes from the woman and the man, or in something of this kind" (ibid.). Although Aristotle's understanding of the role of males and females in procreation departs from our own, he sought empirical evidence from ontogeny to explain how offspring derive physical traits from their parents.

The difference between these ancient thinkers and contemporary biology lies less in the basic concepts of ontogeny and phylogeny than in their theoretical articulation. For Empedocles, a phylogenetic account structurally similar to neo-Darwinian evolutionary theory (random or accidental variation culled by natural selection) was combined with a view of ontogenesis based on somatic inheritance and the transmission of acquired traits. For Aristotle, stress on the functional unity and adaptation of organisms led to concern for mechanisms of inheritance, while phylogeny was either ignored or dismissed in favor of the view that all species form a natural continuum of complexity. Modern biology has, of course, combined phylogenetic views like those of Empedocles with ontogenetic principles akin to those of Aristotle.

Aristotle is therefore not primarily interested in what we call the evolutionary process, focusing instead on the need to explain the ontogeny of structure and behavior—that is, what he calls "the form of the living being" as distinct from its "material nature":

> For the generation is for the sake of the substance, and not this for the sake of the generation. Empedocles, then, was in error when he said that many of the characters presented by animals were merely the results of incidental occurrences during their development; for instance, that the backbone is as it is because it happened to be broken owing to the turning of the foetus in the womb. In so saying he overlooked the fact that propagation implies a creative seed endowed with certain formative properties. Secondly, he neglected another fact, namely, that the parent animal pre-exists not only in account, but actually in time. For man is generated from man; and thus it is because the parent is such and such that the generation of the child is thus and so. (*Parts of Animals*, 1.1.640a [1984:I, 996].)

What has been described as "typological" thinking in Aristotle is actually his insistence on treating the organism as an adapted, functioning offspring of a naturally distinct and identifiable species.

The difference between Empedocles and Aristotle thus points to two quite different ways of understanding species. For Empedocles, as later for Darwin, the accidental causes of speciation mean that the classification of species is a human artifact. In contrast, Aristotle would have agreed with a modern systematist like Mayr, for whom species exist in nature even if the boundaries of the species concept cannot always be defined with mathematical rigor. "Whoever, like Darwin, denies that species are nonarbitrarily defined units of nature not only evades the issue, but fails to find and solve some of the most interesting problems of biology. These problems will be apparent only to the student who attempts to determine species status of natural populations" (Mayr, 1963:29).

Although the foregoing discussion barely introduces the complexity of ancient biology, it indicates three points. First, Aristotle was not unaware of a theory akin to the modern notion of nondirected evolution, but he consciously rejected it as inconsistent with any conceivable mechanism of inheritance that could produce functioning organisms (not to mention such phenomena as recessive traits). Second, Empedocles' theory of the origin of species bears a structural resemblance to contemporary models of punctuated equilibria rather than to Darwin's own view of continuous variation. And finally, Aristotle's tendency to treat existing species as given does not imply the taxonomic fixity often attributed to him under the name "Aristotelian essentialism"; on the contrary, his view of change can be compared to "gradualist" models of evolution, particularly since Aristotle flatly denies the eternity of living beings and stresses the mechanisms of small but continual changes both within and between species (*Physics*, 1.6–9, esp. 8.191b [1984:I, 327]; *Generation and Corruption*, 1.3–5.317a–322a [1984:I, 518–527]; *History of Animals*, 3.9.517a [1984:I, 820–821]; 8.1.588b [1984:I, 922]; *Parts of Animals*, 1.1.639a–b [1984:I, 994–995]; *Generation of Animals*, 2.1.733b–735b [1984:I, 1138–1141]).

Aristotle's Politics: Biological Continuity, Historical Gradualism, and Civic Virtue

Empedocles can be described as an ancient philosopher whose concept of evolutionary time is akin to recent models of punctuated equilibria, whereas Aristotle's view can be compared to contemporary gradualist theories. This makes it possible to use the ancients as evidence of the political implications of different concepts of evolutionary change. Consideration of Aristotle in this light is particularly reasonable because his political analysis is directly related to his biology. Empedocles seems to share the views of other pre-Socratics for whom human society is an unnatural or conventional restraint on individual pleasure (cf. Masters, 1977), whereas Aristotle treats politics as "natural" to human beings.

The link between Aristotle's biology and his political theory is symbolized by the term *Zoon politikon* (political animal), which he uses to describe the human species. Just as each animal species has a "way of life" that influences individual and social behavior, so different human populations have a characteristic way of life with social and political consequences (*Politics*, 1.1–10, esp. 8.1256a–b [1984:II, 1992–1993]). Moreover, each political community can be described as having a regime or constitution of a determinate species (*eidos*): the same concept that Aristotle uses to refer to kinds of animals is used to refer to the kinds of political order found in various human societies (Preus, 1979; Arnhart, 1990).

Aristotle's classification of political regimes is famous: he argues that there are broadly six types of political systems, three consistent with the common good (monarchy, aristocracy, and polity) and three run in the interest of the rulers (tyranny, oligarchy, and democracy). In practice, however, the common political divisions are based on class interests: the wealthy social classes favor oligarchy, while the poor seek democracy (*Politics*, 3.7–9.1279a–1281a [1984:I, 2030–2033]). These distinctions at first seem to be based on the number of individuals in the ruling class, but observable social class distinctions, not an abstract criterion like the number of rulers, are the key to understanding the major kinds of regime as well as their subvarieties (ibid., 4.1–15.1288b–1301a [1984:II, 2045–2066]).

The connection between Aristotle's political theory and his biology is particularly relevant to his concept of social change, for Aristotle is perhaps the most profound gradualist in the Western intellectual tradition. When describing the origin of civilized societies like the city-state (*polis*) of the Greeks, Aristotle summarizes briefly an account that can only be characterized as evolutionary (ibid., 1.1–2.1252a–1253a [1984:II, 1986–1988]); at no time does Aristotle imply that there has been—or will be—a sudden and irreversible quantum change in the human condition. Later, when explicitly analyzing political upheavals, he describes "revolutions" as changes in regime that can be more or less extensive, but in no case are they presented as irreversible (ibid., 5.1–12, esp. 1.1301b [1984:II, 2066–2067]).

Aristotle is thus a thinker whose political theory was consistent with his understanding of biological change and human nature since he takes a gradualist approach to both biological and political change. Substantively, Aristotle's political principles include the acceptance of slavery (at least for some individuals), a preference for the mixed regime in an ancient city-state, the rejection of militarism (and, tacitly, criticism of expansionist empires like that of Alexander the Great), and insistence that political activism is lower in status than scientific inquiry (Strauss, 1964). The virtues are "by nature" and can generally be described as "means" between extremes; the virtuous human being has a character exemplified by rationality, moderation, and concern for the social consequences of individual behavior.

Whatever the differences between one society and another, natural justice exists and forms the basis of political judgment (*Nicomachean Ethics*, 5.7.1134b–1135a [1984:II, 1790–1791]). Although Aristotle was clearly not in favor of tyranny, he could hardly be called a democrat in the modern sense; he stressed the importance of moderation and the idea of a mixed, limited regime dominated by a class of leisured, well-educated, and virtuous gentlemen. Political participation, particularly in a community that permits "ruling and being ruled in turn," is a proper and excellent exercise of the natural faculties of a human being (Strauss, 1964; Masters, 1989a, b; Arnhart, 1990). As we will see, gradualist principles in biological evolution and human history have sometimes been associated with quite different political principles.

Lucretius: Natural Discontinuity, Historical Gradualism, and Private Pleasure

Although Empedocles did not leave behind a comprehensive treatment of politics, other thinkers in pagan antiquity shared similar biological ideas and wrote on politics. The continued interest in Empedocles' theory is demonstrated by an extant Epicurean text—a fragment of Hermarque's *Letters on Empedocles* describing the "genealogy" of human ethics (Goldschmidt, 1977:48–49, 287–297). The most complete version of Epicurean thought remaining to us—Lucretius's *De Rerum Natura* (*On the Nature of Things*)—explicitly praises Empedocles as having made "excellent and divine discoveries" (1.716–733 [1951:48]) and presents an account of organic and human evolution which bears many structural similarities to that of Empedocles (5.772–1457 [1951:194–216]).

For Lucretius, the world as we see it is relatively new, yet fundamental change in the cosmos is imminent; like Empedocles, Lucretius does not attribute the beings we see to a beneficent divine plan (5.324–350[1951:181]). During an early phase of the earth's history, natural causes produced a variety of living forms:

> In those days the earth attempted also to produce a host of monsters, grotesque in build and aspect—hermaphrodites, halfway between the sexes yet cut off from either, creatures bereft of feet or dispossessed of hands, dumb, mouthless brutes, or eyeless and blind, or disabled by the adhesion of their limbs to the trunk, so that they could neither do anything nor go anywhere nor keep out of harm's way nor take what they needed. These and other such *monstrous and misshapen births were created. But all in vain.* Nature debarred them from increase. . . . In those days, again, *many species must*

> *have died out altogether* and failed to reproduce their kind. (Lucretius, 5.837–856 [1951:197]; italics in original.)

For Lucretius as for Empedocles, there was a historical period of wide variation in living forms, including many "hopeful monsters" that were culled out by something akin to what is today called natural selection.

Lucretius explicitly rejects the notion of interspecific chimera on the grounds that "each species develops according to its own kind, and they all guard their specific characters in obedience to the laws of nature" (5.922–924 [1951:199]). Hence the conservation of species once formed is, in this view, guaranteed by the divergence between the nature of each kind of animal. Like Empedocles, Lucretius could be described as a theorist whose concept of time shares the discontinuity found in punctuated equilibrium models.

When we turn to Lucretius's account of humans, however, we find an evolutionary description that can only be called gradualist (Lucretius, 925–1457). As he concludes, "Not only such arts as sea-faring and agriculture, city walls and laws, weapons, roads and clothing, but also without exception the amenities and refinements of life, songs, pictures, and statues, artfully carved and polished, *all were taught gradually by usage and the active mind's experience as men groped their way forward step by step*. So each particular development is brought gradually to the fore by the advance of time, and reason lifts it into the light of day" (Lucretius, 1448–1459 [1952:216]). This gradualist view of human history, according to which all social rules arose through consent, is also presented in the Epicurean "genealogy" of Hermarque (*Letters on Empedocles*, in Goldschmidt, 1977:287–297, esp. 293–294).

Substantively, Lucretius and other Epicureans use the gradual emergence of human political institutions to demonstrate that justice is not a thing in itself, but rather is an agreement based on the mutual interests of individuals. Since the political community is therefore founded on "contract," the natural standard of justice varies from one time and place to another. The natural "end" or goal of human life is not to be found in political life. Hence Lucretius follows Epicurus in insisting that the prudent man will avoid political commitment as well as all other activities that risk his calm and contemplation (Goldschmidt, 1977). These substantive conclusions about human politics are, of course, sharply opposed to the position of Aristotle (Masters, 1977, 1989a).

Two conclusions follow from the juxtaposition of Lucretius's views of the origin of species and of human history. First, concepts of biological change do not logically entail corresponding ideas of the rate of change in human history. And second, even with similar gradualist notions of human history, thinkers like Aristotle and Lucretius diverge sharply in their assessment and judgment of politics. To judge by this hasty look at several ancients, the form of a theory of evolutionary rates of change does not seem to entail specific political opinions.

Modern Biology, Human Evolution, and Politics: Hobbes, Rousseau, and Marx

In the epoch following the Renaissance and before Darwin, we can find thinkers who elaborated nontheological concepts of evolution and thus permit a broader test of the relationship between evolutionary models and political principles. To this end, it is instructive to compare three famous modern political theorists: Thomas Hobbes, Jean-Jacques Rousseau, and Karl Marx.

Hobbes: Discontinuity and Radical Change in Human History

Although Hobbes was a secular thinker who does not seem to accept the biblical account of Creation as literally true, he does not focus on evolutionary change in humans or other animals. Like Aristotle, Hobbes usually takes contemporary animal species as given; when he discusses the "natural condition of mankind" in *Leviathian*, Hobbes refers to a currently existent state of affairs rather than the historical origin of the human condition ("it was never generally so, over all the earth"—1.13 [1962:101]).

Hobbes thus combines a view of the natural fixity or stability of animal species with a highly discontinuous interpretation of human history. For Lucretius, as for many other pagan thinkers, the discovery of language, the arts, and political institutions is presented as a slow, progressive process. Hobbes treats each step as if it were a sudden event. Language, for example, is an "invention" whose origin can be compared to the invention of alphabetic writing and printing (ibid., 1.4 [1962:33]). The founding of the state is a "covenant" that is described as a specific event (ibid., 2.22–23 [1962:129–141]), be it violent conquest (a "commonwealth by acquisition") or mutual consent (a "commonwealth by institution").

Perhaps even more important, Hobbes sees his ideas as having potentially enormous or "revolutionary" effects. His theory is strangely constructed so that its teaching and acceptance are formally necessary to its success. Although all previous societies have collapsed into civil war, the general acceptance of Hobbesian principles will make possible an "eternal" political community that resolves all intractable social conflict (ibid., 2.20 [1962:157–158]; 30 [1962:248–249]: 31 [1962:270]).

The key to such a radical change in the human condition lies in the paradoxical character of Hobbes's teaching. On the surface, Hobbes speaks in favor of an apparently omnipotent "sovereign" who acts for the community untrammeled by constraints caused by institutional checks and balances. But Hobbes's principles include such notions as the subjects' inalienable right to refuse to serve in the military, to resist arrest, and to use force against any government that does not effectively protect the citizen (ibid., 1.14 [1962:108]; 2.18 [1962:137]; 21 [1962:165]; 24 [1962:187]). Moreover, Hobbes's supposedly absolute sovereign is the "representative" of the body politic who is obliged to teach his subjects of their "inalienable" natural right to rebel against him if they do not feel secure in their lives, their liberty, and their property (ibid., 2.18 [1962:137]; 21 [1962:159–168]; 23 [1962:181]). These paradoxes disappear if Hobbes is viewed as a complex philosopher for whom universal adoption of a new way of thinking could totally transform the human condition (Masters, 1980).

For Hobbes, therefore, a theory of relatively fixed animal species is associated with a human history marked by discontinuous changes. The substantive political opinions associated with these attitudes toward time are a vigorous denial of any natural standards of right and wrong, emphasis on the state as the only guarantor of stability and civilized life, and support for a strong central government. One can also infer some of the foundations of laissez-faire capitalism and liberal republicanism from Hobbes's notions that commerce and industry are needed for peaceful prosperity, that monopolies are a serious danger to enterprise, and that the sovereign is the "representative of the people," whose interests must be defended lest they have legitimate grounds to oppose the government.

Like Lucretius, Hobbes holds that the only justice accessible to humans stems from the enforcement of man-made laws; there is no natural fairness or justice. But like Aristotle, Hobbes presents a science of politics designed to provide concrete guidance for political leaders. This combination of principles reflects a view of time differing from those of almost all pagan thinkers: Hobbes—unlike the ancients—based his optimism for the future on the assumption that natural science and technology could conquer natural necessity, material scarcity, and political conflict (Masters, 1977). Even more important than the rate of change for Hobbes is its direction, since he foresees historical progress based on scientific knowledge.

Rousseau: Natural Evolution and Revolutionary Politics

Writing a century after Hobbes, Rousseau explicitly criticized the Hobbesian understanding of human nature as inconsistent with what is known about other animals. Unlike Hobbes, Rousseau insists that an analysis of human nature must be based on an evolutionary theory of our species' origins (*Second Discourse*, Preface [1964:91–97]). Yet like Hobbes, Rousseau describes political institutions as the result of a "social contract" and sees human history as a process characterized by discontinuities which he explicitly calls "revolutions" (ibid., Part 2 [1964:168–173]).

Hobbes had argued that all legal and ethical standards are the result of conventions established among selfish individuals; since humans are by nature competitive and aggressive, Hobbes denies that they naturally cooperate on the basis of rationally defined obligations. Although this criticism of rationalist theories of human cooperation is accepted by Rousseau, the Hobbesian definition of the state of nature is rejected as inconsistent with what is known about other animals:

> Hobbes saw very clearly the defect of all modern definitions of natural right; but the consequences he draws from his own definition show that he takes it in a sense which is no less false. Reasoning upon the principle he establishes, this author ought to have said that since the state of nature is that in which the care of our self-preservation is the least prejudicial to the self-preservation of others, that state was consequently the best suited to peace and the most appropriate for the human race. He says precisely the opposite, because of having improperly included in the savage man's care of self-preservation the need to satisfy a multitude of passions which are the product of society and which have made law necessary. (Ibid., [1964:168–173].)

Since human needs and passions have fundamentally changed over time, an evolutionary approach to "human nature" is needed (ibid., Preface [1964:91–97]).

Rousseau explicitly abandons the biblical account of creation in favor of "hypothetical and conditional reasonings" akin to those "our physicists make every day concerning the formation of the world"; the only approach suitable to an analysis of human nature is one that would be at home in the pagan Athens of Plato and Aristotle (ibid., Exordium [1964:103]). Humans must be understood as the product of an evolutionary process; if current physical attributes of the species are taken as given, it is because of the limitations of knowledge rather than the fixity of the species.

> Important as it may be, in order to judge the natural state of man correctly, to consider him from his origin and examine him, so to speak, in the first embryo of the species,[3] I shall not follow his organic structure through its successive developments. I shall not stop to investigate in the animal system what he could have been at the beginning in order to become at length what he is. I shall not examine whether, as Aristotle thinks, man's elongated nails were not at first hooked claws; whether he was not hairy like a bear; and whether, if he walked on all fours, his gaze, directed toward the earth and confined to a horizon of several paces, did not indicate both the character and limits of his ideas. . . . Comparative anatomy has as yet made too little progress and the observations of naturalists are as yet too uncertain for one to be able to establish the basis of solid reasoning upon such foundations. (Ibid., Part 1 [1964:104–105].)

Even starting from the evolutionary epoch in which humans were already bipeds, it is possible to analyze the species as an "animal"—albeit an animal that can be characterized at this stage as "less strong than some, less agile than others, but all things considered, the most advantageously organized of all" (ibid.).

If one abstracts from social institutions and their effects on humans, the species must originally have been composed of healthy individuals

3. Rousseau explicitly compares ontogeny and phylogeny (compare note 4 below). It is not clear whether Rousseau derived this analogy from Vico, Lucretius, or another source, although Lucretius's *On the Nature of Things* was the model for the *Second Discourse* as a whole (see Rousseau, 1964: editorial note 38; and Masters, 1968). Rousseau's assessment of hominid evolution (phylogeny) differs, however, from his judgment of parallel effects at the individual level (ontogeny): after the evolutionary epoch called "savage society," Rousseau concludes that "all subsequent progress has been in appearance so many steps toward the perfection of the individual, and in fact toward the decrepitude of the species" (1964:151). Unlike those eighteenth-century theologians or naturalists who see in "nature" the proof of a beneficent providence, Rousseau does not assume that the functional "perfection" of individual organic traits represents better "adaptiveness" or harmony with nature at the species level.

because of the operation of what we today call natural selection: "Children, bringing into the world that excellent constitution of their fathers and fortifying it with the same training that produced it, thus acquire all the vigor of which the human species is capable. Nature treats them precisely as the law of Sparta treated the children of citizens: it renders strong and robust those who are well constituted and makes all the others perish" (ibid. [1964:106]). Just as wild animals "have a more robust constitution, more vigor, more strength and courage" than domesticated species (ibid. [1964:110]), so humans who are now weakened by the commodities of civilized societies were once stronger and more self-sufficient than at present. Surrounded by a natural environment of abundance and stability, primitive humans could have survived by substituting agility and flight for sheer strength whenever encountering stronger animals (ibid. [1964:107]).

It was therefore possible for the human species to survive without language, formal social cooperation, and political institutions. But such primitive humans would have been stupid animals without the passions that, for Hobbes, led to a "war of all against all" (ibid. [1964:129–130]). Although conflict could arise, the natural condition of peaceful and independent animals with limited needs is not a Hobbesian state of nature (ibid., note i [1964:195]). In short, humans were originally merely animals, albeit animals capable of changing or "perfecting" their own behavior (ibid., Part 1 [1964:114–115]).

Most commentators have stressed Rousseau's description of savage man in the state of nature as solitary and peaceful: "wandering in the forests, without industry, without speech, without domicile, without war and without liaisons, with no need of his fellow-man, likewise with no desire to harm them, perhaps never even recognizing anyone individually" (ibid. [1964:137]). But from a theoretical perspective, it is the resolutely evolutionary approach to human origins that should attract our attention: not only does Rousseau compare the original humans to living primates (ibid., note j [1964:203–213]), but he explicitly states that crows or monkeys "group together in approximately the same way" as savage men presumably did (ibid., Part 2 [1964:144]). For all practical purposes, a natural condition consisting of primitive hominids living in isolated family groups can be described as "solitary" life (*Essay on the Origin of Languages*, ch. ix [1966:31–34]; *Social Contract*, 1.2 [1978:47–48]; *Second Discourse*, Part 2 [1964:147]).

Rousseau's understanding of the evolutionary process is thus similar to that of Darwin. Writing a century before the publication of *Origin of*

Species, Rousseau took a gradualist view of evolutionary change. Something like natural selection operated on all wild animals; human evolution required "multitudes of centuries" (*Second Discourse*, Part 2 [1964:146]) and is a process in which "the lapse of time compensates for the slight probability of events" because of the "surprising power of very trivial causes when they act without interruption" (ibid., Part 1 [1964:140–141]). Elsewhere Rousseau indicated that the historical changes producing society were "determined by the climate and by the nature of the soil," which can be traced to the inclination of the earth's axis and the resulting seasonal variations in climate (*Essay on the Origin of Languages*, chap. ix [1966:38–39]).

To be sure, Rousseau saw unique principles underlying human history. Whereas other species behave on the basis of natural "instincts," humans are "free," at least in the sense that they can "perfect" their own behavior and thereby change the way they adapt to the world (*Second Discourse*, Part 1 [1964:113–115]). But this difference does not seem to rest on special creation or other extraevolutionary processes; on the contrary, traits like "the shape of the teeth, and the conformation of the intestines," as well as "the number of young" and "conjugal society," are treated as natural adaptations in humans as well as in other animals (ibid., note e [1964:187–188]; note g [1964:191–192]; note l [1964:213–220]).

Rousseau thus elaborated a view of evolution as a gradualist process under the pressure of natural selection. Because he accepted Buffon's observation that the life expectancy of mammals is roughly proportional to their growth phase (ibid., note g [1964:191], citing Buffon's *Natural History*, 4:226–227), Rousseau explained distinct features of human ontogeny without invoking either special creation or natural discontinuities more general than climatic changes or localized events such as earthquakes and volcanic eruptions (*Second Disclosure*, Part 2 [1964:152]).

Despite this evolutionary gradualism, Rousseau's view of human history resembles punctuated equilibria rather than continuous infinitesimal changes. The founding of settled homesites is described as "the epoch of a first revolution, which produced that establishment and differentiation of families and which introduced a sort of property" (ibid. [1964:146]). The division of labor and the emergence of "property" in the precise sense were a "great revolution" produced by the twin discoveries of metallurgy and agriculture (ibid. [1964:151–152]). The establishment of governments and laws was the result of a "social contract" invented by the rich as "the most deliberate project that ever entered the human mind" (ibid. [1964:158–159]). Finally, such governments—originally legitimate because based on the voluntary consent of the governed—are usurped by tyrants under whom "everything is brought back to the sole law of the stronger, and consequently to a new state of nature different from the one with which we began, in that the one was the state of nature in its purity, and this last is the fruit of an excess of corruption" (ibid. [1964:177]).

Rousseau summarizes these transitions as a sequence of revolutionary changes, each of which ushers in an "epoch" or period of relative stability within which changes are gradual and slow. Originally, humans were equal, but the process of human history introduced progressively greater inequalities:

> If we follow the progress of inequality in these different revolutions, we shall find that the establishment of the law and of the right of property was the first stage, the institution of the magistracy the second, and the third and last was the changing of legitimate power into arbitrary power. So that the status of rich and poor was authorized by the first epoch, that of powerful and weak by the second, and by the third that of master and slave, which is the last degree of inequality and the limit to which all the others finally lead, until new revolutions dissolve the government altogether or bring it closer to its legitimate institution. (Ibid. [1964:172].)

Although changes seem possible within each stage or epoch, the general process is an irreversible trend toward inequality, physical decrepitude, and moral corruption (*Letter to Philopolis; Letter to d'Alembert; Social Contract*, 3.10–11 [1978:96–99]).

Although Rousseau argued that natural causes like climate influence politics (*Social Contract*, 2.8–11 [1978:70–76]), his theory of human history has two features sharply divergent from our image of Darwinian processes: first, revolutionary discontinuities rather than gradual change; second, the secular trend toward undesirable outcomes departing more and more from adaptive solutions. Unlike the social Darwinists of the late nineteenth century (Hofstadter, 1955), Rousseau neither transferred his gradualist conception of biological evolution to human history nor imputed selective advantage to the results of the political process. Instead, Rousseau combined a gradualist theory of biological change and a punctuational view of human history in formulating one of the most radical political teachings in the Western tradition.

Having asserted that humans were originally stupid and peaceful animals living in small groups if not in isolation, Rousseau concludes that inequality is "almost null in the state of nature"; if so, he claims,

> moral inequality [i.e., social or legal inequalities of wealth and power] is contrary to natural right whenever it is not combined in the same proportion with physical inequality: a distinction which sufficiently determines what one ought to think in this regard of the sort of inequality that reigns among all civilized people, since it is manifestly against the law of nature, in whatever manner it is defined, that a child command an old man, an imbecile lead a wise man, and a handful of men be glutted with superfluities while the starving multitude lacks necessities. (*Second Discourse*, Part 2 [1964:181–182].)

Hereditary monarchy and inequality of wealth are thus illegitimate in principle. Or, to use the resounding phrase that opens Rousseau's most famous political work: "Man was [and is] born free, and everywhere he is in chains" (*Social Contract*, 1.1 [1978:46]).

Rousseau's substantive conclusions were anathema to the *ancien régime* (which banned his works and condemned his teaching): the only political system worthy of obedience is a community of relatively equal citizens who are free to enact the laws and elect the governing authorities annually. Indeed, only small communities with simple economic systems are likely to be decent and free; since most large-scale societies based on rigid distinctions of social class and centralized political power are inherently corrupt, political revolution is legitimate even where it may be imprudent (Masters, 1968). Given this uncompromising rejection of aristocratic and royal privilege, it should hardly be surprising that Robespierre claimed that Rousseau inspired his revolutionary career and that the Jacobins enshrined his body in the Pantheon along with other fathers of the Revolution of 1789.

For Rousseau, therefore, gradual evolutionary processes in biology did not necessarily entail either historical gradualism or deference to constituted political authority. Few political theorists have articulated a clearer account of human origins in a natural process of gradual change, yet few have been more profoundly radical in challenging existing political institutions. Although conventional histories of the sources of Darwinian theory frequently skip over his thought—perhaps on the arbitrary assumption that, as a political theorist, he is not relevant to the history of science—Rousseau's challenge to the orthodox biblical account of human nature on evolutionary grounds clearly affected the thinking of his contemporaries and was a direct intellectual link between the accounts of pagan antiquity and nineteenth-century biology.

Marx: Darwinian Evolution, Political Revolution, and Progress

Although Karl Marx's concept of revolutionary changes throughout human history is well known, the relationship between this political theory and evolutionary biology is less obvious. It has frequently been said that Marx offered to dedicate *Das Kapital* to Darwin (Feuer, 1977; Mackenzie, 1979:50–52; Tiger, 1980), yet exactly what is the relationship between Marx's concept of political change as a process typically marked by sharp discontinuities and his understanding of biological processes? For example, would Marx have favored contemporary sociobiology, as Lewis Feuer (1977) has claimed, or roundly condemned it, as many Marxists insist?

In presenting the model of punctuated equilibria, Gould and Niles Eldredge mention that one of them learned Marxism "at his father's knee" (Gould and Eldredge, 1977). Is this an indication, as is sometimes charged, that the very concept of punctuated equilibria is infected with the virus of Marxian revolutionary doctrine? Or is it an aside in intellectual history, of no more significance than the claim that Kekule first imagined the ring structure of hydrocarbon atoms during a dream?

Without underestimating the importance of Marx's collaboration with Friedrich Engels, it is probably useful to distinguish between the two; though Engels wrote explicitly on the links between evolution and human history (Engels, 1940, 1942), some scholars of Marxism claim that there are important differences between Marx and Engels. In particular, it is charged that Engels adopted—in later life—more mechanistic interpretations of historical processes than those originally proposed by Marx, with the result that the mature Engels was substantially closer to a neo-Darwinian view of time than was the young Marx (Avineri, 1968).

Like Rousseau, Marx was steeped in the philosophy and science of classical antiquity. Although his doctoral dissertation was on the Greek atomists Democritus and Epicurus, the young Marx was particularly impressed by Lucretius as the model of "a mind free and sharp" willing to describe "Nature without gods, gods without a world" (Mackenzie, 1979:54). This does not mean that Marx simply accepted the view of biological and historical change presented in Lucretius's *On the Nature of Things*. Rather, it reflects Marx's primary objective: to replace the theological account of human existence with a more naturalistic one, ultimately capable of unifying the natural and social sciences (see the note to his doctoral dissertation entitled "Reason and the Proofs of God," in Marx, 1967:64–66).

That Marx set himself the goal of unifying the natural and social sciences is evident in the *Economic and Philosophic Manuscripts of 1844*:

> *One basis* for life and another basis *for science* is a *priori* a lie. The nature which develops in human history—the genesis of human society—is man's real nature; hence nature as it develops through industry, even though in an *estranged form*, is true *anthropological nature*. . . . History itself is a *real part of natural history—of nature developing into man*. Natural science will in time incorporate into itself the science of man, just as the science of man will incorporate into itself natural science: There will be *one science*. (1964:143.)

Human beings have evolved or "developed" by a process of "natural history," and it is merely a question of time until the natural and human sciences are unified.

If Marx saw himself as capable of contributing to this fundamental reunification, it was not primarily in terms of the biological process of evolution. This probably explains Marx's enthusiasm for Darwinian theory, since the specific mechanisms of evolutionary change described by Darwin struck Marx as a vulgar illustration of British bourgeois thought; for Marx, Darwin's view of natural selection was little more than a Hobbesian "war of all against all," conveniently transposing to nature the ideological norm and practical reality of competitive capitalism. Even granted these defects, however, Darwin was invaluable for Marx because his *Origin of Species* could serve, as did Lucretius's *On the Nature of Things*, as an ally against religious dogma.

When Engels says that Marx's "fundamental proposition . . . is destined to do for history what Darwin's theory has done for biology" (Marx and Engels, 1971:136–137), the reference is not to Darwin's gradualist model of rates of change; Marx, like Darwin, saw himself as engaged in a battle to demonstrate that God could no longer be used by any serious thinker to explain either nature or human society: "The question about an alien being, about a being above nature and man—a question which implies the admission of the unreality of nature and of man—has become impossible in practice. *Atheism*, as the denial of this unreality, has no longer any meaning, for atheism is a *negation of God*, and postulates the *existence of man* through this negation; but socialism as socialism no longer stands in need of such a mediation" (Marx, 1964:145–146). By 1844 Marx's concern had already focused on the political goals of a fundamental transformation of the human condition, and it was from this perspective that he sought a materialist, scientifically objective account of human history.

Marx's interest in Darwin thus does not mean that he adopted gradualist assumptions concerning the rate and precise character of the evolutionary process. In contrast to Darwin's view that there is no fundamental difference between humans and other animals, Marx—in this regard a student of Hegel—insists on the distinctness of our species. To be sure, Marx views humans as physical, material beings: "The life of the species, both in man and in animals, consists physically in the fact that man (like the animal) lives on inorganic nature" (1964:112). But the way that humans relate to "inorganic nature" differs fundamentally from any other species.

All animals can be viewed as productive, yet Marx claims that human production is unique: "The whole character of a species—its species character—is contained in the character of its life activity; and free, conscious activity is man's species character. . . . The animal is immediately one with its life activity. It does not distinguish itself from it. It is *its life activity.* Man makes his life activity itself the object of his will and of his consciousness. He has conscious life activity. It is not a determination with which he directly merges. Conscious life activity distinguishes man immediately from animal life activity" (1964:113). Using a Hegelian term, Marx speaks of humans as "a species being"—that is, a being capable of "free activity."

Marx is particularly explicit in distinguishing between humans and other species:

> In creating a *world of objects by his practical activity*, in *his work upon inorganic nature*, man proves himself a conscious species being. . . . Admittedly animals also produce. They build themselves nests, dwellings, like the bees, beavers, ants, etc. But an animal only produces what it immediately needs for itself or its young. It produces one-sidedly, whilst man produces universally. It produces only under the dominion of immediately physical need, whilst man produces even when he is free from physical need and only truly produces in freedom therefrom. An animal produces only itself, whilst man reproduces the whole of nature. An animal's product belongs immediately to its physical body, whilst man freely confronts his product. An animal forms things in accordance with the standard and the need of the species to which it belongs, whilst man knows how to produce in accordance with the standard of every species, and knows how to apply everywhere the inherent standard to the object. (1964:113–114.)

Whereas Rousseau had argued that, at least in its origins, the human species should be understood as another "animal" species, Marx seems to postulate a sharp chasm between humans and other animals.

This is not an isolated passage in the writings of the young Marx. In the *German Ideology*, written with Engels in 1845–46, we find the same radical distinction between humans and other species: "Language is as old as consciousness, language *is* practical consciousness that exists also for other men. . . . Where there exists a relationship, it exists for me: the animal does not enter into *'relations' with anything*, it does not enter into any relation at all" (Marx and Engels, 1970:51). At the earliest human period, "this beginning is as animal as social life itself. . . . It is herd-consciousness, and at this point man is only distinguished from sheep by the fact that with him consciousness takes the place of instinct or that his instinct is a conscious one" (ibid.). How human consciousness arose is not discussed by Marx and Engels; rather, they immediately stress that it permitted humans "further development" of the productive process, leading to the division of labor—or, more precisely, the "division of material and mental labour" (ibid.). "From this moment onwards," humans are radically different from any other species, if only because "from now on consciousness is in a position to emancipate itself from the world" (ibid.:51–52). For all practical purposes, the prehistory of these developments is inaccessible and irrelevant to the science of human history.

Given such a focus on the uniqueness of human beings, it is perhaps not surprising that, beyond condemning the doctrine of divine creation, Marx pays relatively little attention to the precise pattern of hominid evolution. Since he sees modern geology as replacing the biblical teaching of the "creation of the earth" by a "process" that is equivalent to "self-generation," the emergence of species might just as well be the result of spontaneous generation: "*Generatio aequivoca* is the only practical refutation of the theory of creation" (Marx, 1964:144). The attempt to account for the origin of species is ultimately circular, because it is impossible to go beyond the Aristotelian account of parent-offspring resemblance without becoming lost in the "abstraction" of thinking of "man and nature as *non-existent*" (ibid.).

In denying that it is legitimate to ask about the natural origins of human life, Marx imagines a short dialogue with a critic: "You can reply: I do not want to conceive the nothingness of nature, etc. I ask you about *its genesis*, just as I ask the anatomist about the formation of bones, etc. But since for the socialist man the *entire so-called history of the world is nothing but the creation of man through human labor*, nothing but the emergence of nature for man, so he has the visible, irrefutable proof of his *birth through himself*, of the *process of his creation*" (ibid.:145). Whatever might be said concerning other species, humans have essentially created themselves through the process of free, conscious labor.

It follows that theories of biological evolution cannot be relevant to the process of human history. For Marx—as for Teilhard de Chardin—evolution has to be considered as "nature developing into man": "The *social reality* of nature, and *human natural science*, or the *natural science about man*, are identical terms" (ibid.). Humans are capable of—and soon will achieve—a radically humanistic appropriation of all "nature," if only because humans are unlike every other animal species. In a profound sense, the young Marx seems to have adopted a secular version of the Christian doctrine of "special creation," albeit one that views the human species as its own creator (Tucker, 1961).

In human history, rates of change appear to be punctuated by relatively sudden events, explicitly described by Marx as revolutions. In both the *German Ideology* and the *Communist Manifesto*, Marx and Engels give a detailed account of the "revolutionary" transformation of feudal society into the capitalist epoch in which the bourgeoisie bases its wealth and power on control over the industrial mode of production; in the *Manuscripts of 1844* as well as the *Critique of the Gotha Programme* (1875), Marx intimates that the transition from capitalism to communism will pass through a brutal stage of "crude communism" before ushering in the definitive freedom of mankind (1964:133–135).

These transitions are described as abrupt rather than gradual. Apart from a few asides on the likely future of England, Marx generally treats the transformation of capitalist society into communism as a process that will require a political revolution: "In depicting the most general phases of the development of the proletariat, we traced the more or less veiled civil war, raging within existing society, up to the point where that war breaks out into open revolution, and where the violent overthrow of the bourgeoisie lays the foundation for the sway of the proletariat" (Marx and Engels, 1971:101). In short, the general pattern of human history is a sequence of revolutionary upheavals followed by more gradual development within each established form of social relations (see also the preface to the *Contribution to the Critique of Political Economy* [1975:425–426]; *German Ideology*, Part 1 [1970:59, 87–89, 94–95]).

The final transition to communism requires a particularly radical step because it must be worldwide in scope. "Empirically, communism is only possible as the act of the dominant peoples 'all at once' and simultaneously, which presupposes the universal development of productive forces and the world intercourse bound up with communism" (Marx and Engels, 1970:56). This global upheaval is possible because, in the process of establishing a single world market, capitalism destroys all class distinctions except the contradiction between the bourgeoisie and the proletariat.

Marx's view of human origins thus reflects the same discontinuity as does his theory of history. On this basis, he develops the substantive political principles now associated with "communism": progressive change is possible through a revolutionary political movement that takes control of the historical process. Once established, a communist society will move toward the definitive liberation of all mankind in abundance and justice. Whereas all prior history can be described as "natural," since changes occurred in an involuntary manner, the next phase of history will be freely chosen or "voluntary": "*All-round dependence*, this natural form of the *world-historical co-operation of individuals*, will be transformed by this communist revolution into the control and conscious mastery of these powers, which, born of the action of men on one another, have till now overawed and governed men as powers completely alien to them" (ibid.:55). By abolishing the division of labor and private property, communism thus frees humans not only from exploitation by others, but from the domination of natural necessity (Marx, 1964:135–144; Marx and Engels, 1970:52–57).

Varieties of Religious Doctrine and Political Thought

It would appear that every conceivable combination of attitudes toward rates of evolutionary change and political transformation can be identified in the Western intellectual tradition. For Aristotle, both animal and human social events seem marked by continuity and gradualism; for Lucretius, a punctuated theory of natural events is combined with historical gradualism; for Rousseau, a gradualist theory of evolution generates a concept of political revolutions; for Marx explicitly—as for Hobbes implicitly—discontinuities characterize both the origin of the human species and the major changes in human history. And if this variability is not sufficiently paradoxical, of all the thinkers surveyed above, it is probably Marx—the most explicit in adopting historical models akin to punctuated equilibria—who exemplifies the "typological" concept of species "essences" condemned by Mayr and other contemporary biologists.

The temporal model used to explain the origin of human beings thus does not entail any inherent view of history and the political process. This finding is reinforced by comparing the practical principles of other thinkers with gradualist views of human history. For Edmund Burke, revolution was anathema—but this did not prevent him from favoring the cause of the American colonists in their conflict with the Crown or from condemning the blatant exploitation of India by Warren Hastings. For John Stuart Mill, as for many nineteenth- and twentieth-century liberals, Western political society will gradually evolve toward a free and democratic regime in which the naturally best will lead. For Edward Bernstein, in contrast, Marx's doctrine is substantially correct in all respects except for the reliance on revolutionary change; the title of his most famous book, *Evolutionary Socialism*, symbolizes the gradual change that will totally transform the market economy and liberal state. A similar combination of gradualism and socialism characterizes the thought of the English Fabians, not to mention the later writings of Engels himself (Engels, 1940, 1942).

Those seeking rapid or discontinuous change mirror this diversity. From Nietzsche to Hitler, some have proclaimed the need for radical if not brutal transformations of society to prevent the type of society sought by Marx and other Marxists. The Founding Fathers of the American regime, the Jacobins and Girondins in revolutionary France, and elites in the colonial world during the last century have all argued that the goal of improving the human prospect legitimates relatively discontinuous changes.

If concepts of change in biology are so independent of substantive political principles, how can one explain the reciprocal allegations that models of gradualism and punctuated equilibria are politically motivated? To this point, theological doctrines have not been discussed substantively; the biblical concept of divine Creation has lurked in the background, as the target of Hobbes, Rousseau, Marx, and Darwin, but little has been said of this teaching in its own right. Perhaps we will understand the nature of political and scientific debates more fully if we turn to religion.

The account of Genesis could be described as the epitome of a theory of discontinuous change in nature. Not only is each day of Creation a distinct stage in the formation of the world as we know it, but the creation of plants on the third day, sea monsters, fish, and birds on the fifth, and land animals on the sixth seems to exemplify discontinuous change followed by epochs of relative stability. When Adam and Eve eat the apple and are expelled from the Garden of Eden (Gen., 3:1–24), the human condition is transformed; further transformations occur at the time of the flood (Gen., 6:11–8:20) and the tower of Babel (Gen., 11:1–9). Later in the Old Testament, one encounters further instances of what we today might call religious or political change that are equally momentous: the covenant of Abraham (Gen., 17:1–18:15); Moses and the Ten Commandments, establishing the law (Exod., 19:16–35:14); or the anointing of Saul as the first human king of the Hebrews (1 Sam. 10:1–24).

The New Testament is characterized by a similar view of time. In addition to accepting the Creation, the covenant with Abraham, and the Mosaic law as epoch-making transformations of the human condition, at least two new critical changes mark the Christian view of time: the life and death of Jesus, and the Last Judgment (e.g., Rom., 4:18; 5:11–21; 13:11–14). The Judeo-Christian tradition thus generally adopted a "progressive" view of history marked by a series of fundamental changes occurring by divine plan. Such a religious teaching establishes the cultural context within which Western thought since the Renaissance has developed.

Comparisons between Judeo-Christian concept of time and evolutionary theories of punctuated equilibria merely because both share a similar view of discontinuous change in history may seem forced. But reference to pagan antiquity reminds us that religion need not hold this view of time. As Mircea Eliade points out (1959), Greco-Roman paganism—and indeed many other primitive religions as well—conceived of the world as an eternal process of repetitive cycles, in which change did not take the progressive form characterized by the Judeo-Christian approach to history.

This theological difference between pagan antiquity and Christianity helps to explain an important difference between the ancient thinkers discussed above (Empedocles, Aristotle, Lucretius) and such moderns as Hobbes, Marx, and even Darwin. For the pagans, whatever the rate of historical change, there is little expectation that its direction is toward perfection; even when development to "higher" states of civilization is praised, the pagans tend to foresee an impending stage of decline (e.g., Thucydides, *History of the Peloponnesian Wars*). If time is cyclical, change can hardly be permanent, irreversible progress.

Whereas the pagans did not expect humans to "conquer nature," secular modern writers from Sir Francis Bacon and Hobbes to Marx simply assume that history is a progressive movement toward the "conquest" or "control and conscious mastery" of nature (Masters, 1977). This directionality has, of course, been criticized when it is injected into evolutionary theory: biologists have usually rejected the simplistic equation of evolution and improvement, whether in social Darwinist theories of progressive improvement or models of "orthogenesis." But at the same time, many biologists—as citizens—are committed to political or religious doctrines in which the progressive directionality of change is embedded.

Relatively few contemporary biologists have adopted Rousseau's explicitly pessimistic view of human history, and still fewer have openly adopted the Aristotelian view that fundamental human progress is for all practical purposes impossible. Just as Darwin took the conscious direction of selection by animal breeders as a model for natural selection, it would seem that many biologists—however rigorous in rejecting orthogenetic trends in "nature" itself—accept them in the political realm when speaking of human affairs (e.g., Dobzhansky, 1955; Stebbins, 1966).

This observation suggests why it is so often claimed that models of the rate of evolutionary change are inherently ideological, despite the evidence to the contrary. Both gradualism and punctuated equilibria imply views of time that can be associated with the most diverse political principles, but under contemporary circumstances the model of evolutionary change associated with the views of one's rivals is more likely to be attacked than the primordial assumption of "progress." Few contemporaries have concluded from neo-Darwinian models of evolution that all modern political systems are doomed to failure, even though ancient political thought suggests such a conclusion (as is evident in the work of Rousseau, the modern who came closest to the pagans in this regard). For Marxists, particularly if radical, evolutionary gradualism is said to be biased because it can be used to justify reforms that inhibit the revolutionary transformations needed in human life; for traditional "liberals," models of punctuated equilibria seem to be an attempt to justify undesirable revolutionary changes. It is easier to attack one's rivals than to abandon an unquestioned tenet of one's own political faith.

Table 12.1. Theories of biological evolution and human history

	Biological Evolution	
	Gradual	Discontinuous
Human history		
Gradual	Aristotle	Lucretius
Discontinuous	Rousseau	Hobbes Marx

CONCLUSIONS

Examples have been given of diverse theories of biological evolution and human history. Insofar as gradualism and discontinuity represent different approaches to temporal development, either of these two patterns or rates of change can be attributed to biological evolution or to political development. As a result, at least four basic combinations are possible (Table 12.1). Aristotle seems a gradualist in both animal and human matters, whereas Lucretius combines the catastrophism of Empedocles with an evolutionary treatment of human history; among the moderns, Hobbes and Marx imply the existence of radical discontinuities between humans and other animals as well as within human social history, while Rousseau takes a gradualist, evolutionary approach to human origins and a punctuated view of human history. Consideration of other secular and religious thinkers confirms that neither model of change contains an implicit political bias apart from the theoretical and historical context.

It follows that biologists can and should develop theories of evolution without fear that in so doing they are exhibiting intractable bias. Like other domains of scientific inquiry, it is important to allow the scientific community to debate these issues, presenting hypotheses and seeking to falsify them with historical evidence (e.g., Williamson, 1981). But like many other areas in which scientific thought was long inhibited by the unwitting injection of political opinion, evolutionary accounts may have been weakened by the assumption—carried over from political theories that distinguished sharply between evolution and revolution—that only one of the two models could possibly be true. Throughout the debate between exponents of gradualism and of punctuated equilibria, this is one of the more amusing and frustrating realizations.

For example, Ernst Mayr admits that much of his career has been devoted to demolishing the catastrophism of Goldschmidt's theory of "genetic revolutions" and "hopeful monsters"; even though Gould has indicated that his conception of punctuated equilibria is derived from Mayr's understanding of speciation, Mayr himself seems uncomfortable with this development of his work (Hapgood, 1984). It is as if most of the major biological theorists of our day have assumed that gradualism and punctuated equilibria are inconsistent with each other. Yet there is no reason to make such an assumption, particularly if it is true that—to use the term François Jacob borrowed from Claude Lévi-Strauss—nature is a "tinkerer" rather than an engineer (Jacob, 1977).

It is therefore fitting to cite a recent statement based on a careful assessment of the empirical evidence at several sites where the paleontological record of the Cretaceous-Tertiary boundary is reasonably clear:

> Recent developments in research on the terminal-Cretaceous extinctions have been widely viewed in terms of a conflict between gradualistic and catastrophic interpretations. This notion is counterproductive and should be discarded. It seems evident to us that major biotic turnovers occur on two completely different time scales. The classical paleontological view recognizes the importance of gradual turnovers on a time scale of 10^5 to 10^7 years. The evidence of changes of this kind is so strong that their existence cannot be denied by any reasonable person. The novelty resulting from the recent work on the K/T boundary is the recognition that very rapid turnovers can also occur, with characteristic times of 1 to 10^3 years. In accepting the evidence for very rapid turnovers, one need not reject the reality of gradual turnovers. (Alvarez et al., 1984:1136.)

Though some might conclude that the debate has been much ado about nothing, one should probably come to the contrary view: like many scientific debates in the past, once rival views can be stated as empirically falsifiable hypotheses, open controversy makes possible a more precise understanding of nature (Gingerich, 1983; Gould, 1984).

If it is true that both punctuated equilibria and gradualism occur in evolutionary processes, the apparently contradictory teachings of different political and evolutionary theorists would all contain an element of truth. Elsewhere, I have suggested that much the same thing can be said for theories of human nature and sociability (Masters, 1983). With reference to time, as with reference to the natural foundations of cooperation, charges that biological theories are politically motivated seem to refer, at most, to trivial details in the history of science; in principle, an understanding of nature and of human nature seems accessible to those willing to study *sine ira et studio.*

Nevertheless, Marx's thought seems less likely to be consistent with contemporary biology than other political theories, if only because he tended to treat species in a more "essentialist" or typological manner than ancients like Aristotle and Lucretius or moderns like Rousseau. This will seem a paradox for those who assume that Marx's brand of historical materialism represented not only a scientific advance over earlier thought, but a successful attempt to adopt a Darwinian view of time in the study of human life. That such was Marx's own goal has been indicated. Why might he have failed?

As critics often point out, Marx's theory of history has striking parallels in structure to the Christian view of time. These similarities do not relate primarily to the communism of the early church (Acts, 5:1–11), but rather to the pattern of time implicit in the teachings of early Christianity. Whether in Paul's epistles (notably Romans) or such subsequent statements as Augustine's *City of God*, we find remarkable parallels between the Christian and Marxian views of time.

For both, the origin of species as well as the history of human life is characterized by catastrophic changes. For both, a duality (Paul's concept of the conflict between "spirit" and "flesh," Augustine's two cities, or Marx's class conflict) leads to radical change; ultimately, for both, the overall direction of "progress" will usher in a beneficent resolution of human travail. Marx's communist society is on earth, but in other respects it reflects a conquest of nature as definitive as the Last Judgment of Christian doctrine.

This seems to have little enough to do with biological thought. But it has everything to do with some of the controversies in modern politics. Marxism became a secular religion justifying forced investment and rapid industrialization in the non-Western world; ostensibly hostile to Western civilization, Marx's teaching was thus transformed into a nationalistic ideology everywhere inviting imitation of the Western approach to science and technology. More specifically, Marxist thought became the major doctrine for those in the non-Western world who sought material progress on the assumption that modern science provides humans with the tools for controlling natural necessity.

When biologists treat the history of their science, they tend to select evidence that mirrors their preferred model of evolutionary biology. Mayr, critical of Goldschmidt and seeking to establish neo-Darwinian gradualism, treats all past thought as definitely surpassed by the progressive development of biological theories; Gould, in seeking to establish punctuated equilibria as a mode of biological evolution, reasons as if human thought developed by virtually discontinuous leaps. We have already seen, for example, how Mayr misread Aristotle, not to mention other thinkers from the seventeenth to nineteenth centuries, as if to exaggerate the "advance" constituted by Darwinian biology. But Gould's account of Western thought is no less marked by an implicit use of his own model of evolution whenever dealing with the history of ideas, since he leaps quickly from an account of Empedocles and Aristotle to the late eighteenth-century naturalist Charles Bonnet; in so doing, Gould fails to indicate that Bonnet's intuition of a parallel between ontogeny and phylogeny can be traced to Rousseau's *Second Discourse*, itself based on Lucretius's restatement of Empedocles.

Gould's otherwise magnificent *Ontogeny and Phylogeny* presents the history of biology in a way that is akin to a punctuated equilibrium model of change. Doing so, however, requires ignoring texts that are—unlike many of the intermediate stages in the fossil record—readily available for study. Empedocles' views were recorded by Aristotle; similar ideas about the origin of species can be found in Lucretius. Rousseau read both Aristotle and Lucretius and, on the basis of his reading of the ancients, presented a rigorously naturalistic account of human origins as a response to Hobbes (Masters, 1968).

In the *Second Discourse*, which Rousseau modeled on Lucretius, the analogy between human ontogeny and human phylogeny is used explicitly several times.[4] One of Rousseau's contemporaries published a critique of the *Second Discourse* under the pen name "Philopolis": this critic was none other than Charles Bonnet.[5] And Rousseau's reply—the

4. The evolutionary origins of hominid physical structures are the "embryo" of the species. In the pure state of nature, "man remained ever a child"; savage society is "youth" (*la veritable jeunesse du monde*); citizens of legitimate societies are "fathers" and full adults; corrupt civilizations represent the "decrepitude"—"old age" and "death"—of the human species (Rousseau, 1964:137, 151, 86–87; *Social Contract*, 2.8 [1978:71]; 3.10 [1978:96]; Masters, 1979).

5. *Lettre de M. Philopolis au sujet du discours de M. J.-J. Rousseau* . . . in Rousseau, *Oeuvres complètes*, ed. Michel Launay (Paris: Seuil, 1971), 2:270–272. Bonnet criticized Rousseau on theological grounds: "Rousseau's *savage man*, that man he cherishes so warmly, is not at all the man that GOD wanted to create but God created

Letter to Philopolis—specifically expands on the analogical parallels between ontogeny and phylogeny in a most remarkable way.[6] Although there is doubt that the reply was actually sent to Bonnet, there can be no question that Bonnet himself read the *Second Discourse* and saw in it a challenge to his own religious faith. Hence Rousseau would seem to be one of the "missing links" omitted in the historical account of biological ideas presented by Gould.

This detail suggests an unusual property of human thought not noticed by most evolutionary theorists. For a number of contemporary structuralists, human thinking consists of varied combinations and transformations of fundamental elements; hence writers as diverse as Jean Piaget, Jacques Derrida, Michel Foucault, Jacques Lacan, Noam Chomsky, and Lévi-Strauss have suggested a radical alternative to the progressive view of human thought. As Lévi-Strauss has put it, since all cultures elaborate the same structural contrasts, human thought is essentially "closed" at the systemic level; though there are myriad new combinations of elements, the elemental contrasts themselves are finite in number.

Gould's account of past biological thought leads—like that of Mayr—to a tacit presumption that contemporary biology has broken new ground inaccessible to the ancients. Before them both, Darwin implied as much by his treatment and misunderstanding of Aristotle at the outset of *The Origin of Species* (see note 2). In contrast, a structuralist approach could see recent theories as a new combination of principles explored whenever scientific thought is elaborated. If mechanisms studied in contemporary genetics could be used as an analogy for scientific change and development, one could speak of "recombination" as a greater source of theoretical variation than those intellectual "mutations" consisting of entirely new ideas.

---

orang-utans and *monkeys* who are not men" (271). The young Bonnet was a good example of the kind of species "essentialist" who opposed evolutionary thought on religious grounds—and explicitly confronted an evolutionary challenge to his beliefs when he read Rousseau's *Second Discourse* in 1755.

6. Rousseau insists that "according to me, society is natural to the human species the way old age is to the individual"—and then goes on to imagine that a scientist has been able to speed up ontogeny in order to illustrate the danger of actions that might hasten the historical development of the species (*Lettre de J.-J. Rousseau a Philopolis*, Rousseau, *Oeuvres complètes*, 2:273). Unlike most of his contemporaries—and ours—Rousseau did not believe that human history is a process of progressive improvement (Masters, 1968, 1983).

Such a perspective leads to a more modest view of human knowledge than the conventional belief that radical novelties in scientific thought are frequently possible. Far from "progressing" toward truth, science—like myth and ritual in at least this respect—typically explores ever anew the humanly possible ways of knowing. To be sure, science is unlike myth or ritual in its mode of integrating competing hypotheses and subjecting them to falsification. Whereas contrary expectations can sometimes be accepted as consistent in a scientific frame, conflicting religious doctrines are far more difficult to bridge. Hence the empirical finding that the fossil record exhibits both gradualism and punctuated equilibria would reflect the extent to which divergent thinkers have explored—and continue to explore—natural possibilities that have been known, at least dimly, for millennia.

Modern biology may well be an advance over ancient biology in its power to effect changes, in no small part because of our discoveries of the precise mechanisms of evolutionary transmission and, increasingly, genetic control over phenotypical traits. But such advances have been bought dearly: in the process of embarking on research that yields potent biotechnologies, the scientific community has unwittingly communicated the myth of a Baconian "conquest of nature" to the layman and policy maker alike. It may well be that only such implications will suffice to elicit the funds needed for contemporary research. Given the all-too-evident dangers that the resulting technologies will have unwanted consequences, whether by realizing *Brave New World*, by ushering in an age of biochemical warfare, or by simple accident, it is perhaps worth remembering that neither gradualism nor punctuated equilibria really entails a demonstration of the human power to control evolutionary outcomes. That idea is—and always has been—a notion that is ultimately political rather than scientific.

References

Alvarez, W., et al. 1984. Impact theory of mass extinction and the invertebrate fossil record. *Science* 223:1135–1141.

Aristotle. 1984. *The Complete Works of Aristotle*, ed. J. Barnes. 2 vols. Princeton: Princeton University Press.

Arnhart, L. 1990. Aristotle, chimpanzees, and other political animals. *Social Science Information* 19:479–559.

Avineri, S. 1968. *The Social and Political Thought of Karl Marx*. Cambridge: Cambridge University Press.

Bernstein, E. 1961. *Evolutionary Socialism*. New York: Schocken.
Bowler, P. J. 1984. *Evolution: The History of an Idea*. Berkeley: University of California Press.
Caplan, A. M., ed. 1978. *The Sociobiology Debate*. New York: Harper & Row.
Capra, F. 1975. *The Tao of Physics*. Berkeley: Shambala.
Darwin, C. n.d. [1859]. *On the Origin of Species*. 2 vol. Chicago: Rand-McNally.
Delbruck, M. 1971. Aristotle-totle-totle. In J. Monod and E. Boerk, eds., *Of Microbes and Life*. New York: Columbia University Press, pp. 50–55.
——. 1977. How Aristotle discovered DNA. In K. Huang, ed., *Physics in Our World: A Symposium in Honor of Victor F. Weisskopf*. New York: American Institute of Physics, pp. 123–130.
Dobzhansky, T. 1955. *Mankind Evolving*. New Haven: Yale University Press.
Eliade, M. 1959. *Cosmos and History*. New York: Harper.
Engels, F. 1940. *The Dialectics of Nature*. New York: International.
——. 1942. *The Origin of the Family, Private Property, and the State*. New York: International.
Feuer, L. 1977. Marx and Engels as sociobiologists. *Survey* 23:109–136.
Gingerich, P. D. 1983. Rates of evolution: effects of time and temporal scaling. *Science* 222:159–161.
Goldschmidt, V. 1977. *Le doctrine d'Epicure et le droit*. Paris: J. Vrin.
Gould, S. J. 1977. *Ontogeny and Phylogeny*. Cambridge, Mass.: Harvard University Press.
——. 1984. Smooth curve of evolutionary rate: a psychological and mathematical artifact. *Science* 226:994–995.
Gould, S. J., and N. Eldredge. 1977. Punctuated equilibria: the tempo and mode of evolution reconsidered. *Paleobiology* 3:115–151.
Hapgood, F. 1984. The importance of being Ernst. *Science 84* 5:40–46.
Heisenberg, W. 1958. *Physics and Philosophy*. New York: Harper Torchbooks.
Hobbes, T. 1962. *Leviathan*, ed. M. Oakeshott. New York: Collier.
Hofstadter, R. 1955. *Social Darwinism in American Thought*. Boston: Beacon Press.
Jacob, F. 1977. Evolution and tinkering. *Science* 196:1161–1166.
Lévi-Strauss, C. 1962. *La Pensée Sauvage*. Paris: Plon.
Lucretius. 1951. *The Nature of the Universe*, trans. R. E. Latham. Harmondsworth, Eng.: Penguin.
Mackenzie, W. J. M. 1979. *Biological Ideas in Politics*. New York: St. Martin's Press.
Marx, K. 1964. *Economic and Philosophic Manuscripts of 1844*, ed. D. J. Struik. New York: International.
——. 1967. *Writings of the Young Marx on Philosophy and Society*, ed. L. D. Easton and K. H. Guddat. Garden City, N.J.: Anchor Books.
——. 1975. Preface to the *Contribution to the Critique of Political Economy*. In L. Colletti, ed., *Karl Marx: Early Writings*. New York: Vintage.

Marx, K., and Engels, F. 1970. *The German Ideology*, ed. C. J. Arthur. New York: International.
——. 1971. *The Communist Manifesto*, ed. D. J. Struik. New York: International.
Masters, R. D. 1968. *The Political Philosophy of Rousseau*. Princeton: Princeton University Press.
——. 1977. Nature, human nature, and political thought. In J. R. Pennock and J. W. Chapman, eds., *Human Nature in Politics*. New York: New York University Press, pp. 69–110.
——. 1978. Jean-Jacques is alive and well: Rousseau and contemporary sociobiology. *Daedalus* 107:93–105.
——. 1979. Nothing fails like success: development and history in Rousseau's teaching. In J. MacAdam, M. Neumann, and G. Lafrance, eds., *Trent Rousseau Papers*. Ottawa: Ottawa University Press, pp. 357–376.
——. 1980. Hobbes and Locke. In R. Fitzgerald, ed., *Comparing Political Thinkers*. New York: Pergamon, pp. 116–140.
——. 1983. The biological nature of the state. *World Politics* 25:161–193.
——. 1987. Evolutionary biology and natural right. In K. Deutsch and W. Soffer, eds., *The Crisis of Liberal Democracy*. Albany: SUNY Press.
——. 1989a. Classical political philosophy and contemporary biology. In K. Moors, ed., *Politikos*, vol. 1. Pittsburgh: Duquesne University Press, pp. 1–44.
——. 1989b. *The Nature of Politics*, New Haven: Yale University Press.
Mayr, E. 1963. *Animal Species and Evolution*. Cambridge, Mass.: Belknap Press of Harvard University Press.
——. 1974. Teleonomy and teleology. *Boston University Studies in the Philosophy of Science* 14:91–117.
Pittendrigh, C. 1958. Adaptation, natural selection, and behavior. In A. Roe and G. G. Simpson, eds., *Behavior and Evolution*. New Haven: Yale University Press, pp. 390–416.
Preus, A. 1979. *Eidos* as norm in Aristotle's biology. *Nature and System* 1:79–101.
Rousseau, J.-J. 1964. *The First and Second Discourses*, trans. R. D. Masters and J. Masters. New York: St. Martin's Press.
——. 1966. *Essay on the Origin of Languages*, trans. John H. Moran. New York: Frederick Ungar.
——. 1978. *The Social Contract, with Geneva Manuscript and Political Economy*, trans. J. Masters. New York: St. Martin's Press.
Sayers, J. 1982. *Biological Politics*. London: Tavistock.
Stebbins, G. L. 1966. *Processes of Organic Evolution*. Englewood Cliffs, N.J.: Prentice-Hall.
Strauss, L. 1964. *The City and Man*. Chicago: Rand-McNally.
Tiger, L. 1980. Sociobiology and politics. *Hastings Center Report* 10:35–37.
Tucker, R. 1961. *Philosophy and Myth in Karl Marx*. Cambridge: Cambridge University Press.
Williamson, P. G. 1981. Palaeontological documentation of speciation in Cenozoic molluscs from Turkana Basin. *Nature* 293:437–443.

Contributors

KENNETH E. BOULDING is Distinguished Professor of Economics Emeritus and Project Director at the Institute of Behavioral Science, University of Colorado at Boulder. He is the author of *Ecodynamics, The World as a Total System,* and *Conflict and Defense.*

SUSAN CACHEL is Professor of Anthropology at Douglass College, Rutgers University. She has authored works in physical anthropology and hominid evolution.

NILES ELDREDGE is associated with the American Museum of Natural History. He is co-originator (with Stephen Jay Gould) of punctuated equilibrium theory. He has written a popular exposition of this theory, *Time Frames*, as well as numerous articles.

BRIAN A. GLADUE is Director of the Program in Human Sexuality at North Dakota State University. The author of many publications, he has presented results of his research widely.

STEPHEN JAY GOULD, with Niles Eldredge, is co-originator of punctuated equilibrium theory. Currently at Harvard University, he has written numerous books, such as *Wonderful Life* and *The Mismeasure of Man*, and many articles for professional journals.

ANTONI HOFFMAN is a paleontologist with the Institute of Paleontology in Warsaw. He has published several important articles during the punctuated equilibrium debate.

ROGER D. MASTERS is Professor of Government at Dartmouth College. He has written widely on biopolitics in numerous articles and such books as *The Nature of Politics* and *Primate Politics* (coedited with Glendon Schubert).

ERNST MAYR, Harvard University, is one of the foremost figures in the development of the modern synthetic theory of evolution. He is the author of such important books as *Animal Species and Evolution, The Growth of Biological Thought*, and *Toward a New Philosophy of Biology.*

ALLAN MAZUR is Professor of Sociology at Syracuse University. He has written a variety of articles on biosociology.

STEVEN A. PETERSON is Professor of Political Science at Alfred University. The author of *Political Behavior: Patters in Everyday Life*, he has also written many articles on biopolitics.

MICHAEL RUSE is Professor of Philosophy at the University of Guelph. He has authored many books, including *Sociobiology: Sense or Nonsense?, The Philosophy of Biology*, and *Molecules to Men: The Concept of Progress in Biology*, as well as numerous articles for professional publications.

GLENDON SCHUBERT is Professor of Political Science at the University of Hawaii at Manoa. He has written many articles on biopolitics as well as such books as *Evolutionary Politics, Sexual Politics and Political Feminism,* and *Primate Politics* (coedited with Roger D. Masters).

ALBERT SOMIT, a political scientist, is Distinguished Service Professor at Southern Illinois University, where he previously served as President. He has edited *Biology and Politics* and coedited (with Rudolf Wildenmann) *Democracy and Hierarchy.*

STEVEN M. STANLEY is with the Department of Earth and Planetary Sciences at The Johns Hopkins University. He has written several books on paleontology and evolution, such as *The New Evolutionary Timetable* and *Macroevolution.*

Index

Library of Congress Cataloging-in-Publication Data

The Dynamics of evolution : the punctuated equilibrium debate in the natural and social sciences / edited by Albert Somit and Steven A. Peterson.

p. cm.

"An earlier version appeared as a special issue of the Journal of social and biological structures (Academic Press) in July 1989"—Acknowl.

Includes bibliographical references and index.

ISBN 0-8014-2531-X (alk. paper).—ISBN 0-8014-9763-9 (pbk. : alk. paper)

1. Human evolution. 2. Evolution. 3. Sociobiology. 4. Social evolution. I. Somit, Albert. II. Peterson, Steven A.

GN281.4.D96 1992

303.4—dc20 91-55569